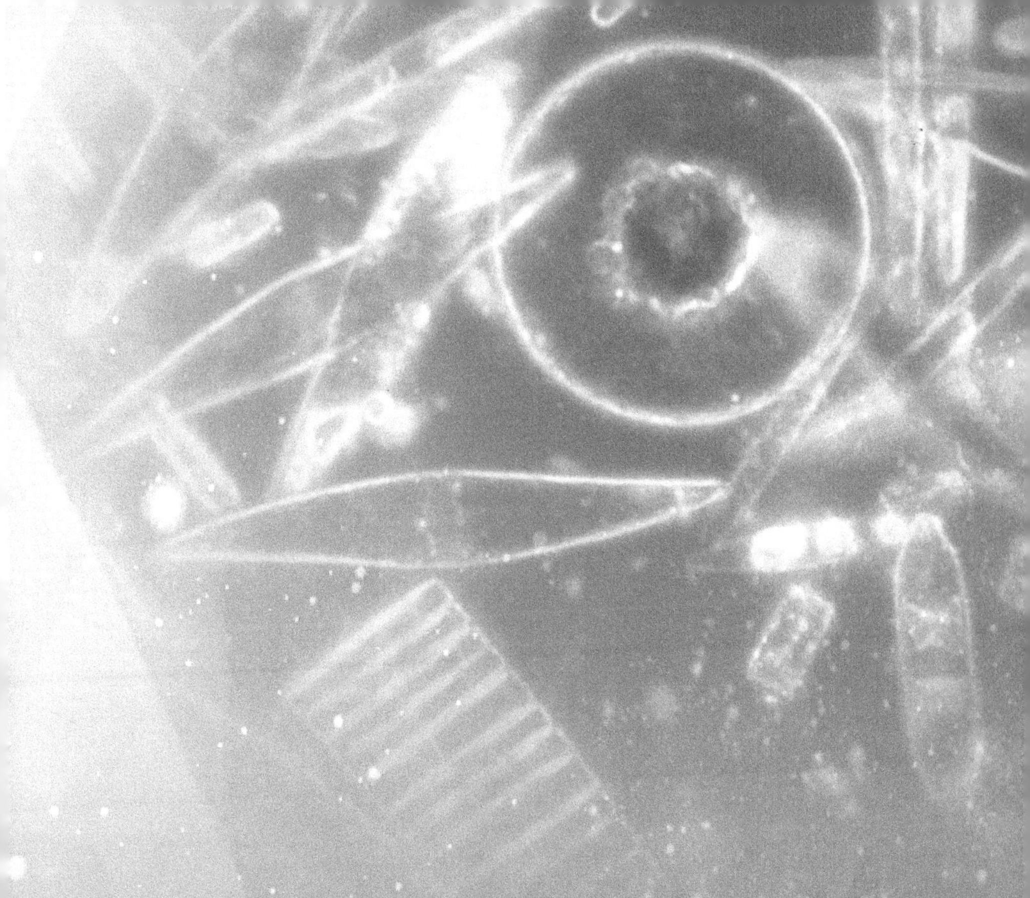

Introduction to Ultrathin Silica Films

Introduction to Ultrathin Silica Films

Silicatene and Others

Shamil Shaikhutdinov

CRC Press
Taylor & Francis Group
Boca Raton London New York

CRC Press is an imprint of the
Taylor & Francis Group, an **informa** business

Introduction to Ultrathin Silica Films: Silicatene and Others

First published 2022 by Jenny Stanford Publishing Pte. Ltd.

Published 2023 by CRC Press
Taylor & Francis Group
6000 Broken Sound Parkway NW, Suite 300
Boca Raton, FL 33487-2742

ISBN 13: 978-981-4877-28-2 (hbk)

Visit the Taylor & Francis Web site at
http://www.taylorandfrancis.com

and the CRC Press Web site at
http://www.crcpress.com

British Library Cataloguing-in-Publication Data
A catalogue record for this book is available from the British Library.

To my parents,

to my family,

to my friends.

Contents

Preface

The idea to write this book came to me after I submitted a review paper for *Advanced Materials* entitled "Ultra-thin silica films on metals: The long and winding road to understanding the atomic structure," which summarized up-to-date information about preparation and structure of thin silica films on metal substrates, including historical excursion into how the atomic structure of what I later took the liberty to name "silicatene" was established. For those who are working like me in the fields related to catalysis, it is not surprising that early works on this topic were primarily driven by attempts to understand surface chemistry of silica and related materials widely used in heterogeneous catalysis. Obviously, chemical reactions on a solid-state catalyst occur on its surface, which is compositionally and structurally very complex in catalysts. It was reasonably thought that making use of many powerful techniques developed in surface science would help to elucidate the atomic structures and the mechanisms of chemical reactions on such catalytic materials if applied to less complex and better defined planar model systems. Since the early days of this so-called "surface science" approach to catalysis, it has been a goal to circumvent the well-known problems related to poorly conducting materials (most of catalytically relevant oxides are insulators or wide-band gap semiconductors) via preparation of well-ordered thin oxide films on metal substrates. Another practical advantage is that such samples can easily be cooled down or heated up, to monitor temperature effects. In addition, the reactions do not suffer from diffusion limitations, typical for porous materials, often obscuring intrinsic reaction properties. On the other hand, the metal support underneath a film may alter the surface composition of the oxide under study, and even stabilize structures which would be unstable if unsupported. All these issues become of particular importance for *ultrathin* films. Although there is no firm definition of "how thick is ultrathin," I use this term for the films of a thickness below which physical and chemical properties deviate from those known for bulk samples and "thick" films. The thickness threshold depends on the material, but usually falls in the sub-nanometer range, which is comparable with the lattice parameters of an oxide. The recent progress in studies of ultrathin oxide films, in particular of transition metals, opened up a new playground for rational design of new-generation heterogeneous catalysts based on the thin-film technology.

While the chemistry-oriented studies of silica within the surface science approach (not only for catalysis, but with respect to dissolution, weathering, etc.) require preparation of relatively thick films hence representing bulk silica, the "materials science"-related disciplines and applications have a long-standing interest in the fabrication of ultrathin crystalline films for further miniaturization of metal-oxide-semiconductor transistors commonly using silica as the gate dielectric. In order to study fundamental properties of such materials, the preparation of *well-defined ultrathin silica* films is a necessity. Moreover, reducing the film thickness, one ultimately ends up with single-layer ("monolayer") films having a two-dimensional (2D) character by definition.

It is needless to say that single-layer materials have recently received enormous attention. The most famous one—graphene—triggered numerous investigations and search for other 2D materials. Silicon and germanium single layers, originally predicted on theoretical grounds, received experimental evidence, in fact, through their successful preparations on metal substrates and named "silicene" and "germanene," respectively. The studies were extended to tin and (very recently) iron layers called "stanene" and "ironene," respectively. Experimental studies carried out in our own laboratories in fruitful collaboration with prominent theoreticians allowed us to establish the atomic structure of the thinnest silica film that only consists of a single layer of corner-sharing SiO_4 tetrahedra. From a purely structural point of view, such a film can be considered as graphene composed of SiO_4 entities instead of the C atoms. Therefore, a single silicate layer deserves its place in the continuously growing family of the truly two-dimensional materials and can be named “silicatene” in analogy to other analogs of graphene, although the suffix “-ene” for these materials may be confusing.

Certainly, those readers who address the structure and properties of silica for the first time, despite it being one of the most abundant chemical compounds on our planet, immediately recognize how much is known about this material and how important silica is in our daily life. Of course, there exist many comprehensive reviews and books about silica related to geoscience, catalysis, and materials science. However, I found reasonable to provide here some basic information about silica compounds, in particular their surface structures, which will be helpful for a better understanding of atomic structure of silica films and interfaces. Note also that the silicate layer constitutes a structural motif in layered (or sheet) silicates existing in abundance in the nature. I found also instructive to introduce the readers to the general considerations applied to the thin film growth, in addition to the basics of experimental tools commonly used in surface science studies of thin films. Dealing with thin silica films, one cannot avoid mentioning the “native” silica films formed by thermal oxidation of Si and SiC substrates. Such information is useful to judge the pros and cons of ultrathin films fabricated on metals, which is the core of this book.

To the best of my knowledge, there are no books on this topic to compare with, and hence to choose one or another scientific issue to highlight. Some sections are written in a free way as for oral lectures, and some text looks more like a scientific manuscript, although I tried to balance. My ultimate goal was to ease newcomers to join this, in my view, fascinating and rapidly growing up but still underrated field of metal-supported ultrathin films of silica and related materials.

I am very grateful to all my colleagues, with whom I worked for many years in the Department of Chemical Physics at the Fritz-Haber Institute of the Max Planck Society, for their tremendous work in the laboratories, in particular to Anibal Boscoboinik, Emre Emmez, Sarp Kaya, Linfei Li, Daniel Löffler, Junling Lu, Qiushi Pan, Dario Stacchiola, Héloïse Tissot, John Uhlrich, Xin Yu, Bing Yang, and Jonas Weissenrieder. I want to thank Hajo Freund, who gave me the opportunity to head the group "Structure and Reactivity," who initiated the project and who was always in the heart of numerous debates and discussions. I was fortunate to collaborate with a great team of theoreticians, especially with Marek Sierka, Tanya Todorova, Radosław Włodarczyk, Itziar Goikoetxea, Frank Fischer, and Joachim Sauer (Humboldt Universität zu Berlin) and Philomena Schlexer, Gianfranco Pacchioni (Università degli Studi di Milano-Bicocca). Without their contribution, this work could never have been completed.

Finally, I want to thank my family, who supported me in this from the first day and provided me the best conditions for writing this book. This support is invaluable.

Shamil Shaikhutdinov
Berlin, January 2020

Chapter 1

Silica and Silicates

Silicon oxides dominate the Si compounds since the Si–O bond is by far the strongest among other Si-containing bonds (except the Si–F bonds which are, in fact, very seldom). Surprisingly, the nature of Si–O bonds in silicas, the discussion about which can be traced back to classical works of Linus Pauling in *The Nature of the Chemical Bond* in 1930s,[1] remains the subject of discussions. What one may certainly say about "the elusive Si–O bond"[2] is that the bond character is intermediate between covalent and ionic.

Silica (SiO_2) is known as a rock-forming mineral and has, therefore, been widely studied in geoscience. Invention of experimental tools based on x-ray diffraction (XRD) more than a century ago, which were immediately applied to natural silica samples, led scientists to a much better understanding of their crystal structures. However, the first studies revealed perhaps a more than expected, structural complexity and diversity of these compositionally very simple compounds. To date, about 50 polymorphs, both occurring in nature and synthesized in laboratories, have been identified. Comprehensive reviews of silica crystal structures can be found in several books like *Silica: Physical Behavior, Geochemistry and Materials Applications*, edited by P.J. Heaney, C.T. Prewitt, and G.V. Gibbs,[3] and *Framework Silicates: Silica Minerals, Feldspathoids and*

Introduction to Ultrathin Silica Films: Silicatene and Others
Shamil Shaikhutdinov

ISBN 978-981-4877-28-2 (Hardcover), 978-1-003-08862-2 (eBook)
www.jennystanford.com

the Zeolites (from the series Rock-forming Minerals) by W. A. Deer, R. A. Howie, W. S. Wise, and J. Zussman,[4] to name a few.

1.1 Silica Polymorphs

A phase diagram of silica is depicted in Fig. 1.1a. Although many crystalline silica polymorphs have been described in the literature, the basic building block is unique, and it is a SiO_4 tetrahedron (Fig. 1.1b). In the great majority of cases, exemplified by quartz, tridymite, and cristobalite, the SiO_4 tetrahedra are connected by sharing corner oxygen atoms such that every Si atom is surrounded by four oxygen atoms, and every O atom binds to two Si atoms. Stishovite is the only known form of silica that shares the rutile crystal structure, in which each Si atom is coordinated by six O atoms (Fig. 1.1c), and each oxygen atom is bonded to three Si atoms. Therefore, the crystal structure contains distorted SiO_6 octahedra sharing edges and corners. Interestingly, the above-mentioned stishovite as well as coesite have been synthesized even before they were discovered in nature, in experiments using very high

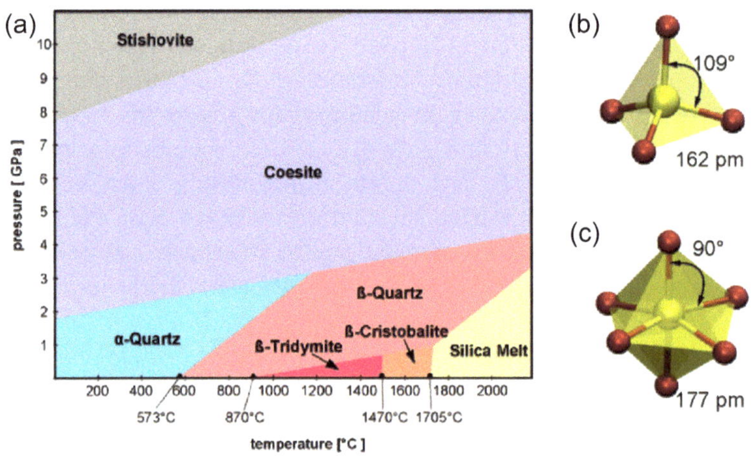

Figure 1.1 (a) Phase diagram of silica (from http://www.quartzpage.de/gen_mod.html). (b) Ideal SiO_4 and SiO_6 building blocks with averaged Si–O bond distances obtained in silica polymorphs indicated.

pressures. It is, therefore, believed that stishovite and coesite are formed on Earth only due to the high-pressure conditions generated, for example, by a meteorite impact. Actually, its presence in the samples is considered as an indicator of metamorphic reactions at extremely high pressures.

Friedrich Liebau, in his famous book *Structural Chemistry of Silicates*,[5] provided a large database of experimentally observed relationships between Si–O distances, oxygen coordination numbers, and Si–O–Si and O–Si–O bond angles both for tetrahedrally and octathedrally coordinated silicate frameworks. Please note that the term *silicate* defines a salt containing an anionic silicon (e.g., SiO_4^{4-} or SiF_6^{2-}). In this sense, pure silica SiO_2 belongs to silicates, although there is no counter-ion in the crystal structures. In tetra-oxosilicates, a mean Si–O bond length is 162 pm, on average, and varies between 157 and 172 pm. Accordingly, an O–Si–O angle varies between 98 and 122° (109°, on average), and Si–O–Si angle between 120 and 180°. Interestingly, the mean Si–O–Si angle in silica polymorphs (147°) considerably deviates from 140° in an ideal tetrahedron. This deviation is considered as indication of some strain while accommodating SiO_4 tetrahedra in the silica framework. In hexa-oxosilicates, the Si-O distances is certainly larger than in tetra-oxosilicates and vary between 170 and 184 pm (the mean value of 177 pm). Note, however, that the database for hexa-oxosilicates is limited as the SiO_6 coordination in silicas is rare.

At atmospheric pressure, the equilibrium sequence of phase transitions, prior to melting at 1705°C, includes α-quartz to β-quartz (through an incommensurate phase at 573°C), then to tridymite (870°C), and finally to β-cristobalite (1470°C). In addition, tridymite is famous for the high number of additional phases which have been reported to exist.[6] α-to-β phase transitions of quartz as well as of cristobalite and tridymite are all reversible and occur relatively fast without any bond breaking via displacive transformations. The transitions between different polymorphs are slow as they include the bond breaking and are referred to as reconstructive transformations.

In addition to experimental studies which allowed to determine crystal structures of silica more precisely, theoretical studies were employed in attempts to generalize the structural relationships

Table 1.1 Structural parameters of most common silica polymorphs

Silica polymorph	Crystal system (space group)	Lattice constants (in Å)
α-quartz	hexagonal $P3_121$ (left hand) $P3_221$ (right hand)	$a = 4.916$ $c = 5.405$
β-quartz	hexagonal $P6_222$ (left hand) $P6_222$ (right hand)	$a = 5.01$ $c = 5.47$
β-tridymite	hexagonal ($P6_3/mmc$)	$a = 5.03$ $c = 8.22$
α-cristobalite	tetragonal ($P4_12_12$)	$a = 4.957$ $c = 6.89$
β-cristobalite	cubic ($Fd3m$)	$a = 7.16$
coesite	monoclinic ($B2/b$)	$a = 7.143$ $b = 12.383$ $c = 7.143$
stishovite	tetragonal ($P4_2/mnm$)	$a = 4.177$ $c = 2.668$

between the silica polymorphs. It was thought that the ordered structures of SiO_2 pertaining to the phase diagram can, in principle, be deduced within the unified picture, specifically from a parent bulk-centered-cubic structure, possessing different fractional occupancies of SiO_2 molecules, through ordering and displacive mechanisms.[7] Since the SiO_4-based structures differ only in the way the tetrahedra are connected, the respective energy differences are rather small, that renders an accurate calculation of the phase equilibria quite difficult. Within the local density approximation used in the density functional theory (DFT) calculations, the atomic volumes of silica polymorphs as well as the bulk modulus can be predicted with a high accuracy. Another approach that uses generalized-gradient approximation leads to less accurate structural parameters. However, gradient corrections were found essential for achieving accurate structural energy differences between the tetrahedrally coordinated phases (major polymorphs) and the octahedrally coordinated phases (high-pressure polymorphs).[8]

Theoretical considerations could even predict new structures of silica. For example, annealing of α-quartz may produce edge-sharing SiO_6 octahedra and SiO_5 cuboids.[9] The diffraction pattern of the new phase agreed reasonably well with that of the intermediate crystalline phase found in silica during the amorphization. Theoretical predictions also suggested that the high-pressure transformation in cristobalite depends not only on the anisotropy of the stress field, but also on the rate of compression, which has experimentally been proven by XRD experiments with α-cristobalite.[10]

Another interesting aspect of structural studies of silica polymorphs refers to the role of disordering in these compounds culminated by silica glasses (see below). This issue has been raised for deeper understanding of the β-cristobalite structure, where Si atoms occupy sites that form a diamond lattice, and O atoms have average positions mid-way between Si atoms. Such a structure suggests the Si–O bond length of 1.55 Å and an Si–O–Si bond angle of 180°, which considerably deviate from typical values for silica and silicate structures (1.62 Å and 145°, respectively), also observed for low-temperature α-cristobalite. Crystal structure refinements of β-cristobalite based on this idealized structure revealed a wide distribution of displacements of the oxygen atoms in directions normal to the Si–Si vector, which suggests either large thermal fluctuations or some degree of disorder in the actual positions of the oxygen atoms. The latter has been considered within two, static disorder and dynamic disorder, models (for historical survey see Ref. [11]). In the static model, β-cristobalite is composed of domains of lower-symmetry phases by orientational disorder. Other model implies a disordered structure created by the dynamic superposition of all rigid unit modes and has received a solid support from the theoretical calculations,[12] predicting that β-cristobalite is composed of randomly distributed low-symmetry α-rings (as in α-cristobalite) and high-symmetry β-rings, constantly changing the ring conformation from one to the other.

It is worth mentioning that the recent years showed a strong increase in numbers of ab initio studies in modelling silica transformations. Certainly, increasing computer power makes it possible to take into accounts also dispersive interactions, which are much weaker than the dominating covalent interactions, but might have a considerable impact on the derived structures. The importance of dispersion corrections in the DFT calculations of silica polymorphs has been examined on several SiO_2 polymorphs having a SiO_4 building block (faujasite, ferrierite, α-cristobalite, α-quartz, and coesite).[13] It has turned out that dispersive interactions in SiO_2

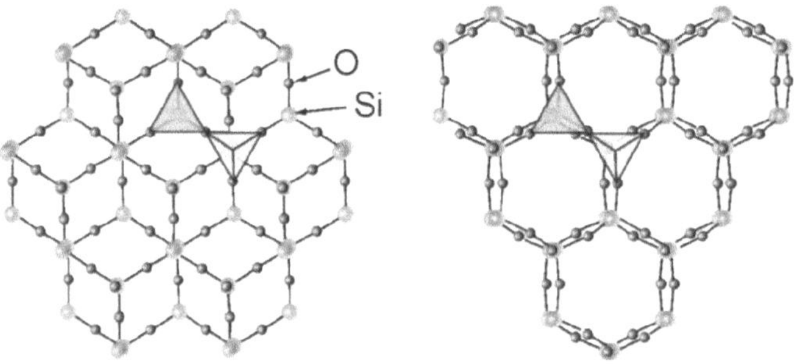

Figure 1.2 Schematic representation of β-cristobalite and β-tridymite structures along the c-axis. The upward-pointing SiO_4 tetrahedra from one layer share oxygen atoms with the downward-pointing tetrahedra in the layer above. In β-cristobalite, the layers are stacked in the ABCABC ... sequence, whereas the layers in β-tridymite are ABAB ... stacked with a mirror plane between the layers.

may not be as weak as previously thought and that more accurate methods including many-body interactions might be necessary in order to achieve a quantitative description of these systems.

In relation to structure of well-ordered silica films, which will be discussed in the following chapters, here we highlight β-tridymite and β-cristobalite, which both exhibit structures based on layers of corner-sharing SiO_4 tetrahedra, with neighboring tetrahedra pointing in opposite directions out of plane as shown in Fig. 1.2. The upward-pointing tetrahedra from one layer share apical oxygen atoms with the downward-pointing tetrahedra in the layer above, thus forming a hexagonal close-packed or face cubic-centered structures, respectively.[14] In β-cristobalite, the layers are stacked in the ABCABC ... sequence, whereas the layers in β-tridymite are ABAB ... stacked with a mirror plane between the layers.

1.2 Silica Glasses

While discussing numerous silica structures, one cannot avoid mentioning the so-called vitreous silica (the other names: fused quartz, fused silica), which is a textbook example of glass. There were diverse definitions of glass in the literature (see, for instance, review by Karmakar[15]). American Standards in 1945 defined glass "as inorganic product of fusion which has cooled to a rigid condition without crystallizing." In a more general sense, the glass stands for every solid that possesses a non-crystalline structure and exhibits a so-called glass transition. The latter, in turn, involves a transformation from a hard and relatively brittle state into a viscous state at increasing temperature. Glass transition occurs always at temperatures below the melting point, which in case of pure silica are about 1200°C and 1700°C, respectively.

Vitreous silica is one of compositionally the simplest forms of glass, and has been subject of interest since early days of solid state physics. Just to mention, the very first international meeting dedicated specifically to the constitution of glass was organized in May 1925. The history of theory of glasses can be found in the nice review by Adrian C. Wright (one of the key persons in the development of pulsed neutron diffraction for structural studies of glasses), who wrote in introduction that although vitreous silica has been extensively investigated, "the greatest surprise, therefore, is just how little is really known, as opposed to merely postulated, about its structure."[16]

One of the very first theories of vitreous silica considered this phase to be an aggregate of small crystallites a few nanometers in size which are connected with each other in a random way. This model is generally attributed to the paper "The Polymorphism and Annealing of Glass" by Alexander A. Lebedev in 1921,[17] who proposed the presence of "minute quartz crystals" in order to explain sharp changes in the physical properties of silica glasses in the temperature region close to the α–β transition of quartz. The so-called crystallite hypothesis was later revisited and eventually rejected by William H. Zachariasen in his famous paper "The Atomic Arrangement in Glass" (1932),[18] which is traditionally cited as the origin of the so-called continuous random network (CRN) theory of glasses, even though (after A.C. Wright[19]) the term *random network* was only introduced a year later by Bertram E. Warren.[20] Basically, in the CRN model SiO_4 units are linked at the vertices by sharing a common or bridging O atom. Although the SiO_4 units themselves are thought to be effectively regular in shape, the mutual

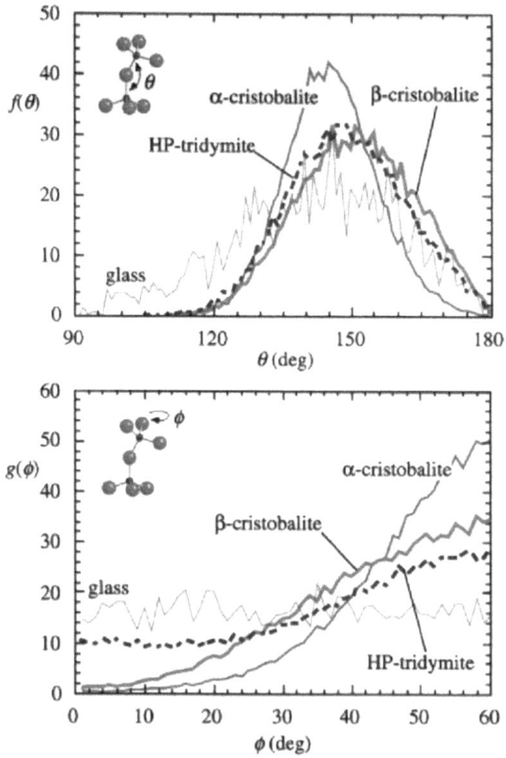

Figure 1.3 Bond angle distribution functions for the Si–O–Si bonds (top panel) and the torsional angles (bottom panel). The angles are defined in the top corners of each diagram. Reproduced with permission from Keen and Dove.[14]

orientation of adjacent units can vary widely throughout the glass in a manner which is apparently random in character, subject only to the restriction that the network be continuously connected.

Comparison of local structures of amorphous and crystalline phases by neutron scattering measurements[14] revealed that the structure of amorphous silica over the length scale 0–10 Å bears close similarities to β-cristobalite and β-tridymite, but is dissimilar to either phases of quartz and α-cristobalite, i.e., of low-temperature phases of silica (Fig. 1.3). The similarity can be explained in terms of the increased flexibility of the two dynamically-disordered

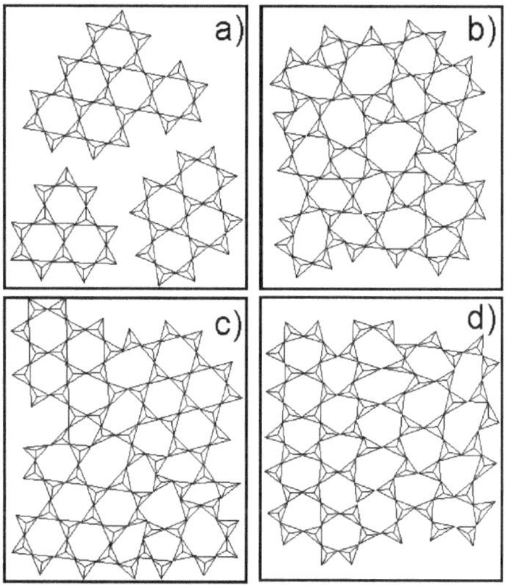

Figure 1.4 Two-dimensional representations of structural models for vitreous silica: (a) crystallite; (b) continuous random network; (c) ordered domains; (d) distorted network of 6-membered rings.

crystalline phases. More recent diffraction experiments combined with Monte Carlo simulations[21] refined a model of silica glass consisting of a fully connected corner-sharing network of SiO_4 tetrahedra.

The CRN model was the subject of several modifications, in particular for doped silica glasses.[22] Some models are depicted in Fig. 1.4. Also, there were attempts to find a compromise between the crystallite and CRN models, which only differ, in essence, by the size of the short-range ordering (15–20 Å in the crystallite theory versus 7–8 Å in the CRN theory).

In addition to experimental studies, theoretical calculations have been carried out to rationalize the atomic structure of vitreous silica. The absence of long-range order obviously limits usage of periodic slab calculations and causes the necessity to include a large number of atoms in the quantum chemistry calculations. In the early ab initio molecular dynamics (MD) simulations,[23] the glass was artificially obtained via quenching a liquid silica. This procedure resembles the experimental preparation of vitreous silica, except that the typical computational cooling rates are about ten orders of magnitude faster than experimental ones. In agreement with experiment, the MD simulations reveal structures of vitreous silica in which almost all Si atoms are bonded to four O atoms and almost all O atoms are bonded to two Si atoms, without long-range ordering. An alternative approach to generate atomistic models of vitreous silica containing up to 300,000 atoms in the network was to use a Monte Carlo scheme with an artificial dynamics consisting of bond transpositions.[24] Basically, all these models showed a good agreement with experimental results, in particular of the electronic density of states.

1.3 Sheet Silicates

As mentioned above, silica belongs to the large family of silicates. Again, for structural and chemical classification of silicates one can refer to the famous book of F. Liebau,[5] although a systematic nomenclature applied to all chemical substances within the IUPAC (International Union of Pure and Applied Chemistry) regulations is missing and remains quite confusing since many minerals are often designated solely by inventor's names, color, morphology, and their combinations. In silicates, Si is coordinated almost exclusively to four oxygen ions and may even exist as isolated SiO_4^{4-} centres, like in olivine with a formula $(Fe, Mg)_2SiO_4$. In a great majority of cases, however, silicates are formed by corner-sharing SiO_4 tetrahedra. Commonly the silicate anions are chains (pyroxenes), double chains (amphiboles), sheets (phyllosilicates), and three-dimensional frameworks (which are all aluminosilicates, except the quartz and its polymorphs, see the previous section). In this sense, silica belongs to the subclass "framework silicates", i.e., forming three-dimensional network of SiO_n polyhedra.

Now, we briefly introduce sheet silicates because of their layered structure which, as you will see in the following Chapters, bears certain similarity to our ultrathin well-ordered silica films grown on metals. The phyllosilicates group is considerably large and includes the serpentine, chlorite, mica, and the clay minerals. In such minerals, the SiO_4 tetrahedra form a so-called tetrahedral

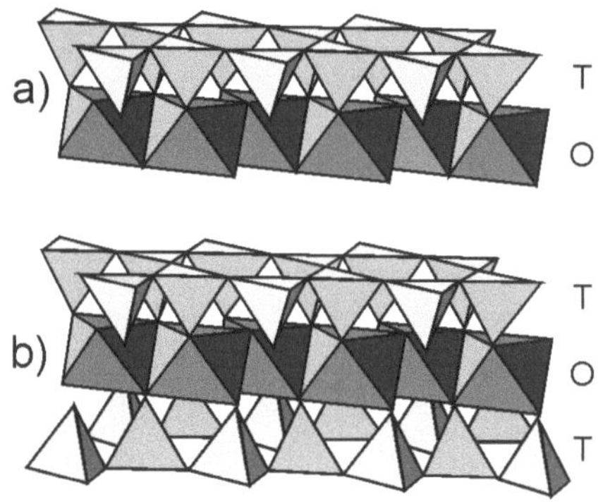

Figure 1.5 Stacking of tetrahedral (T) and octahedral (O) layers in micas and clay minerals. Ideal kaolinite-like (a), and mica-like (b) sheets. Adapted with permission from Liebau.[5]

(hence, denoted T-) layer composed of 6-membered silica rings, with all tetrahedra pointing towards the "octahedral" (O–) layer (see Fig. 1.5a). The latter can be of aluminum hydroxide $Al(OH)_3$ (gibbsite) or magnesium hydroxide $Mg(OH)_3$ (brucite) layer with Al^{3+} and Mg^{2+} ions in octahedral coordination. The T and O layers can be linked together by sharing the apical oxygen ions of the T-layer with the one in the O-layer. In these structures, Mg^{2+} ions occupy all octahedral sites, while Al^{3+} ions only 2 out of every 3, thus giving rise to two groups of sheet silicates: The tri-octahedral silicate, where each O or OH ion is surrounded by three divalent cations and the di-octahedral silicate, where each O or OH ion is surrounded by two trivalent cations. Each T-O sheet is bonded to the top (or bottom) of another T–O sheet by hydrogen bonds. This is the basic structure of industrially important kaolinite $Al_2Si_2O_5(OH)_4$ that belongs to so-called 1:1 clay minerals (one T-layer binds to one O-layer). In a similar way, T–O–T sheets can be formed (Fig. 1.5b) which are weakly connected to each other by van der Waals bonds. The classical example of the 2:1 clay minerals is the talk $Mg_3Si_4O_{10}(OH)_2$, probably best known for its low hardness. Depending on the cation sites occupation in the O-layer and the Al–Si substitution pattern within the T-layer all resulting in the charge imbalance, the T–O–T sheets can be connected by a charge compensating layer of mono- (Na^+, K^+) or divalent (Ca^{2+}) cations in between, which is common characteristics of micas, with the muscovite $KAl_2(AlSi_3O_{10})(F,OH)_2$ being as a best-known example.

In fact, there is a lattice mismatch between the T- and O-layers, which is most clearly manifested itself in the serpentine group of phyllosilicates through their bent crystal structures thus resulting in tubular and fibrous habits. Typical example is the chrysotile (well known as "white asbestos") $Mg_3Si_2O_5(OH)_4$. Solely geometrical arguments suggest that the substitution of cations in one of the layers or in both layers will affect the corresponding lattice constants (because of the different cations radii), thus giving the possibility for different counter ions to be incorporated into the structure. Any strain caused by lattice misfit between two layers may, in principle, be released via several reconstructions such as tetrahedron rotation and tilting and even tetrahedron inversion with concomitant splitting of the O-layer in 2:1 silicates.[5] All these and other factors may explain a large diversity of the clay minerals.

Due to their complex structure and composition, phyllosilicates have only recently become accessible to theoretical calculations. As a consequence, they are studied to a considerably lesser extent as compared to silica polymorphs. Quantum chemical, statistical-mechanical, electrostatic, and crystal chemical studies have been performed in attempts to explore the structure and bonding within and between phyllosilicate layers, the effects of cation substitution on bonding and charge distribution, the structure of water, and the distribution of cations in the interlayer.[25]

To end this section, we mention that the silicate layer as a structural motif can also be observed in other groups of silicates. Solely to illustrate this, Fig. 1.6a depicts the crystal structure of a synthetic silicate $NaKSi_2O_5$. In the basal plane, the structure forms "rings" consisting of *N* tetrahedra, which are commonly referred to as N-membered silica rings, in this case 4-, 6- and 8-membered rings. The stacking of the silicate layers results in a three-dimensional structure in which the alkali cations (Na^+, K^+) reside in the voids between neighboring sheets to compensate the charge localized at non-bridging O atoms.

In principle, the apical oxygen atoms of one T-layer may link to those in another T-layer, thus forming a T-T sheet having a mirror plane. Such structure is common for hexacelsian ($BaAl_2Si_2O_8$) type

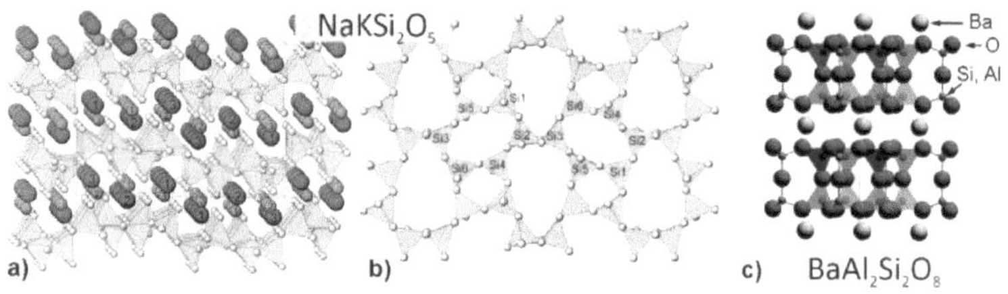

Figure 1.6 (a) Side view of a single layer silicate $NaKSi_2O_5$. Small medium grey spheres represent Na ions; large dark grey spheres correspond to K ions; the numbers differentiate various crystallographic positions. (b) Projection of a silicate sheet perpendicular to the basal plane. Adapted with permission from Rakić and Kahlenberg.[26] (c) Schematic structure of hexacelsian ($BaAl_2Si_2O_8$) consisting of two silicate layers.

phases of feldspars. In these compounds, half of the Si atoms are randomly substituted by Al, and the remaining charge imbalance is compensated by cations between the sheets as shown in Fig. 1.6c.

1.4 Framework Aluminosilicates: Zeolites

Aluminosilicates are the most abundant minerals in the Earth's crust. Mostly represented by feldspar they are constructed by connecting SiO_4 and AlO_4 tetrahedra. The feldspars are remarkably stable, showing the Si(Al)-O-Al(Si) angle close to 140° as an unstrained, ideal tetrahedron. Although these minerals show well-defined compositional stoichiometry, the Al and Si spatial distribution is believed to be random and homogeneous, on average. The conclusion basically follows the so called Loewenstein's rule stating that formation of two four-coordinated Al atoms via Al–O–Al bridges is forbidden.[27] As in the phyllosilicates, the charge imbalance caused by Al^{3+} substitution of Si^{4+} in tetrahedra is compensated by counter ions (e.g., of alkali metals, from Na^+ to Rb^+), which can be accommodated in cavities present in the frame.

In the aluminosilicate family, zeolites seem to be the most famous group thanks to their various technological applications in catalysis, gas separation, and ion exchange, which are based primarily on their microporous (better to say, nanoporous) structure. In zeolites, corner-sharing Si(Al)O_4 tetrahedra, which constitute primary

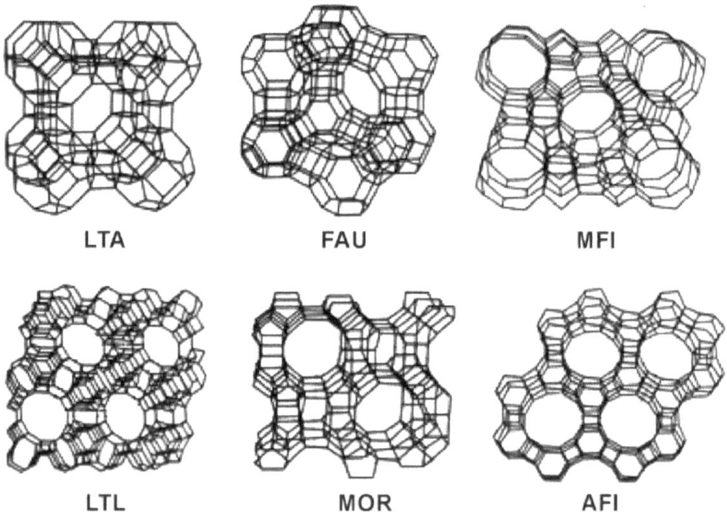

Figure 1.7 Some zeolite framework structures and their acronyms.

building units, form polygons as secondary building units such that the whole framework can be made of only one type of unit repeating in space. Some are depicted in Fig. 1.7. To date, more than 200 zeolite frameworks are known, and over 40 naturally occurring zeolite frameworks have been identified.[28] The number of framework types keeps growing, and the possibilities seem endless with millions of hypothetical structures that have been proposed based on reasonable geometrical constraints.[29]

In contrast to natural feldspars, zeolites may have very low Al:Si ratios. For example, in the synthetic Si-rich ZSM (Zeolite Socony Mobil) zeolites, this value can be as low as 10^{-4}. The name *silicalite* has been introduced in order to highlight the fact that this is practically "all-Si" zeolite in nature.[30] Although the lack of substitutional Al leads to catalytic or ion exchange activity much lower as compared to the regular zeolites, silicalites were found to selectively absorb small organic molecules and behave as a molecular sieve.

Commonly, the distribution of Al atoms in zeolites is rationalized on the basis of the above-mentioned Loewenstein's rule[27] as well as of Dempsey's rule,[31] stating that Al ions tend to locate as far as possible from each other in the framework. It is fair to say, however, that the Al distribution in zeolites remains poorly studied. Since cross-sections of x-ray scattering for Si and Al are very similar, and due to experimental difficulties in obtaining single crystals of zeolites, especially with high Al concentrations, it remains challenging to localize Al atoms by conventional diffraction methods. Synthesis of large crystals of zeolites is essential for refining structures and elucidating the intrinsic properties of these materials. To date, even the "giant" crystals of zeolites falls in the size range below 1 mm.[32] In addition, synthesis conditions may affect the Al distribution. For example, the occupation distribution of Al atoms over the 12 possible tetrahedral sites in the ZSM-5 structure, determined on the basis of Al^{27} nuclear magnetic resonance studies, was found to be non-random and dependent on the Si/Al ratio.[33] Therefore, the substitution of Al atoms in the framework during the crystallization process may be driven kinetically rather than thermodynamically. Also, some theoretical calculations pointed to possible violation of Dempsey's rule.[34] In principle, the equilibrium Al distribution is determined by two factors: The energetics of the Al/Si substitution at an individual site, which tends to populate particular T-sites, and the Al–Al interaction, which at this Si/Al maximizes Al–Al distances. However, according to the recent calculation for ZSM-5,[35] the relationship between the interaction energy and the distance between the two Al atoms deviates from that of expected for merely charge repulsion interaction.

1.5 Electronic Structures of Silica

The continuous interest in the electronic structure of silica is primarily driven by using silica as a dielectric layer in electronic devices. Another related topic deals with optical properties of fused silica which are important for optoelectronics and fiber optic technologies. In early theoretical calculations, the electronic properties of silica were studied within tight-binding theory[36] and Hartree–Fock approximation.[37] The Si–O bond is largely formed by Si $3sp^3$ bonding orbitals and O 2p orbitals. The top of the valence

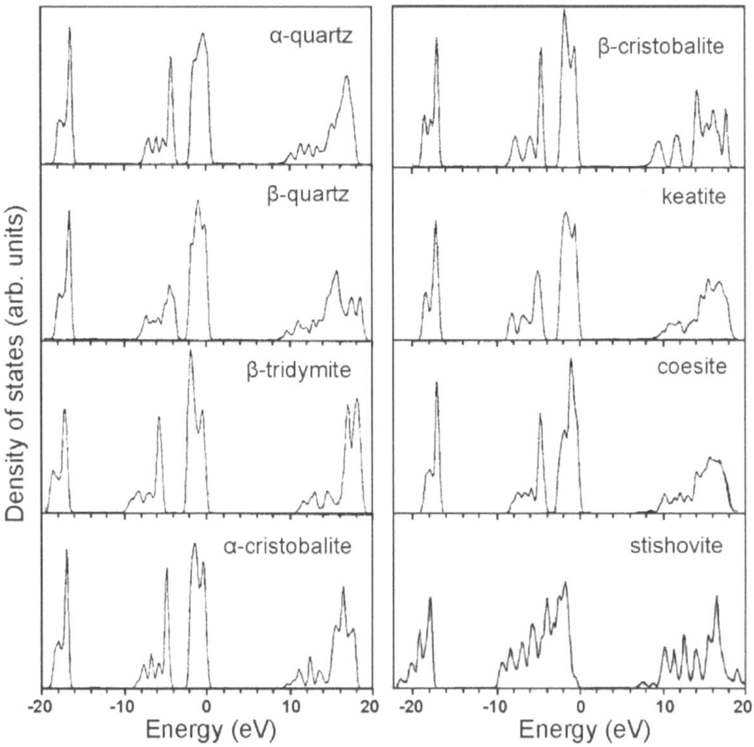

Figure 1.8 Calculated total density of states of several silica polymorphs. Adapted with permission from Li and Ching.[41]

band (VB) in SiO_2 is primarily determined by a narrow band of the O $2p_\pi$ non-bonding orbitals. It has later been proven that also the Si 3d orbitals contribute to the formation of the highest valence states.[38–40]

Figure 1.8 depicts electronic density of states calculated for several crystalline SiO_2 polymorphs using a linear combination of atomic orbitals. Not surprisingly, they look quite similar. However, the band structure of stishovite having edge sharing SiO_4 tetrahedra in the crystal structure considerably deviates from those of tetrahedrally coordinated silicas (i.e., only of corner sharing tetrahedra) and shows more covalency in the bonding character. The calculations revealed a linear correlation between the band gap and the Si–O bond length: The smaller the bond length, the larger the band gap.

To a lesser extent, the gap correlates with the minimum in Si–O–Si angle distribution used instead of the average value. Therefore, the electronic structure of amorphous silica, which actually resembles that of coesite, depends more on the distribution of bond length rather than of bond angle. For α-quartz, for which solid experimental data are mostly available in the literature, the calculated optical absorption and the electron-energy-loss function were found to be in good agreement with existing experimental data.

It should be mentioned, however, that some discrepancy still exists between the theoretically derived and experimentally observed band structures.[42] In particular, the valence band obtained by x-ray absorption and photoelectron spectroscopies did not manifest a gap that separates the top valence band from a more wider bottom valence band constituted from Si 3s,p and O 2p orbitals. A possible explanation originates from many-body electronic interactions usually neglected in the calculations.

To simulate vitreous silica (amorphous, a-SiO_2), a large model has been computed containing 432 SiO_2 units (1296 atoms) and no broken bonds or over- (under-) coordinated Si or O atoms, as shown in Fig. 1.9a.[40] The DFT-optimized structure revealed very small bond length and bond angle distortions. Partial density of states calculated for this model showed that the highest valence band originates from O 2*p* non-bonding orbitals. The lower portion of the upper VB is the mixture of O 2*p* and Si 3*s* and 3*p* orbitals showing the strong Si–O bonding. As in the case of crystalline silica polymorphs, the Si-3*d* orbital participates significantly in the upper portion of the VB in vitreous SiO_2. The calculated electronic structure and optical properties are in fair agreement with experimental ones measured by ultraviolet spectroscopy (Fig. 1.9b).

Besides the dopant or impurity related centers the optical properties of silica are strongly affected by structural imperfections.[43] Even at low concentrations, light transmission properties in optical fibers may be sensitive to the defects because of the large optical path length. The issue is, however, not trivial since definition of defects in amorphous solids is pretty shaky because of the statistical fluctuations which are inherent to the structure, whereas the concept of a defect is easy to define for crystalline materials.

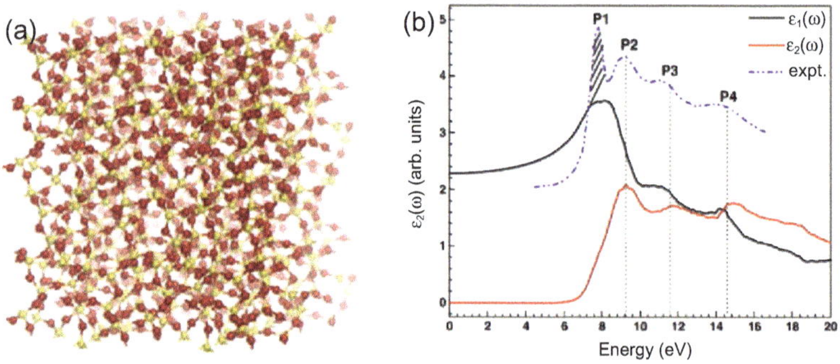

Figure 1.9 (a) Schematics of the model for amorphous SiO_2 (red and yellow circles are for O and Si atoms, respectively). (b) Calculated real and imaginary parts of the dielectric function for a-SiO_2 glass model in comparison with the experimental data. Reprinted with permission from Li and Ching.[40]

The defects in silica glass are usually classified in terms of chemical disorder, such as oxygen-deficient centers, and of physical disorder, such as bond-angle deviations and various kinds of ring structures. Their structural identification is usually performed by absorption and emission spectroscopy in the ultraviolet and visible light region as well as by electron paramagnetic and nuclear magnetic resonance spectroscopies. There are some solid evidences indicating that heavily strained bonds, for example, in three-membered ring structures, affect the fundamental absorption edge.[44] Although published twenty years ago, the book *Defects in SiO_2 and Related Dielectrics: Science and Technology* (edited by G. Pacchioni, L. Skuja, and D. L. Griscom)[45] can still serve as a guide in modern studies of defects in silica.

1.6 Surface Structures

Understanding the surface structures is a key prerequisite for elucidating the mechanisms of chemical reactions on silica and finding structure-reactivity relationships at the molecular level. However, even the first step in any experimental study targeted on this objective, namely the preparation of a non-contaminated, well-defined silica surface, is quite a difficult task in case of silica polymorphs. Another experimental hurdle relates to insulating properties of all silica-based compound that render application of surface-sensitive techniques very difficult (see more in Chapter 2).

Like in early days of crystallography, when the just-discovered x-ray diffraction was almost immediately applied to silica crystals, one of the first "surface science" studies on silica were also performed with diffraction-based technique, namely low-energy electron diffraction (LEED).[46] In that time, LEED was the main tool for characterization of crystal surfaces. It was not surprising that initial LEED patterns of the quartz single crystal surfaces (prepared by etching in HF) all gradually disappeared in time, in the minute's scale, because of sample charging. Surprising was, however, the finding that once disappeared the original structure could only be restored after new etching.

Further studies employing reflection high energy electron diffraction (RHEED)[47,48] revealed irreversible transformations upon reaction with oxygen and argon at elevated temperatures often resulting in surface reconstructions such as (2×2) and (1×4) for (10-11) and (10-10) surfaces, respectively. The reconstructions was revisited by LEED about ten years later.[49] It was found that the α-quartz (01-10) surface, in addition examined by x-ray absorption near edge spectroscopy (XANES) and the electron energy loss spectroscopy (EELS), may exhibit (3×1) and (1×3) structures. The latter was associated either with the bulk α–β transition phase (normally occurring at 573°C; see Fig. 1.1) or the peculiar "surface" phase only present at the surface, not in the silica bulk. Annealing in air at 900°C led to the surface amorphization. Interestingly, these three structures showed different surface electronic properties measured by EELS. The reconstructed surface showed a larger surface band gap than the initial (1×1) surface (9.5 vs. 8.6 eV). The amorphous surface showed the 8.8 eV band gap, i.e., intermediate, although containing two electronic levels associated with surface oxygen vacancies. A more complex reconstruction was observed on the α-quartz(0001) surface upon heating in air above 600°C.[50] Complex LEED pattern, identified as $(\sqrt{84}\times\sqrt{84})R11^\circ$ structure, was associated with the α–β transition phase of quartz, i.e., similarly to the above-mentioned case of the α-quartz (01-10) surface.

Further experiments revealed two electronic transitions at 5.1 and 7.2 eV, which were assigned to neutral oxygen vacancy type defects formed at the surface.[51]

With development of scanning probe microscopies in the early 1980's which revolutionized the field of surface science by achieving spatial resolution down to sub-Ångström level, it was near at hand to image surface structures of silica. However, due to insulating properties of silica, direct imaging of single crystal samples by scanning tunneling microscopy (STM) was thought to be impossible. Therefore, fully unexpected was a short communication[52] reporting STM images of a freshly-cleaved small single crystal of silicalite-1 (Si-rich zeolite ZSM-5). Since STM imaging of this sample was only possible in ambient conditions, not in vacuum, the authors suggested that surface conductivity was provided by a layer of adsorbed water. On the other hand, the tunneling conditions applied for imaging are only used for highly conducting, metallic-like surfaces. This mystery is perhaps the reason for the work is almost non-cited in the literature.

At variance to STM, atomic force microscopy (AFM) does not suffer from electric conductivity constraints. To the best of our knowledge, the first AFM studies of silica were carried out not on silica polymorphs, which are structurally and compositionally more simple, but on easily cleaved (means, naturally clean) mica samples, with the aim to study low friction properties mica is famous for.[53] AFM images recorded in the so called "frictional" mode revealed the periodicity of the hexagonal layer of SiO_4 units that forms the cleavage plane of mica. AFM studies of silicate crystals were then extended towards adsorption of molecules on a zeolite, e.g., clinoptilolite $(Na,K,Ca)_{2-3}Al_3(Al,Si)_2Si_{13}O_{36}\cdot12H_2O$, cut along the (010) plane.[54] In particular, AFM images obtained in various media revealed that molecules could be bound to the surface in different ways: Neutral molecules of *tert*-butanol formed an ordered array, whereas *tert*-butyl ammonium ions formed clusters. These promising results triggered further AFM investigations of many other natural zeolite crystals, such as scolecite, stilbite, faujasite, heulandite, and mordenite, to name a few.[55–58] In essence, the obtained AFM images were in good agreement with models derived from their bulk crystal structures, although really atomic resolution

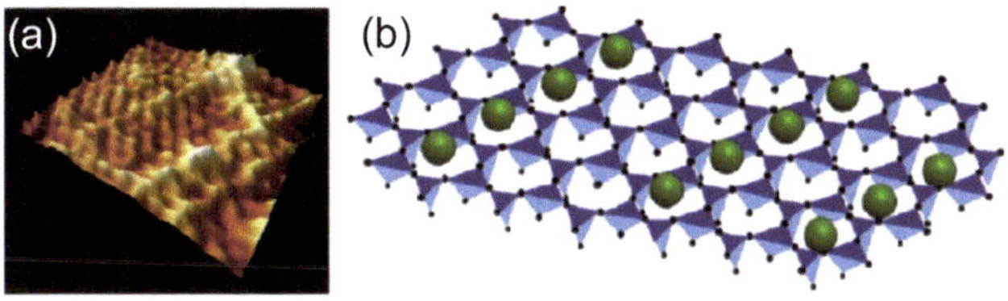

Figure 1.10 AFM image (in 3D view) and schematic model of the montmorillonite surface obtained upon Na–Cs ion exchange. The array of protrusions represents the adsorbed Cs^+ ions (green balls in *b*). Adapted with permission from Araki.[60]

could not be achieved. AFM was also invoked to study feldspars, in this case albite (010) surface $Na_{0.95}Ca_{0.05}Al_{1.05}Si_{2.95}O_8$.[58] Again, the spatial resolution was not sufficient to draw solid conclusions on the atomic structure.

AFM-derived surface morphologies of clay minerals are exemplified by montmorillonite $(Na,Ca)_{0.33}(Al,Mg)_2(Si_4O_{10})(OH)_2{\cdot}nH_2O$ and illite $(K,H_3O)(Al,Mg,Fe)_2(Si,Al)_4O_{10})(OH)_2{\cdot}nH_2O$.[59] High-resolution AFM images revealed a hexagonal array of 6-membered rings expected for the basal planes of (2:1)-type phyllosilicates. The measured periodicity of 5.1 ($\pm$0.3) Å agreed well with the ideal nearest-neighbor spacing of 5.4 Å for hexagonal rings derived from x-ray diffraction data.

Instrumental development of AFM microscopes over the last decades improved significantly both the spatial resolution and understanding of the principles underlying the imaging. For example, a very recent frequency modulation non-contact AFM study in solution allowed to localize ion positions on montmorillonite (sheet silicate) surface upon ion exchange (see Fig. 1.10).[60] In addition, some differences of the amount and the distribution of adsorbed cation between on montmorillonite surface and on muscovite (mica) surface were demonstrated. Last thing to mention is that mica is often used as an imaging standard for calibration of commercial AFM microscopes.

It appears that surface morphology of pure silica polymorphs has been studied for the first time with AFM by Gratz et al.[61] in order to visualize silica dissolution processes. AFM images of the (10-10) and (10-11) surfaces of natural Brazilian quartz revealed

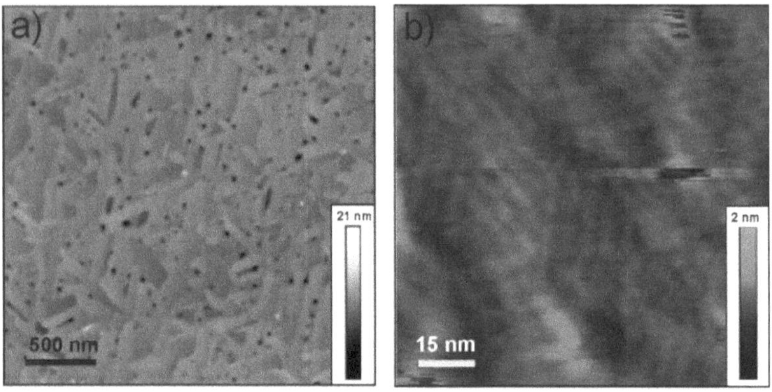

Figure 1.11 AFM images of the α-quartz (0001) surface. (a) Large, atomically flat terraces with an average width of about 200 nm are separated by steps with an average height of 1.9 nm. The terraces are oriented with angles of 60° relative to each other. Etched pits with depths of 2–5 nm are also seen. Adapted with permission from Steurer et al.[62] (b) Three domains of "ribbons" rotated 120° relative to each other. The periodicity within the ribbon is about 5 nm. Adapted with permission from Eder et al.[63]

faceted pits upon treatment in 0.01 M KOH solution at elevated temperatures. In general, the results validated previously suggested "ledge-motion" models of both dissolution and growth of silica crystals.

AFM was employed to inspect the α-quartz (0001) samples (see Figure 1.11a) used for helium atom scattering measurements which revealed a complex diffraction pattern.[62] In addition to diffraction peaks corresponding to the periodicity of the bulk unit cell (4.90 $\pm$ 0.02 Å), weak (2×2) peaks and satellite peaks were observed which pointed out an incommensurate superstructure with a periodicity of about 56 Å. Tentatively, this has been assigned to strain relief. A more recent AFM study[63] showed, indeed, a surface reconstruction consisting of 56 Å wide ribbons, oriented 11° relative to the bulk unit cell (Fig. 1.11b). With the help of molecular dynamics simulations, the ribbons were assigned to a self-organized pattern of nano-Dauphine twins, which were initially considered for the explanation of the "old" LEED results.

Nonetheless, atomic resolution in AFM images of quartz has not yet been achieved. In fact, available information about *surface* structures of silicas at the atomic level remains scarce. This situation brought researchers to focus on computational studies, in particular of α-quartz representing the most stable polymorph at ambient conditions.

Cleaved quartz surfaces are unstable because of the formation of dangling Si-O bonds upon cleavage, and hence the surface must undergo certain reconstruction/relaxation. Several mechanisms of surface reconstruction were considered: (i) formation of two-membered rings from vicinal silyl ($Si^*–O_3$) and siloxy ($O^*–Si–O_3$) radicals; (ii) formation of a pair of tri-coordinated/uni-coordinated oxygen atoms; (iii) formation of three-membered rings; and (iv) transformation of silanone (Si=O) sites into siloxane (Si–O–Si) sites.[64]

The reconstructed quartz (0001) surface exhibits the lowest surface energy, and it is the most stable amongst "dry" surfaces. The stability originates from all the surface Si atoms relaxing to be fully four-coordinated, forming Si–O–Si bridges on the surface as depicted in Fig. 1.12.[65] Interestingly, the reconstruction showed densification of the two uppermost tetrahedral layers, resulting in 3-membered and 6-membered rings that actually do not exist in bulk. The electronic density of states for this surface, albeit

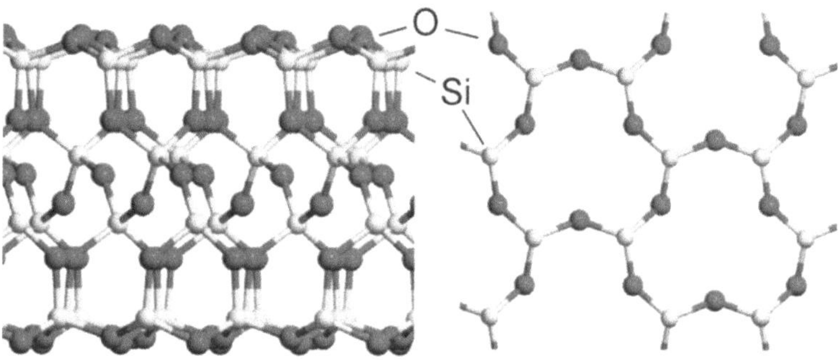

Figure 1.12 Relaxed α-quartz (0001) surface after energy minimization in side (on the left) and top (on the right) views. For clarity, the top view only shows outmost Si and O atoms. Adapted with permission from Goumans et al.[66]

very similar to the bulk one, showed disappearance of the gap between Si-O bonding and O 2p nonbonding states (see previous section), indicating a strong hybridization between them. Note that the "dense" reconstruction of the α-quartz (0001) surface distorts at least the top six layers of atoms. Consequently, the calculated surface energies critically depend on the depth of the reconstructed surface.[66] Apparently, hybrid functionals used in DFT calculations predict energies more accurately. Further structural refinement has recently been reported using Born–Oppenheimer molecular dynamics simulations and DFT.[67]

On the other quartz surfaces, however, the surface Si atoms are four- and three-coordinated, and many O atoms bind to only one Si atom. The resulted species having dangling-bond orbitals are unstable and react with ambient, in particular with water.[65,68] The hydroxylated α-quartz (0001) surface is, in fact, much more stable than the dry surface, that is in agreement with the well-known hydrophilicity of quartz surfaces. According to the DFT calculations for hydrated surfaces the energy increases in the row: (001) $<$ (110) $<$ (102) $<$ (111) $<$ (100) $<$ (112) $<$ (101).[64]

As far as other silica polymorphs are concerned, several DFT studies addressed surface structures of β-cristobalite with respect to their interaction with water.[69,70] Ab initio MD calculations of the dehydroxylation reactions on the β-cristobalite surface guided in the generation of a model of the dehydroxylated surface which features two-membered silica rings as shown in Fig. 1.13.[71] This structural motif was further used for modeling amorphous silica surfaces as well. In particular, such highly strained two-membered rings were suggested to explain spectral features appearing in infrared spectra of dehydroxylated silica glasses.

Overall, we have to admit that truly atomistic picture of silica (and silicate) surfaces largely stem from chemical intuition, albeit supported by theoretical calculations, rather than from solid experimental data. To some extent, such a situation is caused by the well-known difficulties of employing surface sensitive tools to insulating materials. In addition, the preparation of well-defined surfaces of silica compounds either existing in nature or synthesized in laboratories remains non-trivial.

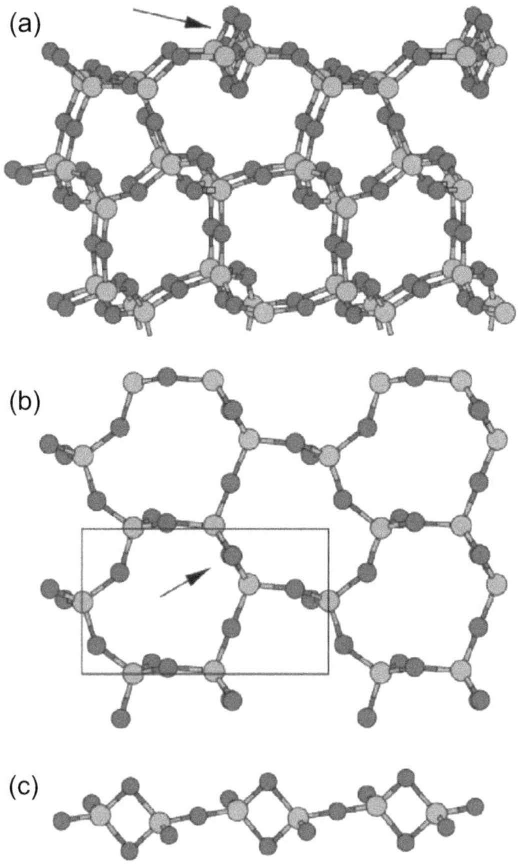

Figure 1.13 Optimized structure of the dehydroxylated (111) surface of β-cristobalite in top (a) and side (b) views. A single two-membered ring (indicated by arrows) is present in the unit cell shown as rectangle. The chain of two-membered rings along the (1–10) direction is schematically shown in panel (c). Adapted with permission from Ceresoli et al.[71]

References

1. Pauling, L., *The Nature of the Chemical Bond and the Structure of Molecules and Crystals: An Introduction to Modern Structural Chemistry*, 3rd ed. 1960, Cornell University Press.
2. Gibbs, G.V., J.W. Downs, and M.B. Boisen, The elusive SiO bond. *Reviews in Mineralogy and Geochemistry*, 1994. **29**(1):331–368.

3. Heaney, P.J., Prewitt, C.T., and Gibbs, G.V., eds. Silica: Physical Behavior, *Geochemistry and Materials Applications*, 1994, Cambridge University Press.
4. Deer, W.A., et al., *Rock-Forming Minerals: Framework Silicates: Silica Minerals, Feldspathoids and the Zeolites*. 2004, London: The Geological Society.
5. Liebau, F., *Structural Chemistry of Silicates: Structure, Bonding, and Classification*. 1985, Springer.
6. Pryde, A.K.A. and M.T. Dove, On the sequence of phase transitions in tridymite. *Physics and Chemistry of Minerals*, 1998. **26**(2):171–179.
7. Dmitriev, V.P., et al., Theory of reconstructive phase transitions between SiO_2 polymorphs. *Physical Review B*, 1998. **58**(18):11911–11921.
8. Demuth, T., et al., Polymorphism in silica studied in the local density and generalized-gradient approximations. *Journal of Physics: Condensed Matter*, 1999. **11**(19):3833.
9. Wentzcovitch, R.M., et al., A new phase and pressure induced amorphization in silica. *Physical Review Letters*, 1998. **80**(10):2149–2152.
10. Dera, P., et al., New insights into the high-pressure polymorphism of SiO_2 cristobalite. *Physics and Chemistry of Minerals*, 2011. **38**(7):517–529.
11. Cope, E. and M. Dove, Evaluation of domain models for β-cristobalite from the pair distribution function. *Journal of Physics: Condensed Matter*, 2010. **22**(12):125401.
12. Yuan, F. and L. Huang, Alpha-beta transformation and disorder in beta-cristobalite silica. *Physical Review B*, 2012. **85**(13):134114.
13. Hay, H., et al., Dispersion effects in SiO_2 polymorphs: an ab initio study. *Physical Review B*, 2015. **92**(14):144111.
14. Keen, D. and M. Dove, Local structures of amorphous and crystalline phases of silica, SiO_2, by neutron total scattering. *Journal of Physics: Condensed Matter*, 1999. **11**(47):9263.
15. Karmakar, B., Fundamental of glass and glass nanocomposites, in *Glass Nanocomposites. Synthesis, Properties and Applications,* B. Karmakar, K. Rademann, and A. Stepanov, eds. 2016, Elsevier: Oxford, p. 389.
16. Wright, A.C., Neutron scattering from vitreous silica. V. The structure of vitreous silica: What have we learned from 60 years of diffraction studies? *Journal of Non-Crystalline Solids*, 1994. **179**(0):84–115.
17. Lebedev, A.A., The polymorphism and annealing of glass. *Trudy Gos. Opt. Inst.*, 1921. **2**:1.

18. Zachariasen, W.H., The atomic arrangement in glass. *Journal of the American Chemical Society*, 1932. **54**(10):3841–3851.
19. Wright, A.C. and M.F. Thorpe, Eighty years of random networks. *Physica Status Solidi B*, 2013. **250**(5):931–936.
20. Warren, B.E., X-ray diffraction of vitreous silica. *Zeitschrift Fur Kristallographie*, 1933. **86**(5/6):349–358.
21. Tucker, M.G., et al., Refinement of the Si–O–Si bond angle distribution in vitreous silica. *Journal of Physics: Condensed Matter*, 2005. **17**(5):S67.
22. Greaves, G.N., Random network models, in *Defects and Disorder in Crystalline and Amorphous Solids*, C.R.A. Catlow, ed. 1994, Springer Netherlands, pp. 87–122.
23. Sarnthein, J., A. Pasquarello, and R. Car, Model of vitreous SiO_2 generated by an ab initio molecular-dynamics quench from the melt. *Physical Review B*, 1995. **52**(17):12690–12695.
24. Vink, R.L.C. and G.T. Barkema, Large well-relaxed models of vitreous silica, coordination numbers, and entropy. *Physical Review B*, 2003. **67**(24):245201.
25. Bleam, W.F., Atomic theories of phyllosilicates: quantum chemistry, statistical mechanics, electrostatic theory, and crystal chemistry. *Reviews of Geophysics*, 1993. **31**(1):51–73.
26. Rakić, S. and V. Kahlenberg, Single crystal structure investigation of twinned $NaKSi_2O_5$ — a novel single layer silicate. *Solid State Sciences*, 2001. **3**(6):659–667.
27. Loewenstein, W., The distribution of aluminum in the tetrahedra of silicates and aluminates. *American Mineralogist*, 1954. **39**(1–2):92–96.
28. Baerlocher, Ch., Meier, W.M. and Olson, D.H., *Atlas of Zeolite Framework Types*, 6th ed. 2007, Elsevier.
29. Foster, M.D. and M.M.J. Treacy. A database of hypothetical zeolite structures. Available from: http://www.hypotheticalzeolites.net.
30. Flanigen, E.M., et al., Silicalite, a new hydrophobic crystalline silica molecular sieve. *Nature*, 1978. **271**(5645):512–516.
31. Dempsey, E., Acid strength and aluminum site reactivity of Y zeolites. *Journal of Catalysis*, 1974. **33**(3):497–499.
32. Lethbridge, Z.A.D., et al., Methods for the synthesis of large crystals of silicate zeolites. *Microporous and Mesoporous Materials*, 2005. **79**(1):339–352.
33. Han, O.H., C.S. Kim, and S.B. Hong, Direct evidence for the nonrandom nature of Al substitution in zeolite ZSM-5: an investigation by ^{27}Al MAS

and MQ MAS NMR. *Angewandte Chemie - International Edition*, 2002. **41**(3):469–472.

34. Schroder, K.P. and J.C. Sauer, Preferred stability of Al-O-Si-O-Al linkages in high silica zeolite catalysts-theoretical predcitionss contrary to Dempsey's rule. *Journal of Physical Chemistry*, 1993. **97**(25):6579–6581.
35. Ruiz-Salvador, A.R., et al., Aluminium distribution in ZSM-5 revisited: the role of Al-Al interactions. *Journal of Solid State Chemistry*, 2013. **198**:330–336.
36. Pantelides, S.T. and W.A. Harrison, Electronic structure, spectra, and properties of 4:2-coordinated materials. I. Crystalline and amorphous SiO_2 and GeO_2. *Physical Review B*, 1976. **13**(6):2667–2691.
37. Schneider, P.M. and W.B. Fowler, Band structure and optical properties of silicon dioxide. *Physical Review Letters*, 1976. **36**(8):425–428.
38. Gnani, E., et al., Band-structure calculations of SiO_2 by means of Hartree-Fock and density-functional techniques. *IEEE Transactions on Electron Devices*, 2000. **47**(10):1795–1803.
39. Xu, Y.-n. and W.Y. Ching, Electronic and optical properties of all polymorphic forms of silicon dioxide. *Physical Review B*, 1991. **44**(20):11048–11059.
40. Li, N. and W.-Y. Ching, Structural, electronic and optical properties of a large random network model of amorphous SiO_2 glass. *Journal of Non-Crystalline Solids*, 2014. **383**:28–32.
41. Li, Y.P. and W.Y. Ching, Band structures of all polycrystalline forms of silicon dioxide. *Physical Review B*, 1985. **31**(4):2172–2179.
42. Nekrashevich, S.S. and V.A. Gritsenko, Electronic structure of silicon dioxide (a review). *Physics of the Solid State*, 2014. **56**(2):207–222.
43. Awazu, K. and H. Kawazoe, Strained Si–O–Si bonds in amorphous SiO_2 materials: a family member of active centers in radio, photo, and chemical responses. *Journal of Applied Physics*, 2003. **94**(10):6243–6262.
44. Tamura, T., et al., First-principles analysis of the optical properties of structural disorder in SiO_2 glass. *Physical Review B*, 2008. **77**(8):085207.
45. Pacchioni, G., Skuja, L. and Griscom, D. *Defects in SiO_2 and Related Dielectrics: Science and Technology*. NATO Science Series II, 2000, Springer.
46. Jánossy, I. and M. Menyhárd, LEED study of quartz crystals. *Surface Science*, 1971. **25**(3):647–649.
47. Sokoll, R. and B. Müller, Veränderung der oberflächenstrukturen von quarzeinkristallen infolge ihrer wechselwirkung mit einfachen gasen

bei hohen temperaturen (I) behandlung mit argon und sauerstoff. *Kristall und Technik*, 1979. **14**(12):1529–1538.

48. Sokoll, R., The use of a computer program for the simulation of RHEED diffraction diagrams in the investigation of structural changes on quartz monocrystal surfaces. *Surface Science*, 1982. **118**(1):165–179.
49. Bart, F., et al., (01-10) α-quartz surface: a LEED, XANES and ELS study. *Surface Science*, 1992. **274**(3):317–328.
50. Bart, F. and M. Gautier, A LEED study of the (0001) a-quartz surface reconstruction. *Surface Science*, 1994. **311**(1–2):L671–L676.
51. Bart, F., et al., Electronic structure of the (0001) and ($10\bar{1}0$) quartz surfaces and of their defects as observed by reflection electron energy loss spectroscopy (REELS). *Surface Science*, 1994. **306**(3):342–358.
52. Jansen, J.C., et al., STM image of silicalite-1 pore structure. *Zeolites*, 1991. **11**(4):306–307.
53. Erlandsson, R., et al., Atomic scale friction between the muscovite mica cleavage plane and a tungsten tip. *The Journal of Chemical Physics*, 1988. **89**(8):5190–5193.
54. Weisenhorn, A.L., et al., Imaging and manipulating molecules on a zeolite surface with an atomic force microscope. *Science*, 1990. **247**(4948):1330–1333.
55. Scandella, L., N. Kruse, and R. Prins, Imaging of zeolite surface structures by atomic force microscopy. *Surface Science*, 1993. **281**(3):L331–L334.
56. MacDougall, J.E., et al., Molecular resolution of zeolite surfaces as imaged by atomic force microscopy. *Zeolites*, 1991. **11**(5):426–427.
57. Yamamoto, S., et al., Surface structure of natural crystals of mordenite as imaged by atomic force microscopy. *Chemical Physics Letters*, 1996. **260**(1):208–214.
58. Hochella, M.F., et al., Atomic structure and morphology of the albite $\{010\}$ surface; an atomic-force microscope and electron diffraction study. *American Mineralogist*, 1990. **75**(7–8):723–730.
59. Hartman, H., et al., Molecular-scale imaging of clay mineral surfaces with the atomic microscope. *Clays and Clay Minerals*, 1990. **38**(4):337–342.
60. Araki, Y., et al., Localization of cesium on montmorillonite surface investigated by frequency modulation atomic force microscopy. *Surface Science*, 2017. **665**:32–36.
61. Gratz, A.J., S. Manne, and P.K. Hansma, Atomic force microscopy of atomic-scale ledges and etch pits formed during dissolution of quartz. *Science*, 1991. **251**(4999):1343–1346.

62. Steurer, W., et al., The structure of the α-quartz (0001) surface investigated using helium atom scattering and atomic force microscopy. *Surface Science*, 2007. **601**(18):4407–4411.
63. Eder, S.D., et al., A giant reconstruction of α-quartz (0001) interpreted as three domains of nano Dauphine twins. *Scientific Reports*, 2015. **5**:14545.
64. Murashov, V.V., Reconstruction of pristine and hydrolyzed quartz surfaces. *The Journal of Physical Chemistry B*, 2005. **109**(9):4144–4151.
65. de Leeuw, N.H., F.M. Higgins, and S.C. Parker, Modeling the surface structure and stability of α-quartz. *The Journal of Physical Chemistry B*, 1999. **103**(8):1270–1277.
66. Goumans, T.P.M., et al., Structure and stability of the (001) α-quartz surface. *Physical Chemistry Chemical Physics*, 2007. **9**(17):2146–2152.
67. Malyi, O.I., V.V. Kulish, and C. Persson, In search of new reconstructions of (001) alpha-quartz surface: a first principles study. *RSC Advances*, 2014. **4**(98):55599–55603.
68. Rignanese, G.M., J.C. Charlier, and X. Gonze, First-principles molecular-dynamics investigation of the hydration mechanisms of the (0001) α-quartz surface. *Physical Chemistry Chemical Physics*, 2004. **6**(8):1920–1925.
69. Rozanska, X., F. Delbecq, and P. Sautet, Reconstruction and stability of β-cristobalite 001, 101, and 111 surfaces during dehydroxylation. *Physical Chemistry Chemical Physics*, 2010. **12**(45):14930–14940.
70. Mahadevan, T.S. and S.H. Garofalini, Dissociative chemisorption of water onto silica surfaces and formation of hydronium ions. *The Journal of Physical Chemistry C*, 2008. **112**(5):1507–1515.
71. Ceresoli, D., et al., Two-membered silicon rings on the dehydroxylated surface of silica. *Physical Review Letters*, 2000. **84**(17):3887–3890.

Chapter 2

Preparation of Thin Oxide Films: Concepts and Toolkits

Atomistic pictures of silica surfaces are often based on theoretical considerations rather than solid experimental results. To some extent, this status quo is caused by difficulties to study electrically insulating materials with experimental techniques based on detecting charged species (e.g., electrons and ions) due to the well-known surface charging effects. In addition, the preparation of the clean surface of silica is not a trivial task. For example, electrochemical etching or other preparations using liquids or aggressive gases can easily contaminate the surface. Clean surfaces prepared by sputtering in ultrahigh vacuum (UHV) often suffer from the compositional non-stoichiometry. To circumvent these problems and to gain more insight into atomic structures and surface properties of silicas, in situ fabrication of a thin silica film on an electrically conducting substrate is a good alternative to the single crystals. Such an approach has been shown to be quite successful for transition metals oxides, and many of those are wide bandgap semiconductors.[1–6] A continuously growing body of studies in this field allows one to formulate general rules about how to prepare a thin film of a target material from scratch. What

Introduction to Ultrathin Silica Films: Silicatene and Others
Shamil Shaikhutdinov

ISBN 978-981-4877-28-2 (Hardcover), 978-1-003-08862-2 (eBook)
www.jennystanford.com

one needs to know first about the film prepared is its chemical composition, thickness, surface orientation, and degree of ordering. This information is an input necessary for a subsequent quest for precise atomic structure of the film and film/substrate interface. The latter plays an important, if not decisive, role in case of *ultrathin* films, such that the substrate may even govern the structure–property relationships. In addition, the structural motif in ultrathin films may deviate from those in the crystals. Knowing the atomic structure, one can explain and even tune functional characteristics of the films, for example, their electric, mechanical, or chemical properties the possible applications of these films are based on. It is fair to say that the quest often succeeds only by combining experimental results with theoretical calculations.

2.1 Experimental Tools for Thin-Film Characterization

It appears that almost every just-invented or relatively "old" but further advanced experimental technique has been employed for silica-based materials. Nowadays, the majority of "surface science" laboratories are equipped with a set of standard, commercially available analytic tools operating in UHV chambers. In the following, we will introduce basic principles of the commonly used techniques and the information they provide, particularly for thin films. Where appropriate, we will illustrate capability of the techniques by showing the results obtained on silica-relevant systems. Note that thickness of ultrathin films is comparable with the average depth of sampling in these techniques. Therefore, characterization of ultrathin films with surface-science tools gives information not only about the surface itself, but the entire film including the interface as well. The most widely used experimental techniques, which will be largely referred to with respect to the structure and properties of supported ultrathin silica films, are

- Electron spectroscopy, e.g., ultraviolet (and x-ray) photoelectron spectroscopy (UPS and XPS), and Auger electron spectroscopy (AES). The methods allow to determine the elemental surface composition, the oxidation states of the constituting elements, and also the valence band structure.

- Low-energy electron diffraction (LEED) and surface x-ray diffraction (SXRD). The methods are primarily used for structures and interfaces showing a long-range order, and provide information about surface symmetry and the unit cell.
- Vibrational spectroscopy, e.g., infrared reflection-absorption spectroscopy (IRAS) and high-resolution electron energy loss spectroscopy (HREELS). With these techniques, both the lattice vibrations (phonons) and the vibrations of adsorbed species can be investigated.
- Scanning tunneling microscopy (STM) and atomic force microscopy (AFM). These seem to be the only techniques allowing to look into surface morphology (topography) with atomic resolution. In addition, tunneling spectroscopy gives information about both the valence and conduction band structures.
- Ion scattering spectroscopy, e.g., low-energy ion surface scattering (LEISS), provides information on the elemental composition of the topmost layer.
- Very often, adsorption of gas molecules at surfaces is structure sensitive, i.e., depends on the crystallographic plane, its atomic structure, and presence of structural defects. In such cases, adsorption of so called "probe" molecules (which molecularly adsorb and desorb intact upon heating) also provides structural information exclusively about the surface. In addition to vibrational spectroscopies, adsorption can be monitored by temperature-programmed desorption (TPD). Planar systems do not suffer from diffusion limitations commonly existing for powder samples and porous materials. In addition, the presence of a metal support facilitates uniform heating of the films and precise control of the surface temperature.

The surface sensitivity of many above-mentioned methods, based on detection of electrons either emitted or scattered from surface, stems from the fact that the inelastic mean free path (IMFP), i.e., the average length over which the electron travels before losing energy (and hence information) in a scattering event, is in the range

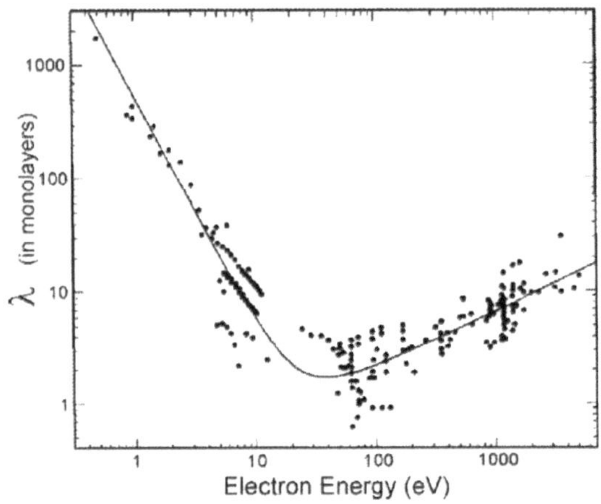

Figure 2.1 Universal curve showing IMFP of elements in atomic layers (λ) as a function of electron kinetic energy. Reproduced with permission from Seah and Dench.[7]

of 10–100 Å, thus only probing surface and sub-surface regions within the "escape" depth. The relation between the IMFP and the kinetic energy of electron follows the so-called universal curve, depicted in Fig. 2.1, which is usually associated with Seah and Dench,[7] who analyzed experimental results available at that time and found empirical fitting equations for the energy dependent electron escape depth in many solids. The curve shows a minimum at around 50 eV and reflects the energy dependence of different mechanisms for electron energy loss. The excitation of phonons and transitions between the electronic bands dominates in the low-energy range. At increasing energies, the coupling to collective modes (e.g., plasmons) becomes important. At high energies, the IMFP increases simply because faster electrons have less time to be scattered. As various scattering mechanisms can be involved, the escape depth depends also on the material under study. Typically, metals exhibit a smaller IMFP as compared to oxides due to additional electron scattering on free electrons. Although theoretical description of the universal curve remains the subject of debate,[8] the general trend definitely holds true, and as such it can be used for analysis of experimental results.

2.2 Low-Energy Electron Diffraction (LEED)

LEED is one of the mostly used techniques for studying crystalline and well-ordered surfaces and interfaces. The principles of electron diffraction are intimately connected to the wave-like nature of elementary particles, first introduced by Louis de Broglie in 1924 and then confirmed experimentally by Davisson and Germer in 1927. However, LEED only became an experimental tool in the 1960s thanks to substantial progress in electron detection methods and development of vacuum systems capable to prepare clean surfaces. In LEED, the incident e-beam is considered as a plane wave with a wavelength depending on the electron energy through the de Broglie wave equation. Accordingly, the 100 eV electrons have the de Broglie wavelength about 1 Å, i.e., comparable with interatomic distances in solids. The method is particularly surface sensitive because of using electron energies close to the minimum in the escape depth (see Fig. 2.1). As all diffraction-based techniques, LEED requires long-range ordering at surface, at least the presence of ordered domains larger than the coherence length of the probing electrons, in the range of 5–10 nm, in order to be detected with conventional apparatus.

If obtained, the LEED pattern immediately tells us about the symmetry and the unit cell of the surface under study. The azimuthal orientation and the size of the unit cell may deviate from those derived by truncation of the crystal lattice at the surface, therefore indicating surface reconstruction. As an example, Fig. 2.2 shows two LEED patterns observed on α-quartz (0001) surface illustrating that the surface underwent complex reconstruction upon high-temperature annealing in air (but measured in vacuum, of course). From these patterns, the relationship between unit vectors on reconstructed (A,B) and pristine (a,b) surfaces can be described either as $\left(\frac{A}{B}\right) = \begin{pmatrix} 10 & 2 \\ -2 & 8 \end{pmatrix} \left(\frac{a}{b}\right)$ in matrix notation or $SiO_2(0001)-(\sqrt{84}\times\sqrt{84})R11^\circ$ in Wood notation, meaning that the ratio between absolute values $|A|$ and $|a|$ is equal to $\sqrt{84}$, and that the unit cells are rotated by 11° with respect to each other.

In principle, the surface under study will exhibit the unreconstructed, i.e., (1×1) pattern, even though the surface atoms were

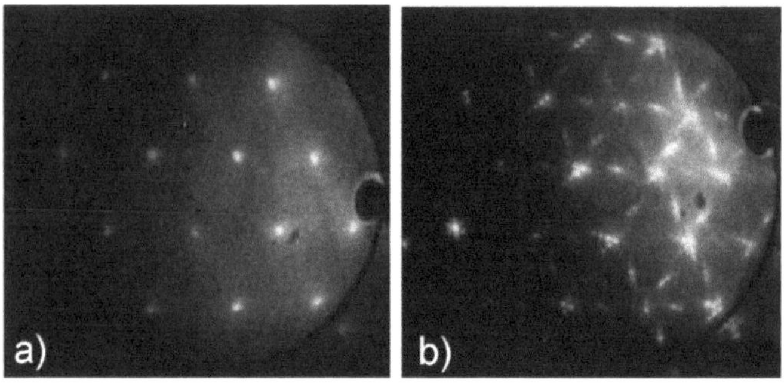

Figure 2.2 LEED patterns observed on a HF-etched-quartz (0001) sample upon annealing in air at 350°C (a) (primary energy is 59 eV) and at 900°C (at 73 eV). Adapted with permission from Bart and Gautier.[9]

relaxed (inwards or outwards of the surface plane). The arrangement of atoms *within* a surface unit cell can only be determined through analysis of so-called I/V curves, which obtained by measuring the intensity (I) of several selected diffraction spots as a function of energy of incident electrons (V) in the wide range of energies, typically between 50 and 300 eV.[10] The experimental curves are then compared to those computed for a particular model structure. The model is tuned in iterative manner until a good fit to the experimental curve is achieved.

In addition, profile analysis of the diffraction spots together with the careful inspection of the background intensity provides valuable information about possible structural defects at surface, such as point vacancies, steps, facets, mosaic, domain boundaries, etc.[11,12] The fine structure of LEED spots anticipated for the most common types of defects observed on hetero-epitaxial films are schematically shown in Fig. 2.3.

2.3 X-Ray Photoelectron Spectroscopy (XPS), Auger Electron Spectroscopy (AES)

Both methods are based on ionization of core levels under high-energy irradiation. In case of XPS, a sample is illuminated with soft *x*-rays (100–1500 eV), whereas for AES the sample is exposed to

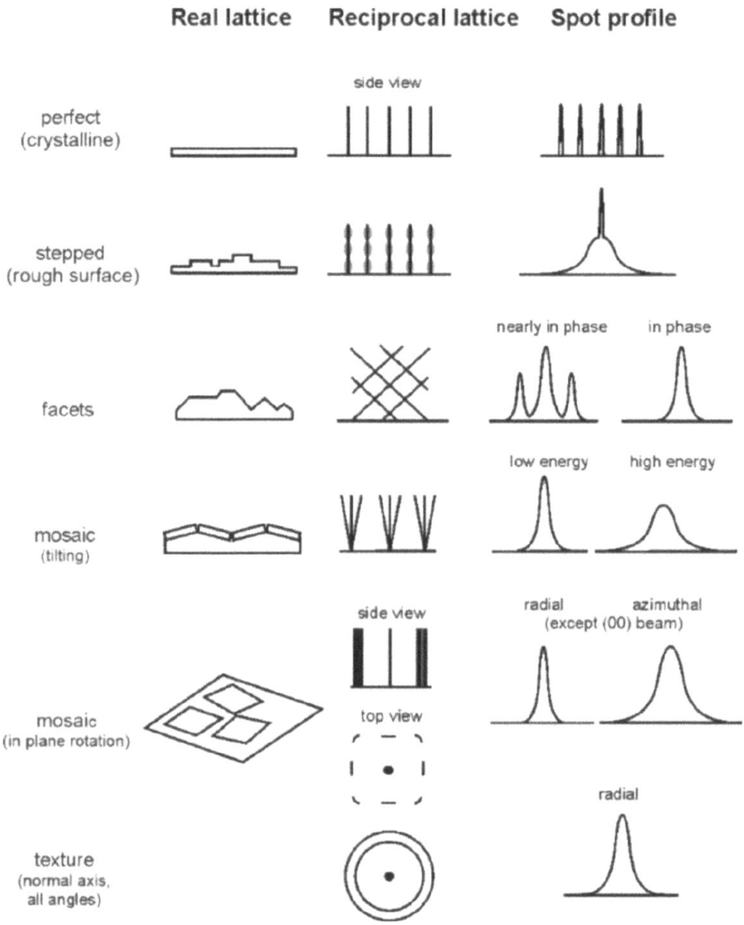

Figure 2.3 The relationship between the real-space morphology and the diffraction pattern for the most common types of defects. Adapted with permission from Henzler.[11]

high-energy electrons (typically, 3 keV). Experimentally measured is the kinetic energy of electrons emitted from the surface which is element specific. Both methods provide a total elemental analysis, except of hydrogen and helium.

Upon illumination of the sample with x-ray photons with energy $h\nu$, electrons are emitted due to photoelectric effect (Fig. 2.2, on the left). The kinetic energies (KE) of electrons, measured by a spectrometer (having the work function ϕ_{sp}), depend on the ionization potential of specific orbitals (i) in the atom via the equation $KE(i) = h\nu - BE(i) - e\phi_{sp}$. The binding energy BE(i) (referenced to the Fermi level) shifts depending on the chemical environment of the ionized atom. This shift provides additional information about the chemical state of the element (see below).

Conventional x-ray sources produce photons at fixed energy, i.e., Mg Kα line at 1253.6 eV (a half-width at maximum is 0.6 eV), Al Kα line at 1486.6 eV (0.85 eV), and (since recently) Ag Lα line at 2984.2 eV (2.6 eV). With monochromatic synchrotron radiation sources, one may tune the primary photon energy, typically between 100 and 1000 eV, which can be chosen such that the corresponding kinetic energy of emitted electrons falls into the range close to the minimum in the universal curve (Fig. 2.1) and thus be most surface sensitive. This is of particular importance in detection of adsorbed (ad-) species. In addition, photoelectron spectra recorded as a function of photon energy allows one to perform depth profile analysis using the fact that λ considerably increases with increasing photon energy above ~100 eV. Therefore, measurements at high energies probe deeper layers. Another approach is based on recording the spectra as a function of electron emission angle (θ) following a simple relationship between the escape depth (d) of detected electrons and IMFP: $d = \lambda\cos\theta$. Therefore, as compared to emission in the direction normal to the surface ($\theta = 0°$), the escape depth is considerably reduced at grazing emissions (typically, $\theta = 60°$).

XPS spectra also probe the electron density of states in the valence band, although the quantum chemical description in this case is more complicated because of so-called "many particles" effect. Moreover, due to a high kinetic energy of corresponding valence photoelectrons measured with laboratory x-ray sources (in the range of 1000 eV), the method becomes less surface sensitive. Therefore, for studying the valence band, it is more appropriate to use a He discharge lamp as a photon source ($h\nu$ is 21.2 eV for He I radiation, and 40.8 eV for He II). Accordingly, the technique is referred to as ultraviolet photoelectron spectroscopy (UPS).

In addition to direct photoelectron emission from the core level, one more process is accompanied by electron emission, which is called Auger process, after Pierre Auger who discovered this

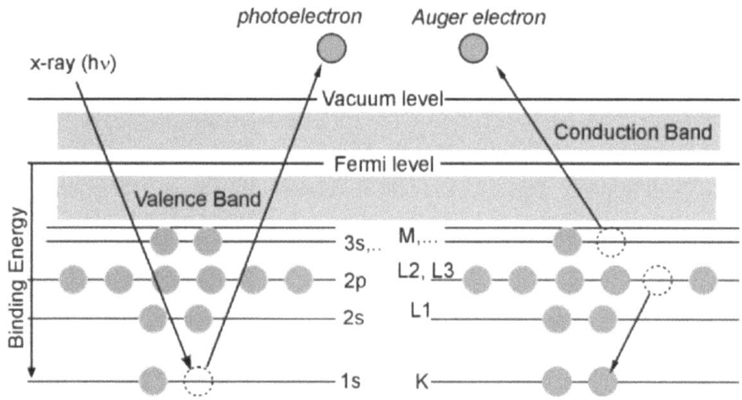

Figure 2.4 Schematic of electron emission processes in XPS (on the left) and AES (on the right).

phenomenon in 1923 (although the effect has first been reported by Lise Meitner in the year before). In this mechanism (see Fig. 2.4, on the right), a core hole left behind the emitted photoelectron on level (K) is filled by an outer shell (L) electron. An amount of energy released during such process, which is equal to the difference in energies of corresponding orbitals, may be transferred to a second outer shell electron (M), which is then emitted and called as Auger electron. In the simplified form, the kinetic energy of emitted electron is $\mathrm{KE} = E(K) - E(L) - E(M)$, where $E(K, L, M)$ are binding energies of the corresponding orbitals. Note that in this process the kinetic energy of the Auger electron is independent of the incident photon (electron) energy, which allows to discriminate the Auger electrons from other emitted electrons by measuring the spectra at different energies. Note that the probability for the Auger process depends on the atomic number and considerably decreases for heavy elements.

As mentioned above, the binding energy of a core level in the element depends on the chemical environment of the ionized atom. This causes some deviation in the BE measured. Although the analysis of the BE shift is not straightforward, it is commonly described in terms of *initial* and *final* state effects. The initial state effects are associated with the electronic state of an atom (ion) *before* ionization has occurred, that, in turn, depends on its valence state, coordination, or specific interaction with surrounding atoms. Accordingly, the final state effects are associated with processes of relaxation and reorganization *after* (or during) the photoelectron emission leading to a hole in the core level. This may involve core hole screening and delocalization.

To separate initial and final state effects in XPS, the so-called Auger parameter has been introduced. Originally defined by Wagner,[13] and later modified by Gaarenstroom and Winograd,[14] the Auger parameter (AP) is sum of the BE of a core level and the KE of the Auger transition that belongs to the decay of the corresponding core hole: AP = BE + KE. The straightforward advantage of using this parameter is that any surface charging induced shifts will be of the same magnitude, but of opposite direction in each of these two components, and therefore they will be cancelled out. The change in AP when comparing the same spectra for a given atom in different environments was thought to relate exclusively to changes in the final state of core-level photoemission processes. The latter statement has been revisited in that only the quantity $\Delta\beta(i) = \Delta E_A(kii) + 2\Delta E_B(i) - \Delta E_B(k)$ directly relates to the final state relaxation contribution $\Delta E_R(i)$ of the binding energy shift $\Delta E_B(i)$.[15] (Here, ΔE_A(kii) is the kinetic energy shift of the Auger transition which corresponds to the decay of a hole state with a hole in level k into a final state with two holes in level i). Although conceptually similar to the Wagner's Auger parameter, the parameter $\Delta\beta(i)$ is based on less drastic approximations, in essence, only on the assumption that $\Delta R(i)$ is determined by Coulomb contributions.

For silica-based compounds, the Si 2p and O 1s core levels are primarily used for identification of the valence state and composition stoichiometry. In principle, one has to take into account the spin-orbit splitting of the Si 2p level into the Si $2p_{3/2}$ and Si $2p_{1/2}$ states separated by about 0.6 eV, although many studies neglect such a small splitting because it is comparable with the energy resolution of conventional spectrometers. Due to substantial surface charging of bulky silica samples, the majority of XPS and AES studies on silica have been performed on thin silica films thermally grown on Si crystals, often used as model for vitreous silica.[16] Besides technological importance of SiO_2/Si interfaces, additional advantage of using a Si substrate is to have an internal

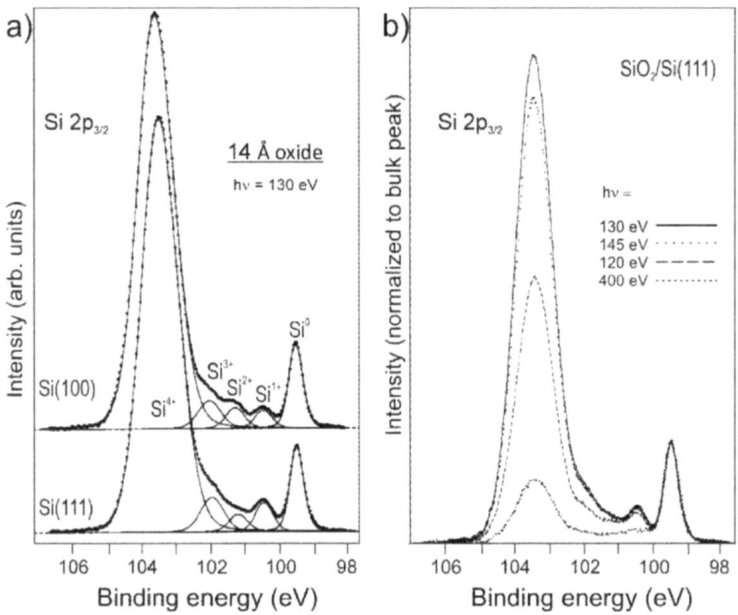

Figure 2.5 (a) Si $2p_{3/2}$ spectra extracted from XPS spectra measured on thin SiO_2 films, grown on Si(100) and Si(111), and deconvoluted into several Si^{n+} states. (b) Normalized spectra for SiO_2/Si(111) as a function of photon energy as indicated. The spectra are replotted from Himpsel et al.[17]

reference for energy calibration. To illustrate this, Figure 2.5a displays Si $2p_{3/2}$ spectra obtained on thin silica films formed by oxidation of the Si(001) and Si(111) surfaces in pure oxygen. The spectra additionally benefit from using tunable low photon energy provided by synchrotron facilities to probe primarily the surface states, whereas with conventional x-ray sources the Si 2p level photoelectrons possess the kinetic energy above 1 keV and hence more bulk-sensitive.

Five different states of Si are simultaneously observed in the spectra shown in Fig. 2.5, i.e., starting from Si^0 in a substrate (at 99.5 eV) up to Si^{4+} in the SiO_2 film. The intermediate peaks are equally separated by about 1 eV and can, therefore, be assigned to Si^{n+} species ($n = 1 - 3$), in addition suggesting that the initial state effects associated with the oxidation state of Si dominate in the observed BE shifts. With the progressive film growth, the highest BE signal gains in intensity and saturates, while all other signals attenuate and ultimately disappear, thus suggesting SiO_2 stoichiometry in the thick films. Another set of data presented in Fig. 2.5b illustrates the photon energy dependence measured in order to get information about the Si^{n+} depth profile. The SiO^{4+}/Si^0 intensity ratio maximizes at about 130 eV due to a minimum in the escape depth for the range of kinetic energies studied.

Analysis of Auger parameter in silica compounds is rather scarce. Partially, it is because the kinetic energy of Auger Si(KLL) electrons falls into the range of ~1600 eV that is above excitation energies provided by conventional x-ray sources. For this purpose, one can use a continuous x-ray light provided by Bremsstrahlung radiation. Such measurements reported in Ref. [18] for nine inspected silicates having different SiO_4 tetrahedra arrangements (i.e., monomeric, dimeric, chains, sheets, and 3D-network; see Chapter 1) showed that Auger parameter AP = BE(K) + KE(KLL) was almost constant (1711.7 ± 0.2 eV), although BE of the Si 2p level shifted substantially, i.e., from 101.3 to 103.4 eV. It was suggested that the variation of the Si 2p energies correlates with electrostatic force of the Si–O bonds in these phases calculated from the Pauling rule. (Note that in these measurements sample charging was minimized by using additional ("flood") electron gun which compensates the photoelectric current.)

The O 1s core level in metal oxides is less sensitive to the atomic structure. In great majority of XPS studies on silica surfaces, the O 1s signal showed, in essence, a single peak (if, of course, the surface is not hydrated; see Chapter 8). Nonetheless, the energy separation between the Si 2p and the O 1s levels may also be useful for surface characterization and comparative studies. In particular, the value 429.9 ± 0.2 eV has been reported[19] for vitreous silica thick films. In contrast to silica polymorphs having only Si–O–Si type of bonds, silicate minerals may show several oxygen species in spectra. For example, in pyroxenes (single chain silicates, with a general formula $M^{2+}Si_2O_6$), two oxygen species were observed in the high-resolution spectra, which were assigned to bridging oxygen (i.e., Si–O–Si, like in quartz) and non-bridging oxygen (i.e., Si–O–M^{2+}, like in olivine, in which SiO_4 tetrahedra are isolated and not connected to each other).[20]

In principle, both XPS and AES techniques allow to determine the thickness of oxide films, that is, in fact, not a trivial task in case of *ultrathin* films.[21] The thickness evaluation by XPS is based on relatively simple formalism which describes attenuation of a signal from a substrate according to a well-known Beer–Lambert exponential law, and growth of a signal from a film which ultimately saturates as the thickness exceeds the escape depth of photoelectrons. The intensity of each signal in spectra is determined by the photoionization cross sections, the escape depths, and the density of corresponding atoms in oxide and support. In case of a silica film grown on Si to the thickness d_{oxide}, the intensity ratio of the Si peaks in SiO_2 and Si is given by

$$\frac{I_{\text{SiO}_2}}{I_{\text{Si}}} = \frac{I^{\infty}_{\text{SiO}_2}}{I^{\infty}_{\text{Si}}} \frac{\{1 - \exp[-d_{\text{oxide}} / \lambda_{\text{SiO}_2}(E_{Si}) \cos\theta]\}}{\exp[-d_{\text{oxide}}/\lambda_{\text{SiO}_2}(E_{\text{Si}}) \cos\theta]}$$

where I^{∞}_{SiO2} and I^{∞}_{Si} are the signal intensities measured with the same apparatus and setting parameters for a clean Si substrate and a very thick SiO_2 film, respectively. The parameter $\lambda_{\text{SiO}_2}(E_{\text{Si}})$ is the IMFP value for the Si 2p photoelectrons in SiO_2, and θ is the angle of emission of the detected electrons with respect to the surface normal. If the escape depth in silica is known, then the film thickness can be calculated by the equation

$$d_{\text{oxide}} = \lambda_{\text{SiO}_2}(E_{\text{Si}}) \cos\theta \; \ln(1 + R_{\text{exp}}/R_{\text{o}})$$

where R_{exp} is the measured ratio $I_{\text{SiO}_2}/I_{\text{Si}}$ and R_0 is the ratio $I^{\infty}_{\text{SiO}_2}/I^{\infty}_{\text{Si}}$. Alternatively, one can determine $\lambda_{\text{SiO}_2}(E_{\text{Si}})$ by measuring spectra on the films of known thickness measured, for example, by cross sectional electron microscopy or ellipsometry. This approach was tested on thermally oxidized silica films[22] by measuring XPS spectra with different x-ray sources as a function of the film thickness, which was, in turn, determined by ellipsometry in air and varied in the range of 2–20 nm. The results yielded the escape depth 26 ($\pm$2) Å for the Si 2p level and of 21 ($\pm$5) Å for the O 1s level by using the Al Kα (1487 eV) radiation. However, good agreement has only been obtained for the relatively thick films. For nominal thicknesses below 2 nm, considerable discrepancy was obtained, which has been attributed to substantial non-stoichiometry of the oxide layer at the interface to the Si substrate. Thorough analysis performed by Himpsel et al.[17] on the films prepared in situ under well-controlled conditions led to the λ values about 7 and 4 Å for SiO_2 and Si, respectively, for low photon energies (100–200 eV). These values are close to those previously found from AES analysis of SiO_2/Si films, combined with high energy ion backscattering and nuclear microanalysis determination of the film thicknesses, which gave 6.5 Å for the escape depth in SiO_2.[23] Most recent evaluation of large dataset of experimental results using conventional x-ray sources gave λ_{SiO_2} (Mg source) $= 2.8 \pm 0.5$ nm and λ_{SiO_2} (Al source) $= 3.1$ nm $\pm$ 0.3 nm.[21]

In principle, similar formalism can be applied to the case of hetero-epitaxial growth of a metal oxide film on a substrate of another metal. The analysis is more complicated because the energy dependent attenuation lengths of photoelectrons from the two materials should be taken into account. This gives:[24]

$$d = \lambda_{\text{sub}} \cos\theta \ln(R/R_0 + \alpha)$$

$$\text{and} \quad \alpha = \exp(d/\cos\theta \times (1/\lambda_{\text{sub}} - 1/\lambda_{\text{film}}))$$

where $R = I_{\text{film}}/I_{\text{sub}}$ is the relative peak intensity ratio measured for the film of the thickness d, and R_0 is the ratio measured on the bulk material (or a very thick film) and a bare substrate. Obviously, if the attenuation lengths are similar, α approaches unity, and the equation reduces to the case of homoepitaxial growth of SiO_2 on Si.

2.4 Infrared Reflection Absorption Spectroscopy (IRAS)

Although the technique is mostly used for studying adsorption and surface reactions, it became very informative for structural characterization of ultrathin oxide films, particularly if the films are grown on metallic substrates. Reflection of IR light from a metal surface covered by a thin insulating film is treated on the basis of macroscopic classical electrodynamics using Maxwell's equations for the interface with complex dielectric functions $\varepsilon^* = n + ik$ (Fig. 2.6a). The analysis is simplified by assuming the film thickness (d) much smaller than the IR wavelength, i.e., 1–10 μm.

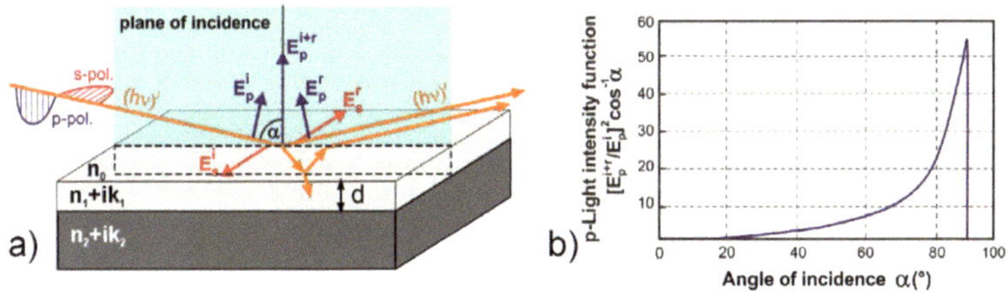

Figure 2.6 (a) A three-phase model of interaction of IR light with a thin film (or adsorbate layer) on a support. Orientation of s- and p-polarized light upon reflection and the electric fields of incident beam ($\mathbf{E}^i$), reflected beam ($\mathbf{E}^r$), and net electric field upon reflection ($\mathbf{E}^{i+r}$). (b) Surface intensity function of reflected light versus angle of incidence (α).

The reflectivity of the s-polarized light becomes negligibly small (a phase shift for the s-component of the electric field at surface is about 180°), and only p-polarized component of the light reflects from surface. The phase shift for the p-light depends on the angle of incidence (α) so that the reflection maximizes at grazing incidence as depicted in Fig. 2.6b. In modern IRAS setups, the angle α is set between 80°and 85° with respect to the surface normal.[25]

The presence of a metal substrate results in a surface selection rule stating that IR bands associated with vibrations resulting in an oscillating dipole moment parallel to the surface will be absent, and only vibrations giving rise to a dynamic dipole moment perpendicular to the surface will be observed.[26,27] One of the reasons is the 180° phase shift for the electric field component parallel to the surface. The out-of-phase electric field vectors of the incident and reflected waves cancel each other, and therefore no electric field can exist to interact with vibrations parallel to a metallic surface. The other reason is that the metal surface creates image charges. Their oscillations sum up in the direction normal to the surface, but cancel out in the surface plane. Generally speaking, with IRAS one probes all vibrations having a non-zero component in the normal direction.

The IRAS setups implemented in modern UHV analytical chambers typically cover frequencies in the range between 600 and 4000 cm^{-1} that allows investigation both of molecular vibrations of adsorbates and lattice vibrations (phonons). The phonons are

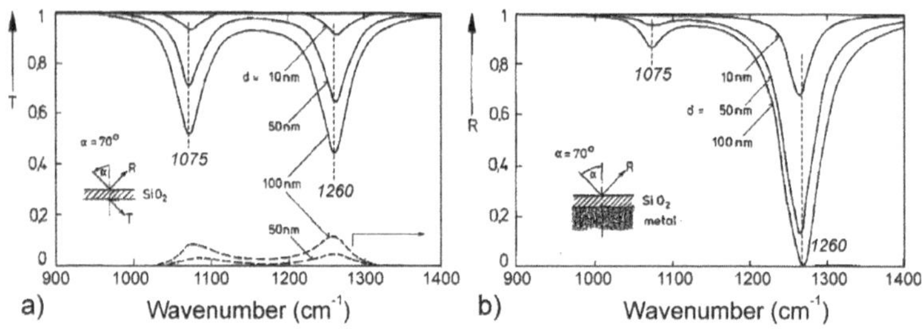

Figure 2.7 Transmission (T) and reflection (R) spectra calculated for thin SiO_2 films of various thicknesses (as indicated) upon illumination with p-polarized light at angle $= 70°$: (a) unsupported films; (b) films supported onto a metal substrate. Adapted with permission from Harbecke et al.[29]

labeled "longitudinal" if the vibration occurs along the direction of the wave propagation. Accordingly, the "transverse" phonons are originated from vibrations perpendicular to this direction. In addition, phonons are branched into "acoustic" phonons, i.e., with a wavelength much larger than the interatomic distances, and "optical" phonons which can be excited by radiation in the IR region. Electromagnetic waves, which are transverse in nature, cannot interact with longitudinal phonons in an infinite crystal. It was thought, therefore, that the longitudinal optic (LO) phonons cannot be measured by IR spectroscopy in the transmission mode. However, this is, in fact, possible thanks to the effect predicted by Berreman,[28] who showed that, upon illumination with the p-polarized light not normal to the surface, the absorption bands will be observed at the frequencies characteristic of LO modes in infinite crystal. The effect is most pronounced for thicknesses smaller than the "Berreman thickness"[29]:

$$d_{\mathrm{B}} = (\lambda/2\pi)(\cos\alpha/\sin^2\alpha)(\mathrm{Im}(-1/\varepsilon^*))^{-1}_{\mathrm{max}}$$

At this thickness the absorption is nearly unity at the LO frequency, and almost zero at the TO resonance. For silica films, the calculated d_{B} values fall in the range of 50–100 nm. Figure 2.7 shows transmittance calculated for SiO_2 films, unsupported (a) and supported on a metal (b), as a function of the film thickness. Obviously, such bands are easier to detect in reflection mode from a crystal film deposited onto a metallic substrate.

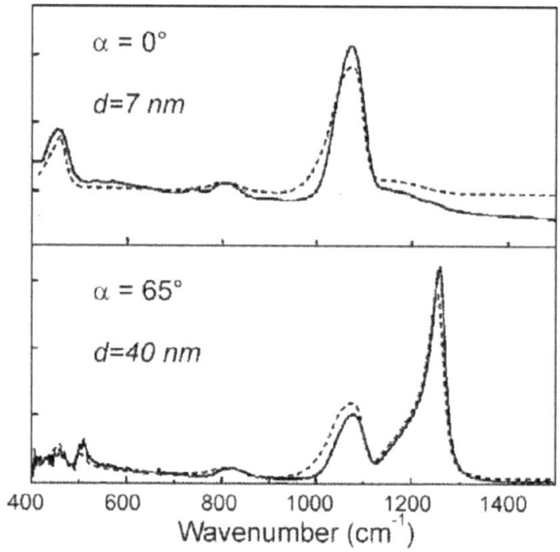

Figure 2.8 Experimentally measured (solid lines) and calculated (dashed lines) absorbance spectra obtained with p-polarized light at normal ($\alpha = 0°$) and oblique ($\alpha = 65°$) incidence on SiO_2/Si films. Adapted with permission from Martinet and Devine.[30]

Comparison of experimental spectra displayed in Fig. 2.8 measured for amorphous silica films on Si at two angles of incidence (0° and 65°) shows nice agreement with predictions based on the Berreman effect: The 1257 cm^{-1} signal only appears at oblique incidence, while the 1075 cm^{-1} band attenuates.

Although classical oscillator model may explain the main adsorption bands in silica, explanation of the "extra" lines and shoulders in spectra required incorporation of long-range interaction between the SiO_4 tetrahedral units resulting in TO–LO pairs.[31] Moreover, careful analysis found in addition to the three LO–TO splitting located at 507(LO) - 457(TO), 820(LO) - 810(TO), and 1256(LO) - 1076 (TO) cm^{-1} one more mode which exhibited inverted LO-TO splitting at 1160(LO) - 1200(TO) cm^{-1},[32] as the independent-oscillator model failed to yield a consistent value of mode strength for the optically active oxygen asymmetric stretch (AS_1) TO mode at 1076 cm^{-1} (in-phase motion of oxygen atoms). To explain the results, disorder-induced mechanical coupling between the

AS_1 mode and the relatively optically inactive oxygen asymmetric stretch (AS_2) mode (out-of-phase motion of adjacent oxygen atoms) was implemented in the oscillator model.[32] Apparently, the coupling increases for amorphous silica as compared to quartz. Note, however, that multiple reflection effects have to be carefully examined in spectra measured on thin films.

2.5 High-Resolution Electron Energy Loss Spectroscopy (HREELS)

The technique is based on measuring the energy loss of monochromatic electrons upon their inelastic interaction with a surface resulting in energy and momentum transfer to surface excitations. Typically, the losses are in the range between 1 meV and 1 eV. Possible interactions are mainly the excitations of vibrations of adsorbate molecules, vibrations of the upper layers of the substrate (surface phonons), and vibrations of the electrons in the substrate or in films or islands on the substrate surface (plasmons).

The surface phonons are commonly referred to as "Fuchs–Kliewer phonons" after Fuchs and Kliewer, who theoretically investigated optical modes of vibration in ionic crystals in the mid-1960s.[33] Basically, these modes come as solutions of the Maxwell's and material equations. Experimentally, really high-resolution spectra were only reported five years later by Ibach for ZnO crystals measured in specular reflection.[34] Since that, HREELS has been realized as a powerful tool for studies of surface vibrational modes in crystalline materials.[35]

There are two principal regimes of HREEL spectroscopy, namely the dipole scattering and impact scattering. In the *dipole scattering*, the method is based on the same fundamental photon absorption mechanism as IR spectroscopy: The incident electron is treated similarly to an electromagnetic (infrared) wave that interacts with oscillating dipoles at the surface. The interaction between the probe electron and the sample is dominated by long-range Coulomb fields. The incident electron scatters inelastically and excites vibrations in the dipole structure. Since the momentum transfer is small, the electrons are primarily scattered in the specular direction.

The response of the solid sample surface to the external electric field of the probe electron can be described in terms of the surface dielectric response with frequency of the external electric field. Since the dielectric theory is a macroscopic approach, describing the sample in terms of bulk parameters, its application to ultrathin films is not straightforward. In case of metal supported films, the same selection rules apply as for IRAS spectroscopy: Only vibrations having a component perpendicular to the surface can be detected. This is because: (i) no electric field exists to interact with dipoles parallel to the surface; (ii) the dynamic dipole moment vibrating along the metal surface is cancelled out by its image dipole. Accordingly, dynamic dipole moment in the normal direction is reinforced by that of its image dipole.

In the *impact scattering* regime, electron is scattered as a particle upon collision with atomic core potentials of surface modulated by nuclear vibrations. Contrary to dipolar scattering, it is not limited to small momentum transfer. The electrons scatter away from the specular direction over a wide range of angles, which depends on surface dipole amplitude/direction (symmetry of vibrational mode) and electron energy (momentum). The theory of impact scattering is complex and remains poorly-developed as the theoretical treatment requires a quantum mechanical formalism. Note also a much lower cross-section in this regime as compared to dipole scattering, thus resulting in weak loss features. On the other hand, electron momentum neither parallel nor normal to the surface is conserved in the process. Therefore, the metal-surface selection rules do not apply to impact scattering, and one can even detect vibrational modes with dynamic dipoles parallel to the surface plane.

As an example of HRRELS applied to silica related materials, Fig. 2.9 displays a spectrum obtained on synthetic quartz (0001), which seems to be the only successful HRRELS study of silica polymorphs. The spectrum could be recorded with a high resolution thanks to using of an additional electron gun to avoid sample charging that obviously destabilizes the surface potential. The three main peaks are located at 498, 798, and 1176 cm^{-1}, which are, in fact, close to the IR absorption bands, if measured at normal

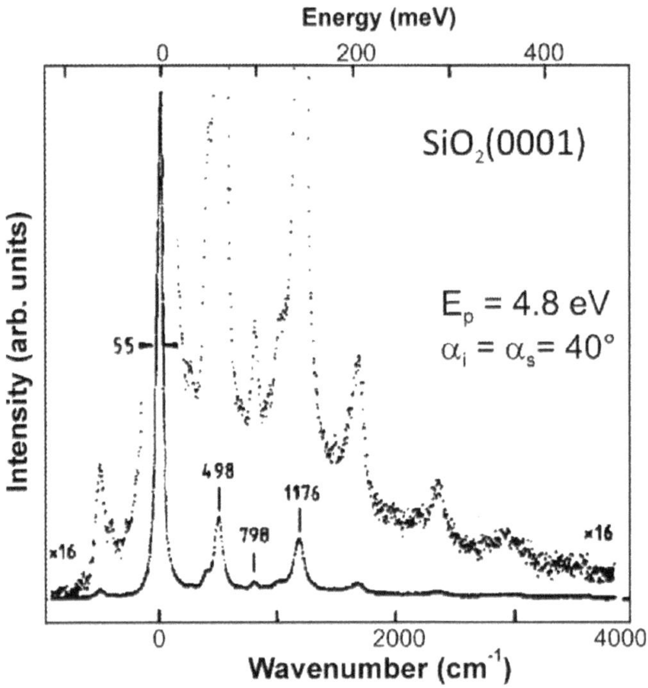

Figure 2.9 High-resolution energy loss spectrum measured on a synthetic quartz (0001) surface in specular direction. The angle of incidence is 40°. Adapted with permission from Thiry et al.[36]

incidence (see top spectrum in Fig. 2.8). The comparison shows close similarities in phonon spectra observed by the two techniques. One may note, however, that HREELS has the advantage of exploring small energy losses, i.e., below 500 cm^{-1}, practically inaccessible to conventional IRAS setups having IR-transparent windows (usually, KBr) with a cut-off at about 600 cm^{-1}. In addition, HREELS shows a better sensitivity with respect to adsorbed molecules. On the other hand, IRAS provides a much better resolution (typically 2–4 cm^{-1}).

Solely to illustrate complementarity of HREELS and IRAS techniques, in particular for studying ultrathin oxide films, Fig. 2.10 compares both spectra obtained on a "monolayer" film of alumina formed by oxidation of the NiAl(110) surface. The prominent losses at 420 cm^{-1}, at 620–660 cm^{-1}, and most intense at 860–870 cm^{-1}

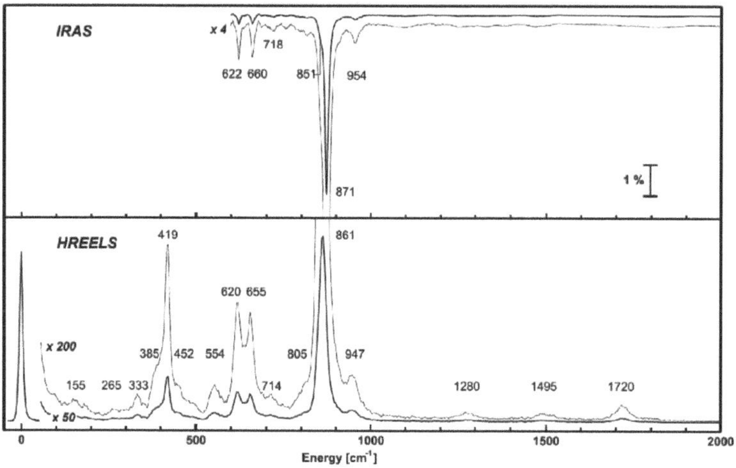

Figure 2.10 Direct comparison of IRAS and HREELS spectra obtained on an ultrathin alumina film grown on NiAl(110). Reproduced with permission from Frank et al.[37]

are assigned to the Fuchs–Kliewer phonons. For a dielectric film on a metallic substrate, loss signals near the LO frequencies of the corresponding bulk dielectric are expected similar to the Berreman effect in IRAS. Although a difference between these two techniques exists in terms of momentum transfer (which is negligible for photons and relatively large for electron scattering) spectral features are observed at similar energies.

2.6 Scanning Tunneling Microscopy (STM)

Apparently, STM is the only technique that provides information about surface topography of conducting materials with atomic resolution. The technique is based on the quantum tunneling effect between the metallic tip and the surface separated by a gap of a few Ångströms. Within the semiclassical approximation, the tunneling current (I) between two electrodes that arise upon applying small voltage (V) is given by

$$I \propto \int dE \rho_s(r_{tip}, E)\rho_t(E - eV)M(z, E, V),$$

where ρ_t and ρ_s are local density of states (DOS) in the tip and the sample, respectively. Tunneling probability $M = \exp(-2z/h\sqrt{(2m(\phi - E)})$ depends on tunneling gap (z), and the work function $\phi = \frac{1}{2}(\phi_t + \phi_s)$ which is averaged of both tip and sample. Due to exponential dependence for M on energy, the current is dominated by the electronic states near the Fermi level.

The most common mode in STM imaging is a "constant current" mode. In this mode, one scans the surface with a tip at a fixed sample bias voltage. An electrical feedback loop controls the voltage on the piezo-driver (used to approach and retract the tip from the surface) to keep the tunneling current constant. By recording the voltage applied to the piezo as a function of coordinate in surface plane, the surface topography is mapped. In the so-called "constant height" mode, only used for atomically flat surfaces, the sample is scanned at high speed with the feedback switched off, and variations in the tunneling current are recorded. In general, STM images reflect spatial distribution of electronic states which are, of course, linked to the arrangement of atoms at surface. Accordingly, the image contrast depends on the applied tunneling parameters, i.e., bias magnitude and its polarity, and tunneling current.

As an example, Fig. 2.11 displays high-resolution STM images of the Si(111)-(7×7) reconstructed surface obtained with modern microscopes. Note that it was this surface, for the first time atomically resolved in the early 1980s by Gerd Binnig et al.,[38] that demonstrated tremendous capabilities of STM for surface characterization.

Since STM microscope is based on measuring net electric current, whereas many oxides are wide band semiconductors or insulators like silica, STM imaging of such materials is only possible either by doping or partial reduction (via annealing in UHV at elevated temperatures) to make specimens having, at least, surface conductivity. As alternative approach, the material under study can be fabricated as a thin film on a conducting substrate to the thickness which allows electrons to pass through the film to result in a net electric current, in the order of a few pico-amperes, at least. In this case, STM imaging is even possible at bias magnitudes lower than the oxide bandgap.

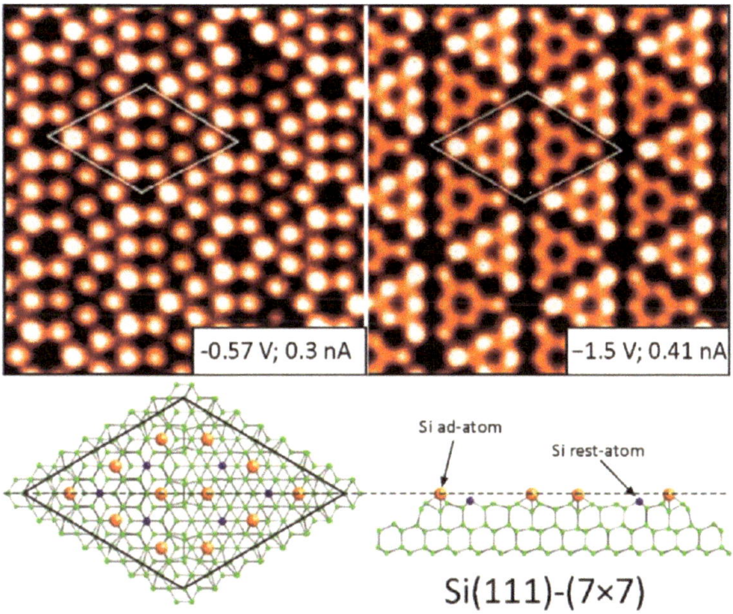

Figure 2.11 STM images of the Si(111)-(7 × 7) surface at tunneling bias and current as indicated. Shown below is the theoretical model in the top view (on the left) and in the side view across the dash line (on the right). Adapted with permission from Wang et al.[39]

A complex potential diagram of tunnel junction for the case of a thin oxide film supported on a metallic substrate is schematically shown in Fig. 2.12. Depending on the work functions the charge redistribution first occurs between a metal support and an oxide film as a result of alignment of the Fermi levels in the two materials. In addition, due to low conductivity, the film cannot fully screen the electric field from the tip, which may, therefore, penetrate into the bulk, thus leading to the so-called band bending (ΔE), the well-known phenomenon in semiconductors. The extent of band bending depends on the film thickness and intrinsic dielectric properties of the oxide. All these effects must be taken into account while interpreting STM images on thin insulating films.

In principle, at low voltages smaller than the band gap electronic states of the oxide are not much involved in the tunneling process, and the film is almost "transparent", albeit it alters the topography

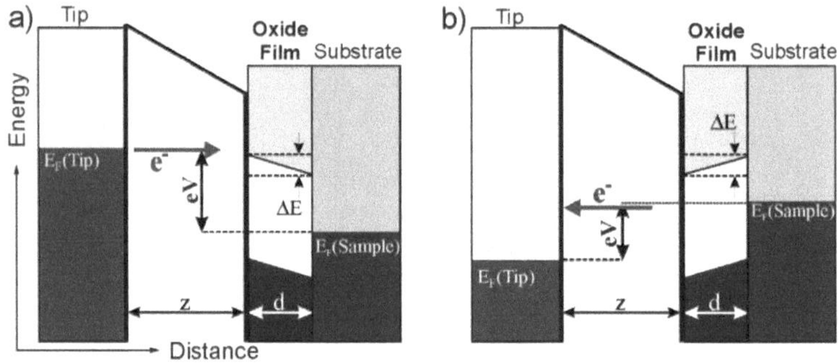

Figure 2.12 Potential diagram of an STM junction containing an insulating (oxide) film supported by a metal substrate. Due to the tip-induced field, the bands of the film shift to higher energies with increasing positive bias (a) and to lower energies at negative bias (b) applied to the sample. Adapted with permission from Nilius.[40]

imaged. As voltage increases above the bandgap, the electrons can tunnel directly into (or out) the empty (occupied) states in oxide. The difference between two regimes is illustrated in Fig. 2.13 showing two STM images of ultrathin alumina film grown on NiAl(110).[41] At the bias voltage around 4 V, the image shows the oxide island with an apparent height of 3.5 Å with respect

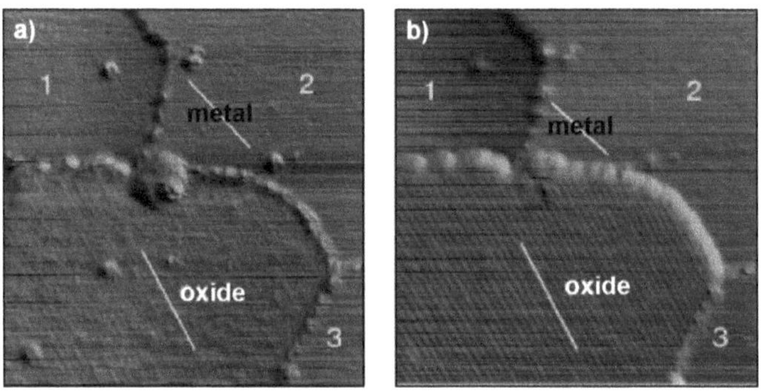

Figure 2.13 STM images of ultrathin alumina film on NiAl(110) recorded at two different bias voltages (0.4 V (a) and 4 V (b)), whereas the tunneling current was set to 2 nA for both. The image size is 50 nm × 50 nm. Numbers 1, 2, and 3 indicate three terraces of a metal support. Reproduced with permission from Bertrams et al.[41]

to the metal support. At lower biases, however, the island height approaches zero as if the oxide were transparent for tunneling.

The band structure of the oxide films can directly be addressed by sweeping tunneling voltage and measuring the current at the fixed tip–sample distance, resulting in tunneling spectra, which gives information about both, conduction and valence, bands simultaneously, whereas many electron spectroscopies have access only to one of those. On the other hand, a considerable electric field between the tip and the sample may affect the electronic structure, which has also to be taken into account.

2.7 Electron Microscopy

Since the invention of a transmission electron microscope (TEM) in the early 1930s, the method had become widely used for characterization of solid materials. A high spatial resolution is achieved thanks to the small de Broglie wavelength of electrons accelerated to high energies, typically 100–300 keV, and modern high-precision electronic optics and detectors. In addition, electron microscopes make use of high energy electron diffraction as well as energy dispersive x-ray (EDX) and/or electron energy loss spectroscopy (EELS), which allow to determine the chemical composition in the area of interest with nanometer resolution.

With the introduction of aberration-corrected optics, new opportunities have been opened up, even for direct visualization of single atoms.[42,43] Further advantage became possible by using high-angle annular dark field (HAADF) scanning electron microscopy, the capability of which was first demonstrated for heavy elements like uranium and thorium.[44,45] In this mode, signal is collected from electrons incoherently scattered (Rutherford-scattered) to high angles. The images can directly be interpreted in terms of positions of atoms or atomic columns. Since the contribution of an atom to the intensity in the recorded image is roughly proportional to the square of its atomic number, contrast variations help to identify elements in the atomic structure.

For studying interfaces and thin films, cross-sectional images are used. This needs, however, special preparations of specimens

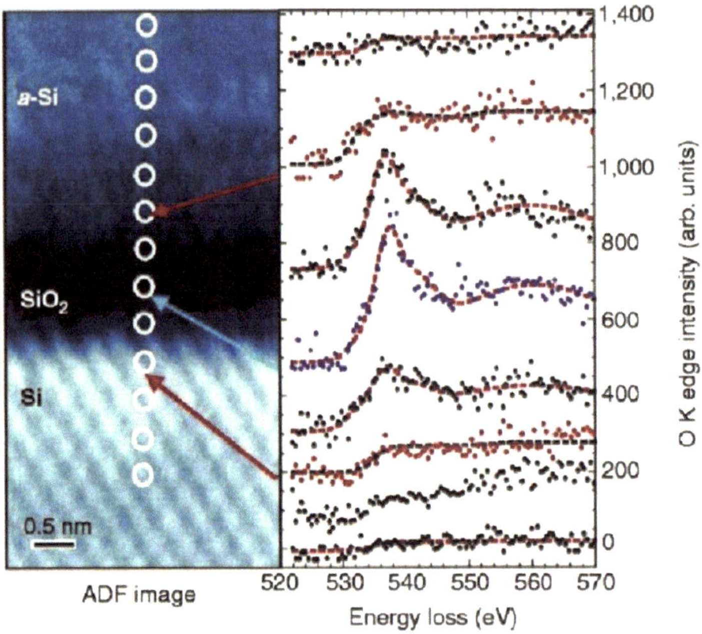

Figure 2.14 Left panel: ADF SEM image of a gate stack (c-Si/SiO_2/a-Si) containing a thin gate oxide. Right panel: Background-corrected EELS O K edge spectra recorded point-by-point as indicated. Reproduced with permission from Muller et al.[48]

including mechanical thinning, dimpling, and ion milling.[46,47] Figure 2.14 shows an example of the ADF SEM image of a stack consisting of crystalline and polycrystalline Si with a thin gate oxide in between. The image is a projection of about hundreds of atoms in the path of the beam, so no structural information can be obtained from the amorphous region where the atoms are not aligned along the beam direction. On the right panel, EELS O K edge spectra are shown recorded point-by-point across the interface, which allow to address the electronic structure in the silica film.

Although TEM is, no doubt, a very powerful tool for structural characterization of solids, information about *surface* structures is difficult to obtain as it requires careful precautions about specimen preparation that do not contaminate the surface. In addition, it requires much more elaborative analysis of the image contrast in combination with image simulations.

Final remark: Most of the above-listed techniques can only be applied in vacuum, typically below 10^{-6} mbar. Currently, many efforts are being made for development of methods which can operate in gas atmosphere and even in liquids. In particular, sum frequency generation (SFG), polarization-modulation IRAS (PM IRAS), near-ambient pressure XPS (NAP XPS), high-pressure STM (HP STM), and "environmental" SEM techniques allow one to perform studies under technologically relevant conditions.[49–51]

2.8 Film Growth: General Considerations

Aiming at preparation of a thin oxide film, one would immediately suggest direct oxidation of a respective metal substrate as the first thing to try. However, there are not so many examples in the literature when this preparation results in *well-ordered* films.[6] A crystalline $Cr_2O_3(0001)$ film was found to grow on Cr(110). Crystalline NiO(001) films were observed on Ni(001), although the surface appeared highly defective. Molecular oxygen O_2, atomic oxygen O, ozon O_3, and nitrogen oxides (NO, NO_2) can all be used as the oxidizing agents.

Another approach involves oxidation of bimetallic or alloyed (intermetallic) surfaces where one of the constituting elements that has a higher affinity for oxygen will be preferentially oxidized, thus forming oxide layer (Fig. 2.15). Some of the best examples of the crystalline films fabricated using this approach are alumina films grown on NiAl(110), NiAl(111) and $Ni_3Al(111)$ single crystals.[52–54]

In the great majority of cases, however, thin oxide films are deposited onto another metal substrate using various deposition methods. The preparation often includes additional post-oxidation and/or thermal treatments.

According to the Bauer's classification,[55] there are three principal modes of the film growth depicted in Fig. 2.16:

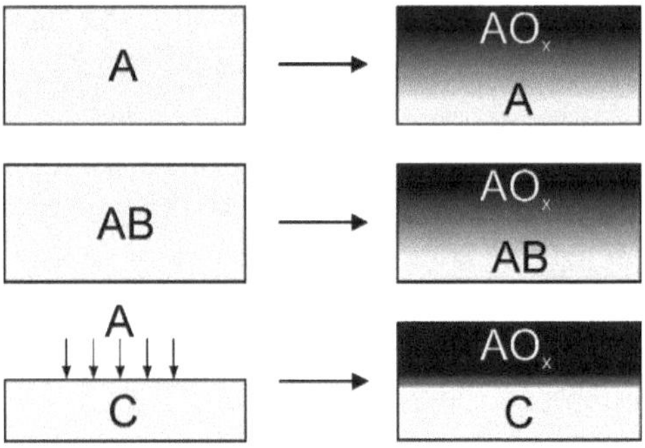

Figure 2.15 Schematic presentation of the oxide film fabrication on metal substrates.

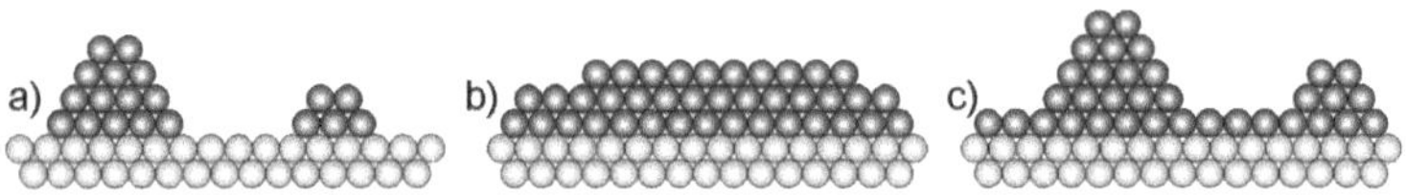

Figure 2.16 Growth modes: (a) the three-dimensional (3D) island mode; (b) the layer-by-layer mode; (c) 3D island-on-wetting-layer mode.

1. The three-dimensional (3D) island mode (also referred to as Vollmer-Weber (V-W) mode)
2. The layer-by-layer mode (or Frank–van der Merve (F-M) mode)
3. 3D island-on-wetting-layer mode (or Stransky–Krastanov (S-K) mode)

Which one dominates the process depends on the relation between surface free energies (γ) of the substrate, of the film, and the film/substrate interface. The layer-by-layer growth occurs when

$$\Delta\gamma = \gamma_{\text{film}} - \gamma_{\text{substrate}} + \gamma_{\text{interface}} < 0.$$

In the opposite case ($\Delta\gamma > 0$), the 3D growth takes place from the onset. In the intermediate regime, $\Delta\gamma$ is smaller than zero until a certain thickness of the film that grows in layer-by-layer mode, above which it grows via the V-W mode on top of the preformed film.

To predict the growth mode, one first needs to know the surface free energies of corresponding materials, which can be measured experimentally or even calculated from the first principles. Usually, surface energies of metals are higher than of their oxides. More difficult, albeit more important is to determine the interfacial energy as it depends on many parameters which can be grouped into "chemical" and "structural" ones. The structural parameters primarily refer to the lattice symmetry and the lattice constants. For example, to fabricate an oxide film in the (001) orientation, one would certainly consider a metal substrate exposing the (001) surface or another one of square symmetry rather than the (111) or hexagonal (0001) surface. In addition, the difference in the lattice constants (**a**) of a support and an oxide (the lattice mismatch $= (\mathbf{a}_{\text{film}} - \mathbf{a}_{\text{substrate}})/\mathbf{a}_{\text{substrate}}$) must be as small as possible, since the large mismatch will cause energetically costly stress or strain deformations at the interface, and hence affect the quality of the prepared films. The corresponding tension and/or compression at the interface can be released in different ways such as dislocations, coincidence superstructures, film buckling, and their combinations. For example, LEED analysis of thin MgO(001) films grown on Ag(001) together with STM imaging revealed the mosaic structure (see Fig. 2.3 for description) because MgO(001) has a larger (by 3.5%) lattice constant and must therefore be compressed to fit the Ag(001) surface. The formation of coincidence superstructures giving rise to a Moiré pattern is a typical situation for the ultrathin films with a relatively large mismatch. The famous example is an FeO(111) monolayer on Pt(111). Here, eight lattices of FeO approximately coincide with nine lattices of Pt along the main directions (Fig. 2.17). In addition, the layer may slightly rotate, by about 10°, depending on preparations. However, further growth of FeO in a layer-by-layer mode becomes energetically unfavorable. Instead, islands of Fe_3O_4(111) grows on top of the FeO(111) layer, thus exemplifying the S-K mode. The islands coalesce until the entire surface is covered by Fe_3O_4(111) at increasing film thickness.

Intuitively, a substrate with the square or rectangular surface unit cell is hardly suited for the film growth in hexagonal or trigonal orientation, and vice versa. However, such reasonable expectations are not always found in experiment. For example, iron oxide films (FeO and Fe_3O_4) were found to grow in the (111) orientation both on Pt(111) and Pt(001) substrates,[3,56] and some experimental "tricks" are required to push Fe_3O_4 films to grow in the (001) orientation

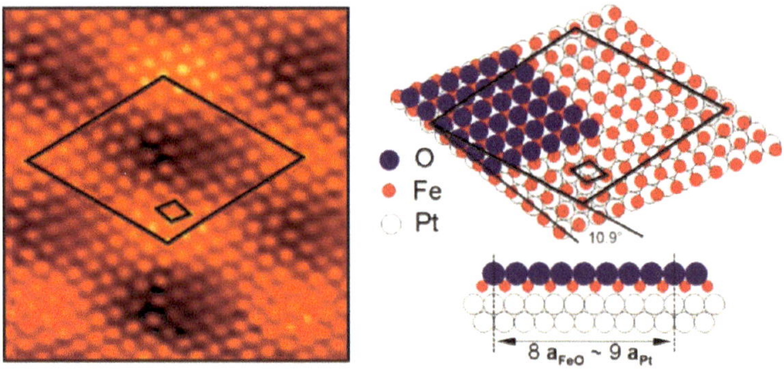

Figure 2.17 High-resolution STM image and structural model of an FeO(111) "monolayer" film on Pt(111). The rhombs indicate unite cells of FeO(111) and of coincidence superstructure.

on Pt(001).[57] It, therefore, appears that chemical interactions at the interface contribute much to the epitaxial relationships. In another example, ultrathin films of MnO and NiO showed (001)-like stacking arrangement even grown on the Pt(111) surface.[58] Apparently, the oxide/support interaction is weak in this case so that the film grows along its most stable crystallographic direction, which is the (001) one for oxides having the rock salt crystal structure like MnO and NiO studied.

Another factor that may influence the interface energy between oxide film and metal support is the affinity of the latter to oxygen. For example, MgO(001) films readily grow on an relatively inert Ag(001) substrate in such a way that the O atoms in oxide sit on top of the Ag atoms. Meanwhile, for MgO(001) films on Mo(001), the O-Mo bonds are much stronger, thus leading to a considerably shorter interlayer distance between oxide and metal at the interface, i.e., from 2.73 to 2.15 Å, for Ag and Mo supports, respectively.[40]

Miscibility of an oxide to be grown and of a metal substrate (or of its oxide, if forms) may also play a considerable role in the resulted film structure. It should be pointed out, however, that thermodynamic predictions of whether two materials are intermixing or not, usually consider their bulk properties, often tabulated in phase diagrams. If they are, indeed, miscible, the formation of mixed oxide phase is highly possible at the interface, which may even extend to the surface. If moderate, the effect is limited to the interface only and can be ignored for surface structures of relatively thick films. However, migration of metal support atoms to the surface should always be examined carefully. For example, growth of $Fe_3O_4(001)$ films on the MgO(001) surfaces at high temperatures resulted in Mg diffusion and formation of an $Mg_xFe_{3-x}O_4$ layer, which turned out to be even smoother than the original substrate surface.[2]

In essence, the above-discussed general considerations are sufficient for choosing a proper substrate for the preparation of a target oxide film, if there are no additional criteria to be taken into account, either relevant to the functional properties (for instance, catalysis or spintronics) or compatibility with existing technologies.

2.9 Deposition Methods

There are plenty of deposition techniques existing in the modern laboratories which remain the subject to continuous development. The techniques differ in a way how the material is brought onto a substrate, i.e., deposited "physically" or "chemically". In turn, the physical deposition methods are differentiated by the mechanism the deposited material is generated. This can be physical vapor deposition (PVD), pulsed laser deposition, magnetron and/or ion sputtering, to name a few. In contrast, chemical vapor deposition (CVD) involves volatile precursor compounds which adsorb and then decompose on a substrate.

PVD means a direct deposition of metal atoms from the heated to high temperatures pure metal source. This can be performed either in vacuum or in oxygen, at pressures low enough for the mean free path of evaporated atoms to be compatible with the experimental setup geometry. In some variations, a metal substrate is pre-covered with oxygen to ease oxidation of incoming metal atoms and also to minimize the intermixing with a substrate. Some of the PVD preparations are depicted in Fig. 2.18. In case of large lattice mismatch between a film and a support, combination of oxygen pre-adsorption and deposition at low temperatures such that surface diffusion is suppressed, may somewhat prevent aggregation into

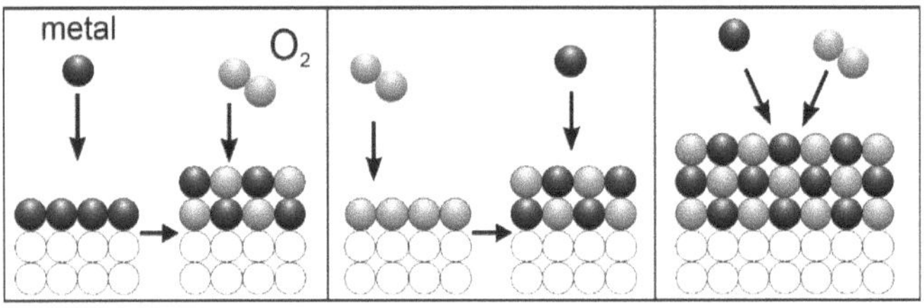

Figure 2.18 Oxide film preparation schemes using metal vapor deposition: metal deposition in UHV followed by oxidation in oxygen or another oxidizing agent; oxygen chemisorption (oxidation) of a metal substrate followed by metal deposition in vacuum (or oxygen); metal deposition in oxygen.

3D particles and hence favor the 2D film growth. For making the prepared films well-ordered, final steps usually include post-annealing, either in UHV or in oxygen.

Molecular beam epitaxy (MBE), originally invented and developed for fabrication of semiconductor films, in combination with in situ monitoring of the growth process, for example, by reflection high-energy electron diffraction is now widely used for complex oxide systems.[59] In MBE, low energy ($\sim$1 eV) thermal atomic beams for each of the constituent metals are directed towards a substrate together with oxidizing agent. The kinetics of epitaxial film growth, along with the requirements for proper phase formation, typically requires deposition at elevated temperatures. Basically, the choice depends on whether metal oxidation on the surface occurs at a rate higher than that of film growth. If so, this would minimize the formation of sub-oxides, for metals that exhibit multiple oxidation states, or even metallic clusters.

Pulsed-laser deposition (PLD) became a widely used approach after its success in growing high temperature superconductor films.[60,61] In this technique, a pulsed-laser is focused onto a target of the material to be deposited. At sufficiently high laser energy, each pulse vaporizes (ablates) some amounts of the material, which is then ejected from the target in a forward-directed plume. Although no background oxygen is required in PLD of oxides, it is recommended to introduce oxygen while deposition. In contrast to PVD and MBE, kinetic energies of ionized and neutral species can reach several hundred electronvolts that may result in the uncontrollable defect formation in the growing film. In addition to its simplicity in practical realization, PLD is thought to have advantage of stoichiometric transfer of complex material from the target, but this seems to be case sensitive. In sputter depositions,[62] a glow discharge plasma is generated in front of the target which removes (sputters) surface atoms which are then condensed onto a substrate.

For large area deposition and deposition onto porous supports, CVD is apparently better suited because of using volatile metal-containing molecules, whose adsorption onto a support does not suffer from shadowing effects as in case of PVD using directed flow. CVD makes use of direct sublimation of solid precursor into vacuum or in the carrier (typically, inert) gas. Some preparations simply involve passing the carrier gas through the precursor solution (bubbling). The technique yields a uniform deposition even on a non-planar, porous substrate, and can be used at elevated oxygen pressures, if needed. On the other hand, the CVD method suffers from the lack of control of cleanliness arising from the stripping of the ligands in the precursor and its thermal decomposition at surface. Another important issue for CVD of oxide materials is identification of the appropriate metal-organic precursors.

Another deposition method which received tremendous interest worldwide is referred to as of atomic layer deposition (ALD). Interestingly, the principle of ALD has been discovered twice: in the 1960s under the name "molecular layering" in the Soviet Union, and in the 1970s under the name "atomic layer epitaxy" (ALE) in Finland.[63] It includes sequential, via alternating pulses, deposition of chemical precursors exhibiting self-limiting growth at monolayer coverage. The general trends in ALD applied to inorganic films can be found in many comprehensive reviews (for instance, Ref. [64]) and will be not discussed here.

Finally, thin films can be deposited onto a planar substrate from precursor solution. Again, the process does not suffer from diffusion-related limitations as for porous supports. However, sedimentation of precursor onto a flat support followed by drying in ambient conditions often leads to a non-uniform material distribution across the surface. To circumvent this problem and prepare homogeneous films, one can make use of a spin coating procedure, i.e., putting a few droplets of the solution onto the rotating substrate to uniformly spread the material by centrifugal force. Another approach consists of dipping the substrate into the solution and pulling it slowly out, which is referred to as a dip-coating method.

References

1. Franchy, R., Growth of thin, crystalline oxide, nitride and oxynitride films on metal and metal alloy surfaces. *Surface Science Reports*, 2000. **38**(6–8):195–294.
2. Chambers, S.A., Epitaxial growth and properties of thin film oxides. *Surface Science Reports*, 2000. **39**(5):105–180.
3. Weiss, W. and W. Ranke, Surface chemistry and catalysis on well-defined epitaxial iron-oxide layers. *Progress in Surface Science*, 2002. **70**(1–3):1–151.
4. Surnev, S., M.G. Ramsey, and F.P. Netzer, Vanadium oxide surface studies. *Progress in Surface Science*, 2003. **73**(4–8):117–165.
5. Wu, Q.-H., A. Fortunelli, and G. Granozzi, Preparation, characterisation and structure of Ti and Al ultrathin oxide films on metals. *International Reviews in Physical Chemistry*, 2009. **28**(4):517–576.
6. Kuhlenbeck, H., S. Shaikhutdinov, and H.-J. Freund, Well-ordered transition metal oxide layers in model catalysis: a series of case studies. *Chemical Reviews*, 2013. **113**(6):3986–4034.
7. Seah, M.P. and W.A. Dench, Quantitative electron spectroscopy of surfaces: a standard data base for electron inelastic mean free paths in solids. *Surface and Interface Analysis*, 1979. **1**(1):2–11.
8. Werner, W.S.M., Electron transport in solids for quantitative surface analysis. *Surface and Interface Analysis*, 2001. **31**(3):141–176.
9. Bart, F. and M. Gautier, A LEED study of the (0001) α-quartz surface reconstruction. *Surface Science*, 1994. **311**(1–2):L671–L676.
10. Moritz, W., J. Landskron, and M. Deschauer, Perspectives for surface structure analysis with low energy electron diffraction. *Surface Science*, 2009. **603**(10):1306–1314.
11. Henzler, M., Capabilities of leed for defect analysis. *Surface Review and Letters*, 1997. **04**(03):489–500.

12. Wollschläger, J., Morphology and defect characterization of epitaxial oxide films. *Defect and Diffusion Forum*, 1998. **164**:37–56.
13. Wagner, C.D., Chemical shifts of Auger lines, and the Auger parameter. *Faraday Discussions of the Chemical Society*, 1975. **60**(0):291–300.
14. Gaarenstroom, S.W. and N. Winograd, Initial and final state effects in the ESCA spectra of cadmium and silver oxides. *The Journal of Chemical Physics*, 1977. **67**(8):3500–3506.
15. Hohlneicher, G., H. Pulm, and H.-J. Freund, On the separation of initial and final state effects in photoelectron spectroscopy using an extension of the Auger-parameter concept. *Journal of Electron Spectroscopy and Related Phenomena*, 1985. **37**(2):209–224.
16. Hirose, K., et al., Photoelectron spectroscopy studies of SiO_2/Si interfaces. *Progress in Surface Science*, 2007. **82**(1):3–54.
17. Himpsel, F.J., et al., Microscopic structure of the SiO_2/Si interface. *Physical Review B*, 1988. **38**(9):6084–6096.
18. Okada, K., Y. Kameshima, and A. Yasumori, Chemical shifts of silicon X-ray photoelectron spectra by polymerization structures of silicates. *Journal of the American Ceramic Society*, 1998. **81**(7):1970–1972.
19. Grunthaner, F.J., et al., Local atomic and electronic structure of oxide/GaAs and SiO_2/Si interfaces using high-resolution XPS. *Journal of Vacuum Science & Technology*, 1979. **16**(5):1443–1453.
20. Zakaznova-Herzog, V.P., et al., High resolution core and valence band XPS spectra of non-conductor pyroxenes. *Surface Science*, 2006. **600**(16):3175–3186.
21. Seah, M.P., et al., Critical review of the current status of thickness measurements for ultrathin SiO_2 on Si Part V: Results of a CCQM pilot study. *Surface and Interface Analysis*, 2004. **36**(9):1269–1303.
22. Flitsch, R. and S.I. Raider, Electron mean escape depths from x-ray photoelectron spectra of thermally oxidized silicon dioxide films on silicon. *Journal of Vacuum Science & Technology*, 1975. **12**(1):305–308.
23. Derrie, J. and M. Commandré, SiO_2 ultra thin film growth kinetics as investigated by surface techniques. *Surface Science*, 1982. **118**(1):32–46.
24. Kim, K.J., et al., Determination of the absolute thickness of ultrathin Al_2O_3 overlayers on Si(100) substrate. *Analytical Chemistry*, 2009. **81**(20):8519–8522.
25. Greenler, R.G., Design of a reflection–absorption experiment for studying the ir spectrum of molecules adsorbed on a metal surface. *Journal of Vacuum Science & Technology*, 1975. **12**(6):1410–1417.

26. Hoffmann, F.M., Infrared reflection-absorption spectroscopy of adsorbed molecules. *Surface Science Reports*, 1983. **3**(2–3):107–192.
27. Hollins, P. and J. Pritchard, Infrared studies of chemisorbed layers on single crystals. *Progress in Surface Science*, 1985. **19**(4):275–349.
28. Berreman, D.W., Infrared absorption at longitudinal optic frequency in cubic crystal films. *Physical Review*, 1963. **130**(6):2193–2198.
29. Harbecke, B., B. Heinz, and P. Grosse, Optical properties of thin films and the Berreman effect. *Applied Physics A*, 1985. **38**(4):263–267.
30. Martinet, C. and R.A.B. Devine, Analysis of the vibrational mode spectra of amorphous SiO_2 films. *Journal of Applied Physics*, 1995. **77**(9):4343–4348.
31. Galeener, F.L. and G. Lucovsky, Longitudinal Optical Vibrations in Glasses: GeO_2 and SiO_2. *Physical Review Letters*, 1976. **37**(22):1474–1478.
32. Kirk, C.T., Quantitative analysis of the effect of disorder-induced mode coupling on infrared absorption in silica. *Physical Review B*, 1988. **38**(2):1255–1273.
33. Fuchs, R. and K.L. Kliewer, Optical modes of vibration in an ionic crystal slab. *Physical Review*, 1965. **140**(6A):A2076–A2088.
34. Ibach, H., Optical surface phonons in zinc oxide detected by slow-electron spectroscopy. *Physical Review Letters*, 1970. **24**(25):1416–1418.
35. Avouris, P. and J. Demuth, Electron energy loss spectroscopy in the study of surfaces. *Annual Review of Physical Chemistry*, 1984. **35**(1):49–73.
36. Thiry, P.A., et al., Vibrational study of the SiO_2/Si interface by high resolution electron energy loss spectroscopy. *Journal of Vacuum Science & Technology B: Microelectronics and Nanometer Structures*, 1985. **3**(4):1118–1121.
37. Frank, M., et al., Phonons of clean and metal-modified oxide films: an infrared and HREELS study. *Surface Science*, 2001. **492**(3):270–284.
38. Binnig, G., et al., 7 x 7 reconstruction on Si(111) resolved in real space. *Physical Review Letters*, 1983. **50**(2):120–123.
39. Wang, Y.L., et al., Tip size effect on the appearance of a STM image for complex surfaces: theory versus experiment for Si (111) - (7 x 7). *Physical Review B*, 2004. **70**(7):073312.
40. Nilius, N., Properties of oxide thin films and their adsorption behavior studied by scanning tunneling microscopy and conductance spectroscopy. *Surface Science Reports*, 2009. **64**(12):595–659.

41. Bertrams, T., A. Brodde, and H. Neddermeyer, Tunneling through an epitaxial oxide film: Al_2O_3 on NiAl(110). *Journal of Vacuum Science & Technology B: Microelectronics and Nanometer Structures Processing, Measurement, and Phenomena*, 1994. **12**(3):2122–2124.
42. Liu, J., Advanced electron microscopy of metal–support interactions in supported metal catalysts. *ChemCatChem*, 2011. **3**(6):934–948.
43. Pennycook, S.J., et al., Materials advances through aberration-corrected electron microscopy. *MRS Bulletin*, 2011. **31**(1):36–43.
44. Crewe, A.V., J. Wall, and J. Langmore, Visibility of single atoms. *Science*, 1970. **168**(3937):1338–1340.
45. Krivanek, O.L., et al., Atom-by-atom structural and chemical analysis by annular dark-field electron microscopy. *Nature*, 2010. **464**(7288):571–574.
46. Weaver, L., Cross-section TEM sample preparation of multilayer and poorly adhering films. *Microscopy Research and Technique*, 1997. **36**(5):368–371.
47. Liu, Y., et al., A cross-sectional TEM sample preparation method for films deposited on metallic substrates. *Materials Characterization*, 2007. **58**(7):666–669.
48. Muller, D.A., et al., The electronic structure at the atomic scale of ultrathin gate oxides. *Nature*, 1999. **399**(6738):758–761.
49. Rupprechter, G., A surface science approach to ambient pressure catalytic reactions. *Catalysis Today*, 2007. **126**(1–2):3–17.
50. Hendriksen, B.M., S. Bobaru, and J.M. Frenken, Looking at heterogeneous catalysis at atmospheric pressure using tunnel vision. *Topics in Catalysis*, 2005. **36**(1–4):43–54.
51. Salmeron, M. and R. Schlögl, Ambient pressure photoelectron spectroscopy: a new tool for surface science and nanotechnology. *Surface Science Reports*, 2008. **63**(4):169–199.
52. Jaeger, R.M., et al., Formation of a well-ordered aluminium oxide overlayer by oxidation of NiAl(110). *Surface Science*, 1991. **259**(3):235–252.
53. Franchy, R., J. Masuch, and P. Gassmann, The oxidation of the NiAl(111) surface. *Applied Surface Science*, 1996. **93**(4):317–327.
54. Becker, C., et al., Oxygen adsorption and oxide formation on Ni3Al(111). *Journal of Vacuum Science & Technology A*, 1998. **16**(3):1000–1005.
55. Bauer, E., Phänomenologische Theorie der Kristallabscheidung an Oberflächen. I, in *Zeitschrift für Kristallographie*. 1958. p. 372.

56. Shaikhutdinov, S., M. Ritter, and W. Weiss, Hexagonal heterolayers on a square lattice: a combined STM and LEED study of FeO(111) on Pt(100). *Physical Review B*, 2000. **62**(11):7535–7541.
57. Davis, E.M., et al., Growth of Fe_3O_4(001) thin films on Pt(100): tuning surface termination with an Fe buffer layer. *Surface Science*, 2015. **636**(0):42–46.
58. Hagendorf, C., et al., Growth, atomic structure, and vibrational properties of MnO ultrathin films on Pt(111). *Physical Review B*, 2008. **77**(7):075406.
59. Chambers, S.A., Epitaxial growth and properties of doped transition metal and complex oxide films. *Advanced Materials*, 2010. **22**(2):219–248.
60. Norton, D.P., Synthesis and properties of epitaxial electronic oxide thin-film materials. *Materials Science and Engineering: R: Reports*, 2004. **43**(5–6):139–247.
61. Martin, L.W., Y.H. Chu, and R. Ramesh, Advances in the growth and characterization of magnetic, ferroelectric, and multiferroic oxide thin films. *Materials Science and Engineering: R: Reports*, 2010. **68**(4–6):89–133.
62. Kelly, P.J. and R.D. Arnell, Magnetron sputtering: a review of recent developments and applications. *Vacuum*, 2000. **56**(3):159–172.
63. Puurunen, R.L., A short history of atomic layer deposition: Tuomo Suntola's atomic layer epitaxy. *Chemical Vapor Deposition*, 2014. **20**(10–11–12):332–344.
64. Miikkulainen, V., et al., Crystallinity of inorganic films grown by atomic layer deposition: overview and general trends. *Journal of Applied Physics*, 2013. **113**(2):101.

Chapter 3

Thin Silica Films on Si and SiC

The most straightforward way to prepare a thin silica film is thermal oxidation of pure Si substrates and Si-containing alloys, since Si will most likely segregate at surface and be oxidized. Motivated by the key role of the SiO_2/Si interface in microelectronics, oxidation of the Si and SiC surfaces was investigated in an enormous number of studies on the mechanism of oxide formation, and structural and electronic properties of thin SiO_2 films and its interfaces.[1] These systems are the focus of this chapter.

3.1 Thermal Oxidation of Si Surfaces

Silicon crystallizes in a diamond cubic crystal structure (Fig. 3.1a). Among different surfaces, the (111) and (100) surfaces are the most studied, and corresponding single crystals are manufactured on industrial scale. Moreover, the Si(111) surface is one of the most explored systems in the "surface science" community because of a complex (7×7) reconstruction this surface undergoes upon heating in ultrahigh vacuum (UHV), which has attracted many research groups to understand its atomic structure (Fig. 3.1c; see also

Introduction to Ultrathin Silica Films: Silicatene and Others
Shamil Shaikhutdinov

ISBN 978-981-4877-28-2 (Hardcover), 978-1-003-08862-2 (eBook)
www.jennystanford.com

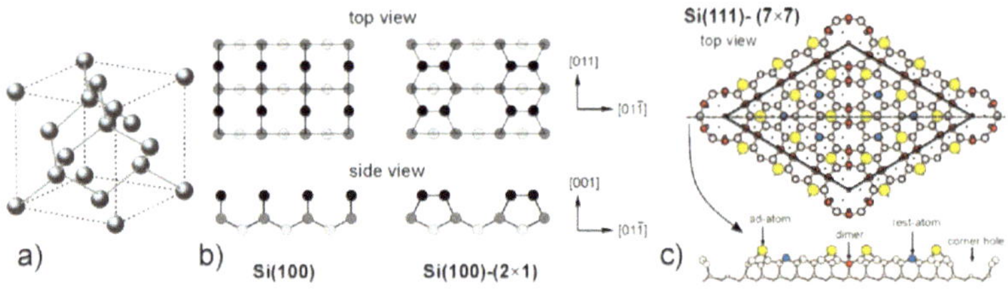

Figure 3.1 Crystal structure of Si (a), and schematic representation of unreconstructed and (2×1)-reconstructed Si(100) surfaces (b), and Si(111)-(7×7) surface (c).

Fig. 2.11 in Chapter 2). The Si(100) surface is thermodynamically more stable, although it exhibits a (2×1) reconstruction under certain conditions (Fig. 3.1b).

It is needless to say that oxidation of metallic surfaces has long been discussed in the literature with respect to corrosion and the formation of a passivating oxide film. In one of the first theories developed by Wagner in the 1930s,[2] the oxide film growth obeys a parabolic kinetics (i.e., thickness (L) depends on time (t) as $L \approx t^{1/2}$) under the following assumptions: (i) rapid interfacial reactions to create fixed bulk concentrations at the two interfaces (i.e., oxide/gas and metal/oxide), which yield concentration gradients; (ii) diffusion of charged species through the oxide film under these gradients; (iii) a zero divergence for each current in the interior of the film; and (iv) zero net charge transport through the film. The process was thought to be primarily controlled by diffusion across the growing oxide film. However, this is only valid for relatively high temperatures and sufficiently thick films so that the concentrations of ions and electrons are equal throughout most of its thickness. Actually, as was later outlined by Cabrera and Mott,[3] at each boundary there will be layers where they are not equal, and where a space charge is set up, giving rise to an electric double layer. The space charge region depends on many parameters, including the temperature, the dielectric constant of the oxide, and the concentration of the dissolved ions. In turn, this causes the electric field that affects the ion diffusion. In case of very thin films, usually formed by oxidation at low temperatures, the field may be so strong that the drift velocity of the ions is no longer proportional to it. The rate of oxidation at these strong fields is determined mainly by the rate at which ions escape from the metal. At temperatures below the critical temperature, the film grows rapidly to some thickness and then essentially stops growing. In the case of Si, early experiments suggested a critical temperature to be around 500°C.[4]

Later, Deal and Grove argued that any deviation from parabolic oxidation is not necessarily to be associated with space charge effects.[5] Focusing solely on the thermal oxidation of Si, they provided a mathematical model that describes the growth of a silica oxide layer to occur on the interface between the oxide and the metallic substrate rather than at the oxide/gas interface. Basically, it includes diffusion of interstitial oxidizing species through the film already formed and the reaction at the SiO_x/Si interface following a first-order kinetics. It should be noted, however, that all these phenomenological theories were developed for uniformly covered and ordered oxide films. That is not obvious and hence had to be proven. Note also that the early experimental results must be taken with some precautions, as most of them are based on experiments performed either in air or in low vacuum conditions, which hardly avoid the spontaneous formation of the "native" oxide layer, which, in turn, may alter the kinetics.

The experimental results obtained for Si oxidation in dry oxygen agreed well with the Deal–Grove theory, but only for thick oxide films prepared at high temperatures (800–1000°C).[6,7] The results for thinner oxide films (3–14 nm) grown on n-doped Si(100) at oxygen pressures of 0.25–2 Torr pointed to the presence of a thin interfacial region about 1.5 nm in thickness.[7] In addition, the oxidation kinetics was found to critically depend on Si surface plane as well as oxygen pressure.[8] For example, wet oxidation at atmospheric pressure at 785°C performed on (100)-, (110)-, and (111)-oriented Si wafers resulted in the initial oxidation rate order (110)>(111)>(100).[9] As the film growth proceeds further, the oxidation rate of the (111) plane surpasses that of the (110) plane, thus resulting in the order (111)>(110)>(100). Similar behavior was observed for dry oxidation.

Further studies provided a substantial body of evidence that the growth of ultrathin oxide films is quite different from thicker films, thus requiring certain modification of the theoretical models.

For example, to explain the initial rapid oxidation of Si in dry oxygen, it was suggested that a fraction of oxidizing species diffuses into the Si substrate.[10] The reaction occurs over this oxygen-diffused zone instead of occurring just at the interface. As the oxide thickness builds up, the concentration of the oxidizing species at the Si/SiO_2 interface decreases due to the increase in the diffusion resistance offered by the growing oxide film. Another model for observed deviations from the Deal–Grove predictions considered the transport of oxygen atoms through the silica network.[11] In other models, a distinct thin oxide layer was proposed separating the Si surface from the amorphous silica.[12] This "reactive" layer was thought to be stabilized by stress generated with the underlying silicon during oxidation at the interface. The layer was supposed to be impermeable to molecular oxygen diffusion after the first sub-oxide formation. Further analysis of experimental data, especially with the help of isotopic experiments,[13] highlighted the important role of the reactive layer adjacent to the Si surface in the conversion of interstitial oxygen molecules into network oxygens. The deviations from the Deal–Grove model for the thin films were explained by modification of oxygen diffusion, probably from structure changes rather than mere compression. However, further oxygen isotopic tracer studies[14] indicated that the main reactions take place directly at the Si/SiO_2 interface, thus supporting the original Deal–Grove model. It appears that only a combination of interfacial, near-interfacial, and surface reactions can properly describe Si oxidation in the initial stages, resulting in the formation of very thin films.

Comparative x-ray photoelectron spectroscopy (XPS) study of more than 50 differently prepared samples of different orientation and doping level, grown in air with or without wet chemical etching, or by anodic oxidation, revealed that a transition region always exists between the Si substrate and the SiO_2 film[15] Si^{3+} species may even be found up to 3 nm in the bulk SiO_2, whereas the Si^{+} and Si^{2+} states were mostly localized at the SiO_2/Si interface. Comparative study of samples oxidized in dry O_2 revealed the distribution depending on surface orientation: Si^{+} cations dominated at the SiO_2/Si(111) interface, whereas Si^{2+} was found in excess at the SiO_2/Si(100) interface.[16]

For thermal oxidation of Si substrates, some critical conditions seem to exist such as oxygen pressure and substrate temperature at which oxide film can only be grown. Basically, at high temperatures and low oxygen pressures, reactive etching of Si occurs, which is accompanied by production of volatile SiO species. When the incoming oxygen flux becomes higher than the rate of SiO desorption, surface oxidation sets in, thus resulting in a thin oxide layer.

Clearly, the formation mechanism and the atomic structure of the interfacial layer in ultrathin silica films formed on Si single crystals could only be elucidated by studying under UHV conditions, providing the best control of the surface. Combined Auger electron spectroscopy (AES) and electron energy loss spectroscopy (EELS) studies of oxygen interaction with the clean Si(111)-(7$\times$7) surface suggested that the SiO_2/Si interface is not abrupt, and a transition layer about 5 Å in thickness exists between the Si substrate and the silica film.[17] For comparison, a combined high-resolution electron energy loss spectroscopy (HREELS) and XPS study of oxygen adsorption and oxidation of the Si(100)-(2$\times$1) surface at temperatures between 300 and 1000 K suggested a rather abrupt Si/SiO_2 interface.[18]

Figure 3.2a displays Si $2p_{3/2}$ high-resolution photoelectron spectra obtained with synchrotron radiation on ultrathin silica films (estimated nominal thickness $\approx$ 5 Å) prepared by oxidation of the clean Si(001) and Si(111) surfaces in 2×10^{-5} Torr of O_2 at 750°C for 20 seconds.[19] In addition to Si^0 signal from a substrate, four differently oxidized Si^{n+} ($n = 1$–4) species are present in the film. Note that the Si^{3+} contribution is enhanced at this photon energy ($h\nu = 130$ eV) relative to Si^{1+} and Si^{2+} partially due to a cross-section resonance of Si^{3+} at this photon energy. Nevertheless, an oxide overlayer on the Si(111) substrate has less Si^{2+} and more Si^{+} than the Si(100) substrate. At increasing thickness by a factor of 2, the Si^{4+} component considerably gains in intensity (see Fig. 2.5 in Chapter 2). However, the interface is not atomically abrupt that manifests itself through the high density of intermediate-oxidation states.

It is fair to say that the majority of surface science studies of SiO_2/Si films in the early 1980s have been carried out primarily with

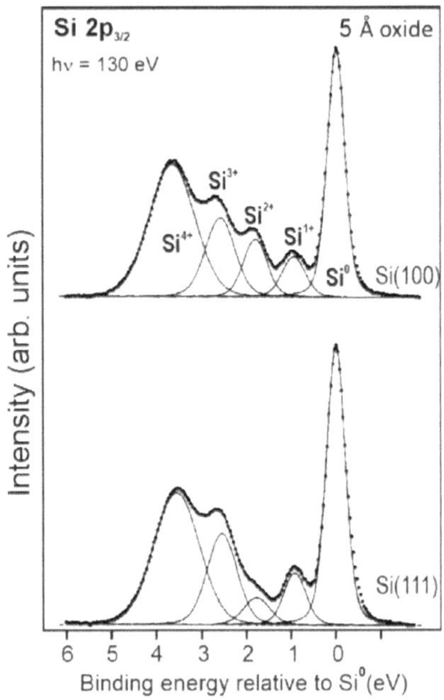

Figure 3.2 Si $2p_{3/2}$ photoelectron spectra from ultrathin silica overlayers on Si(100) and Si(111) surfaces. The spectra deconvolution into several Si^{n+} states after background subtraction. Adapted with permission from ref. 19.

XPS. For a more comprehensive review, the readers are referred to ref. 20.

With a help of scanning tunneling microscopy (STM), it became possible to directly monitor Si surface etching process with spatial resolution at the nanometer scale. In particular, STM images of the initially clean Si(111)-(7×7) surface were recorded in situ at 950 K in 10^{-7} mbar of O_2.[21] The results supported the previously suggested, so-called "step flow" mechanism for removal of the topmost, defect-free Si layer and formation of monolayer deep pits in the defective (7×7) layers in the further reaction process. The follow-up in situ STM study[22] revealed the distinct transition from etching to oxide growth occurring at higher oxygen pressures as displayed in Fig. 3.3. STM images revealed that oxide growth begins by heterogeneous nucleation of small oxide clusters (~2 nm in size) at steps and domain boundaries, simultaneously with

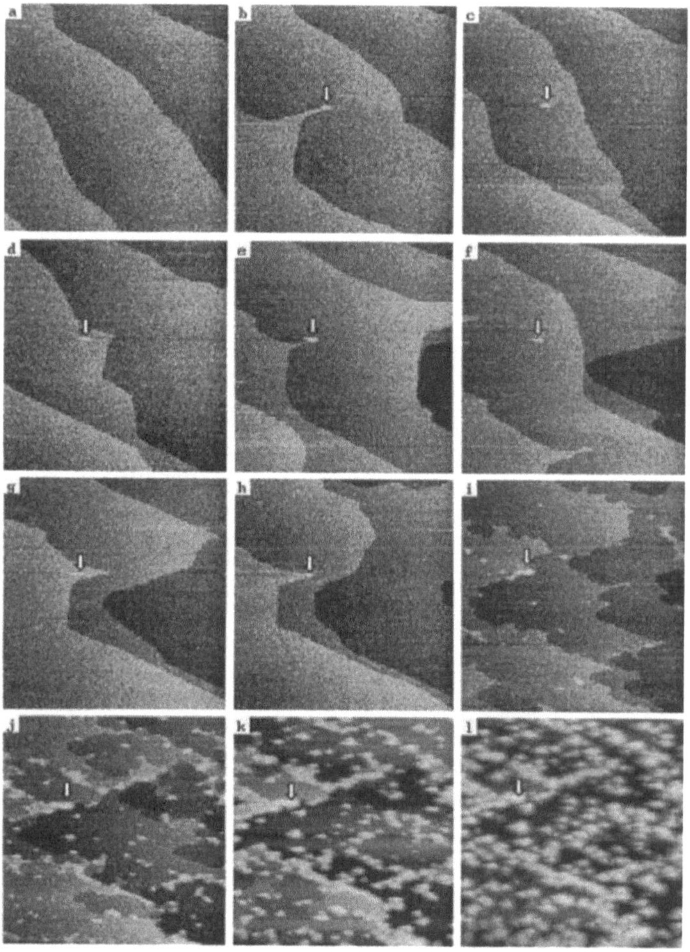

Figure 3.3 Series of STM images (250×250 nm) recorded at 1050 K illustrating the structural evolution of the Si(111)-(7×7) surface during interaction with O_2 at stepwise increased pressures: (a) $<2\times10^{-8}$ Pa; (b) 2×10^{-5} Pa; (c–g) 3×10^{-5} Pa; (h–j) 4×10^{-5} Pa; (k) 5×10^{-5}Pa; and (l) 7×10^{-5} Pa. (The arrow marks local contamination acting as a pinning center.) Reproduced with permission from ref. 22.

the ongoing etching process. Apparently, the clusters stabilize steps against etching and act as a barrier for moving steps in the step flow process. After removal of the first Si bilayer, the process is effectively stopped and further reaction proceeds via slow growth of the oxide covered areas and simultaneous Si removal in between those areas. Finally, the entire surface becomes covered with small oxide clusters.

Based on complementary STM and XPS results, the initial stages of oxide growth on Si(111)-(7×7) depend on temperature.[23] While the high-temperature (600°C) oxidation showed oxide islanding, small oxide clusters uniformly covering the surface were formed at room temperature. In fair agreement with these findings, the density functional theory (DFT) calculations (on Si(001) surface),[24] showed that oxidation at high temperature propagates in the direction normal to the surface rather than in the lateral direction, even in the sub-monolayer coverage regime. It, therefore, appears that the formation of an atomically thin and a uniform oxide layer by thermal oxidation can only be achieved at relatively low oxygen pressures and temperatures.

From early works performed in the 1990s up to date, numerous STM as well as atomic force microscopy (AFM) studies of thermally grown thin silica films on Si wafers could not yet image the surface structure of the film with atomic resolution. Although the native silica films basically follow the morphology of a Si substrate underneath,[25] these films were never atomically flat and exhibited high corrugation, probably reflecting the amorphous nature of silica formed.

In the course of Si thermal oxidation studies, it was realized that the Si(001) surface is better suited than Si(111) for the formation of abrupt and ordered SiO_2/Si interfaces, which are important for the fabrication of the *crystalline* SiO_2 films. A high-resolution transmission electron microscopy (HRTEM) study of thin films thermally grown on the Si(001) surface (prepared by additional Si vapor deposition onto a Si(001) single crystal) revealed an atomically flat interface, which was assigned to a crystalline transition layer about 5 Å in thickness presumably having a tridymite-like structure.[26] In addition, as mentioned above, XPS results pointed to a lower density of intermediate-oxidation states for Si(001) compared to Si(111).[19] Further works using reflection high-energy electron diffraction (RHEED) measurements combined with HRTEM of thin films on a Si(001) single crystal, which was cut with their surface normal at 5° from the (001) axis, revealed good epitaxy of an oxide phase along the (1–10) ledges.[27] The crystal parameters did not fit any known phases of silica, indicating a strained "abnormal" oxide layer that forms during the initial growth of thin films.

On the basis of x-ray diffraction (XRD) patterns of SiO_2/Si(001) films, it was proposed that the films might have silica crystallites dispersed within.[28,29] The crystallites were located not only at the interface, but also inside the amorphous film, preserving an epitaxial relation among the crystallites. However, the proportion of such crystallites having the cristobalite-like structure was approximately a few percent of the entire film. Cristobalite is often considered as the most likely candidate for an epitaxial oxide on Si, since it is structurally similar to Si and can be obtained by the insertion of O atoms between the Si–Si bonds in crystalline Si. On the other hand, its lattice parameter is 40% larger. An epitaxial relationship between Si and cristobalite is therefore difficult to achieve.

A few studies reported ordering at SiO_2/Si(100) interfaces when the films were grown on passivated Si(100)-(1×1) surfaces produced by a special wet chemical cleaning.[30,31] A combination of data obtained by ion beam channeling and nuclear resonance revealed a registry of Si atoms in SiO_2 relative to the Si substrate, which has been attributed to the cristobalite epitaxial phase. However, the structurally ordered phase transforms into an amorphous structure approximately 2 nm from the interface. Beyond this thickness, the oxide was always found in the amorphous state.

Despite numerous HRTEM studies of SiO_2/Si interfaces, only with the help of a spherical aberration correction it became possible to investigate the SiO_2/Si(100) interfaces with truly atomic resolution.[32] Careful image simulations suggested one to two atomic layers of cristobalite at the interface, referred to as "a metastable crystalline SiO_2 layer."[33] Figure 3.4 compares the experimental

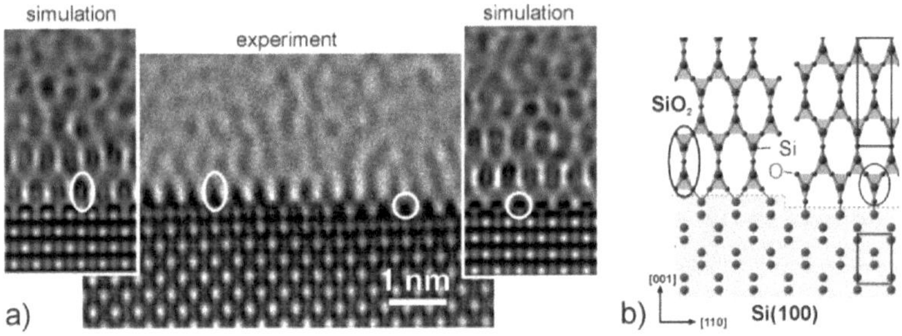

Figure 3.4 (a) Comparison of simulated and experimental TEM images. (b) Pseudo-cristobalite structure model of the interface between SiO_2 and Si(100). The two rectangles show the sizes of unit cells of the Si substrate and pseudo-cristobalite. The circles and ovals in (a) correspond to single and double SiO_4 tetrahedra marked in (b), respectively. Adapted with permission from ref. 33.

cross-sectional TEM image with image simulations, for which a pseudo-cristobalite structure was used for modeling the interface between a SiO_2 film and a Si(001) substrate.

Being sensitive to the details of Si–O bonding configuration, infrared spectroscopy was intensively used to study the SiO_2/Si interfaces.[34–39] Figure 3.5a,b shows a series of IR spectra obtained at normal and grazing incidence on a thermally grown silica film on Si(001), which was then gradually thinned by dilute HF solution down to 0.6 nm (the average film thickness was estimated with XPS). First of all, the spectra clearly demonstrate the Berreman effect, making the longitudinal mode in silica (at around 1250 cm^{-1}) only visible in spectra if measured at grazing incidence (see Chapter 2). Secondly, both longitudinal and transverse optical (LO and TO, respectively) bands considerably red-shift with decreasing film thickness. Such a shift has been reported in many other studies, and Fig. 3.5c summarizes experimental observations to show the general trend.

Analysis of several different factors potentially causing the thickness effect on the IR band position led to the observation that neither image charge effects nor increased stress nor density near the interface and surface/interface roughness were dominant factors.[35] In fact, the data were best fitted by introducing a substantial component of sub-stoichiometric oxide within about 6 Å from the Si surface, in full agreement with conclusions made on the basis of XPS results mentioned above. It seems plausible, however, that not only sub-stoichiometry, but also interface-related strain contributes to the band shift.[39]

Interestingly, the LO and TO frequencies measured on the *ordered* SiO_2/Si(001) interface are slightly higher throughout the thickness analyzed (2.5 nm) than those measured with the same setup for "conventionally" grown, disordered films in the same range of thicknesses. The comparison also revealed that better-ordered films exhibit a more constant frequency across a 1 nm thickness in the interfacial oxide region near the Si substrate.

These experimental findings were theoretically investigated using a scheme for calculating both the LO and TO spectra of surfaces and interfaces from first principles.[40] Applied to the previously suggested structural model for the SiO_2/Si(100) interface,[41] IR spectra were calculated, which basically reproduced the experimentally observed shifts shown as a solid line in Fig. 3.5c. Based on the calculations, the effect is assigned to the lengthening of the Si–O bonds in the sub-stoichiometric oxide at the interface.

Not surprisingly, numerous theoretical studies have been performed in order to establish the atomic structure of the SiO_2/Si(001) interface. Among most common silica polymorphs considered

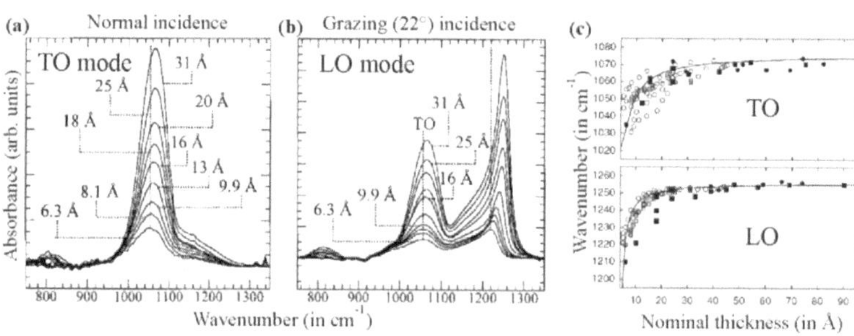

Figure 3.5 IRAS spectra obtained at normal (a) and grazing (b) incidence on thermally grown silica films on Si(001) uniformly thinned from ~3 nm down to 0.6 nm as indicated. Adapted with permission from ref. 35 (c) Position of the TO and LO peaks of the Si(100)/SiO_2 interface vs. oxide thickness obtained in several works. Solid lines are theoretical predictions (see ref. 40).

for theoretical modeling and examined by molecular dynamics calculations, tridymite was found to be the best ordered and energetically the most stable, albeit showing the highest strain in the Si layer.[42] In all interfaces studied, except β-cristobalite, the crystalline order was preserved with defects in the form of dislocated oxygen atoms, whereas the interface with β-cristobalite led, in essence, to a disordered arrangement of atoms.

In addition to thermal Si oxidation, which uses molecular oxygen as oxidizing agent, the formation of ultrathin silica films on a Si(001) wafer has been observed with highly concentrated nitric acid (HNO_3) vapor at 100–200°C. Depending on the exposure time and temperature, silica films were obtained with thickness, measured by XPS and cross-sectional TEM, ranging from 0.6 to 2 nm.[43] Focusing primarily on electric properties of such films, it was found that the leakage current density was lower by one order of magnitude than that for films of the similar thickness grown with O_2. The effect has been attributed to combination of several reasons such as: (i) the atomically flat SiO_2/Si interface and uniform thickness of the ultrathin SiO_2 layer; (ii) the low concentration of sub-oxide species and the low interface state density; and (iii) the high atomic density of the SiO_2 layer, which leads to a high band discontinuity energy at the SiO_2/Si interface.

It is generally believed that functional properties of a crystalline film are superior to the amorphous or highly defective films. Certainly, crystalline phases and the interfaces they form can effectively be studied with experimental and theoretical techniques which are more efficient for periodic than disordered systems. However, thin silica films thermally grown on Si(111) and Si(001) seem all to be amorphous in nature, regardless of the film thickness and Si surface orientation. It is worth mentioning, however, that many other metal oxide films grown on Si substrates do exhibit crystalline structures.[44] In case of very thin films, the film quality to a larger extent depends on the atomic structure of the SiO_2/Si interface.[45] The interface is a key not only to the controlled growth of well-ordered silica films, but it dominates electron transport and hence functional properties of the metal-oxide-semiconductor (MOS) based electronic devices.[46,47]

3.2 Thin Silica Films on SiC

In contrast to Si, silicon carbide (SiC) exhibits more than two hundred crystal modifications (polytypes). The smallest structural units are SiC_4 and/or CSi_4 tetrahedra with sp^3-hybridized silicon and carbon atoms, i.e., each Si is bonded to four C atoms, and each C is bonded to four Si atoms. (For comparison, in SiO_2 each Si is bonded to four O atoms, but O is only bonded to two Si atoms). The Si–C and Si–Si (C–C) distances in SiC are 1.89 and 3.08 Å, respectively. Two interconnected SiC_4 and CSi_4 tetrahedra may form either a hexagonal or a cubic cite, as depicted in Fig. 3.6a. The Si–C bilayers, always hexagonal, may stack differently, that is in an ABAB...sequence like in hexagonal wurtzite crystal structure or in an ABCABC... sequence like in a cubic zincblende structure. Basically, the SiC polytypes differ in the number of distinct basic units and the way the units are connected with each other. Due to a large number of polytypes, the crystal nomenclature is rather complex. It includes a combination of a number and a letter. The number specifies the total number of layers in a unit cell, and the letter denotes the lattice type, either hexagonal (H), cubic (C), or rhombohedral (R) (some are depicted in Fig. 3.6). Just to mention, the 2H–SiC polytype corresponds to the pure hexagonal close packing, i.e., in the ABAB... sequence, and 3C–SiC corresponds to the cubic close packing, i.e., in the ABCABC... sequence. In all cases, the SiC surface may be terminated either with Si or C in the topmost layer.

As in the case of pure Si substrates, a silica film can directly be grown on SiC surfaces in oxygen atmosphere at elevated temperatures. The oxidation of SiC, which is even more complex than of Si, received considerable attention since 1990's, after the considerable development of SiO_2/SiC systems which were thought to be directly applied in MOS-based electronic devices.[48] The formation of a high quality insulating oxide layer on SiC is a key requirement for electronic devices based on this material.[49] However, the electronic properties of SiO_2/SiC interfaces indicated a much higher density of electrically active centers than for the films grown on Si substrates. In fact, only a few SiC polytypes received particular interest

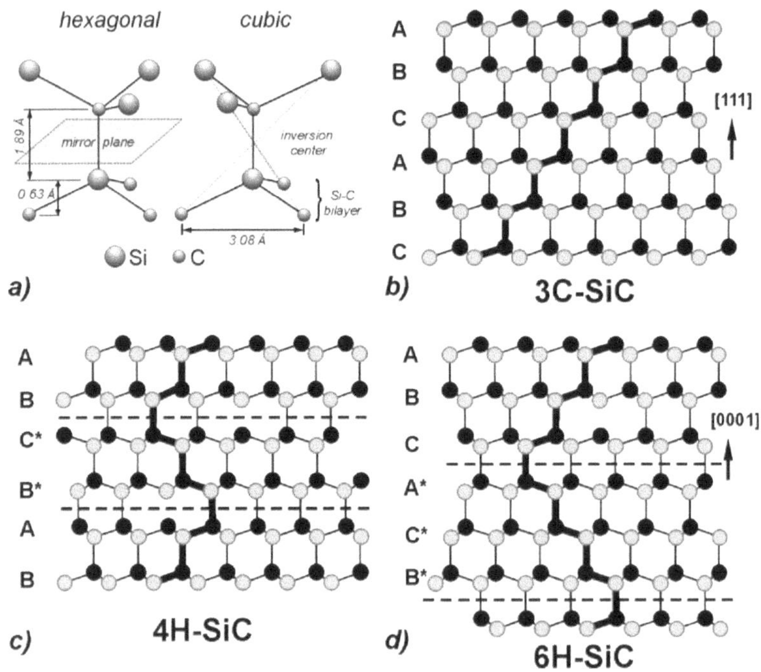

Figure 3.6 Structural units (a) and side views of selected SiC polytypes (b–d). In 3C-phase, the SiC bilayer is regularly stacked throughout the crystal as shown by a bold zigzag line, whereas the stacking sequence changes every two and three bilayers in the 4*H* and 6*H* structures, respectively. Capital letters A, B and C indicate SiC bilayers with three different stacking ways, having the same translational symmetry. A* and others indicate a bilayer which is related by a 180° in-plane rotation with respect to the bilayer A and others.

with respect to their potential applications in electronics.[50] The hexagonal 6H- and 4H-phases are primarily considered thanks to good quality single crystalline wafers became available on market. In addition, these polytypes exhibit considerably wider band gaps, i.e., 3 eV (for 6H-) and 3.3 eV (for 4H-) as compared to 2.4 eV observed on the 3C–SiC phase. The structures are schematically shown in Fig. 3.6 (b–d).

As discussed in the previous section, thin silica films grown on Si single crystals are amorphous, that can at first glance be rationalized in terms of a large lattice mismatch between SiO_2 and Si surfaces. In this respect, the lattice parameters of SiC and SiO_2 match fairly well. Depending on the degree of "hexagonality," the mismatch may even be below 5%. Certainly, this fact standing alone favors good epitaxy between SiO_2 and SiC. On the other hand, ab initio calculations comparing SiO_2/Si(001) and SiO_2/SiC(001) interfaces showed that the SiC surfaces are hardly suitable for the abrupt interface, and the formation of sub-oxides at the interface seems inevitable, although the conclusion was drawn solely on geometric arguments.[51]

In general, oxidation kinetics of SiC surfaces showed similar behavior to Si and was originally described within the Deal–Grove model applied to oxidation of pure Si, although oxygen has a lower solubility in SiC than in Si. In fact, the oxidation rate depends on the polytype and surface termination (Si– or C–) of the most stable basal planes.[49,52] Usually, the Si-terminated surfaces showed considerably slower oxidation rate than the C-terminated ones. The difference was linked to the different amounts of the interfacial oxycarbides $Si_4C_{4-x}O_4 (x<2)$ probably affecting the film growth.[53]

Thermal oxidation of SiC practically always results in amorphous silica films. Nonetheless, the films can be crystallized, at least partially, into β-cristobalite-like structure upon annealing at very high temperatures, that is not possible on Si just because of its lower melting point temperature.[54] Note, however, that such semi-crystalline films were usually prepared in hot furnaces in dry air or high oxygen pressures and had thicknesses in the range of several hundred nanometers and can hardly be considered as truly thin films.

For the first time, the formation of well-ordered silica films on SiC has been reported by Heinz and co-workers[55,56] for hexagonal SiC(0001) substrates. The preparation began with a sacrificial oxidation followed by HF etching. After that the sample was etched in hydrogen microwave plasma at around 800°C. Not surprisingly, upon transfer the sample to the analysis UHV chamber through air, the surface became oxidized. Surprising was, however, that the sample immediately displayed a $(\sqrt{3}\times\sqrt{3})R30°$ LEED pattern (Fig. 3.7a), and the long-range ordering was even observed after having the sample exposed to air for several hours. The resulted oxide film could only be decomposed after UHV annealing above 1000°C.

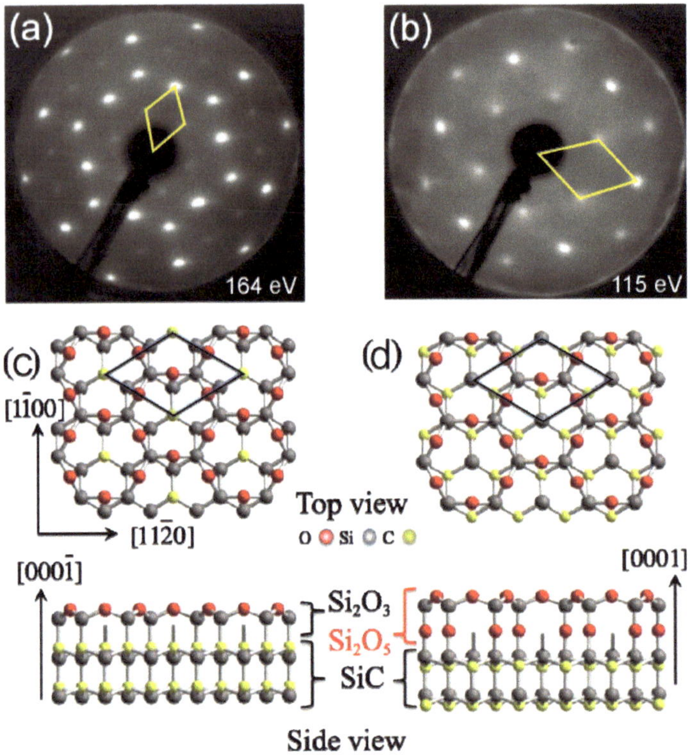

Figure 3.7 LEED patterns obtained on the hydrogen passivated 6H–SiC(000-1) (a), and 4H–SiC(0001) (b) surfaces (at the electron energies as indicated) upon exposure to air at room temperature. (c,d) Top and side views of the silica films formed. The $(\sqrt{3}\times\sqrt{3})R30^\circ$ unit cells are indicated by rhombs. Note that dangling bonds, represented by short vertical bars in the side views, exist on the top atoms of the SiC surfaces for both structures. Adapted with permission from refs. 55 and 58.

Analysis of I/V LEED curves measured on such films yielded the best fit (reliability *R*-factor of 0.14) for a model containing a Si_2O_3 monolayer on top of a bulk-like SiC(000-1) substrate as shown in Fig. 3.7c. Interestingly, the same $(\sqrt{3}\times\sqrt{3})R30^\circ$ structure was observed on the Si-terminated SiC(0001) surface, which was, however, better described as a Si_2O_5 layer (*R*-factor is 0.2) depicted in Fig. 3.7d. Further DFT study[57] showed that both the Si_2O_3 and the Si_2O_5 models have roughly the same formation energy in case of the C-terminated SiC(000-1) surface, whereas the Si_2O_5 model is energetically much more stable on the Si-terminated SiC(0001) surface. The structural parameters calculated by DFT were in excellent agreement with the LEED results.

The "as prepared" films on SiC are usually contaminated, primarily with hydrocarbons as found by XPS studies performed with synchrotron radiation at different photon energies thus allowing to address depth distribution for elemental composition.[59] The surface impurities desorb upon UHV annealing at 650°C simultaneously causing an upward band bending by about 1 eV. In addition, a new weak feature appeared about 0.4 eV above the valence band edge which is typical for a dangling bond state. Indeed, one-third of the carbon atoms in the topmost SiC bulk bilayer have a dangling bond in the structural model shown in Fig. 3.7a. During oxidation performed in air these bonds are most likely saturated by adventitious adsorbates and therefore only show up in XPS spectra after annealing.

It should be mentioned that numerous XPS studies were carried out on different SiC polytypes in order to address initial stages of oxidation and growth of silica films (see, for instance, refs. 60–62). For example, oxidation of the clean 4H–SiC surfaces at increasing oxygen exposures up to 10^6 L (1 L $= 10^{-6}$ Torr$\times$s) was monitored at room temperature and 800°C for comparison. Beyond the Si electronic states in a SiC substrate, spectra revealed two additional states of Si, assigned to Si^{+} and Si^{4+} in films prepared on the Si-terminated surface, and Si^{2+} and Si^{4+} in films on the C-terminated surface (see Fig. 3.8a,b). For comparison, Fig. 3.8c reproduces Si 2p spectra obtained at the same photon energy for a crystalline $(\sqrt{3}\times\sqrt{3})R30°$ layer prepared in air (see Fig. 3.7). On the basis of the intensity ratio of oxide and substrate related signals, oxidation of the clean SiC surface at elevated temperatures in the 10^{-3} mbar range of O_2 pressures (used for samples shown in Figs. 3.7a,b) seems to propagate considerably deeper as compared to the H-etched SiC surface exposed to air at 300 K used for the sample in Fig. 3.7c.

Even today, twenty years later, the mechanism of the ordered film formation on SiC in air remains puzzling. Under preparation conditions used, i.e., treatment either in hydrogen plasma or in 1

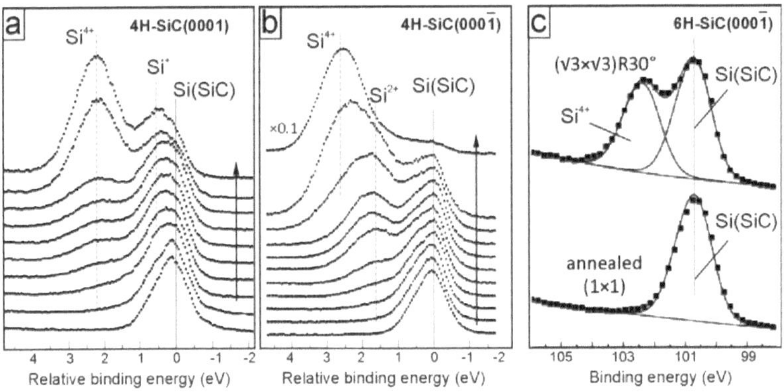

Figure 3.8 Photoelectron spectra of the vacuum-cleaned 4H–SiC(0001) (a) and 4H–SiC(000-1) (b) surfaces exposed to O_2 (10^{-7}–10^{-3} mbar) at 800°C at increasing exposure (from bottom to top). Adapted with permission from ref. 62. For comparison, panel (c) shows spectra for a ($\sqrt{3}\times\sqrt{3}$)R30° silica film formed in air on 6H–SiC(000-1) and the same film annealed to 1100°C in UHV, from ref. 59. The photon energy in all spectra is 130 eV.

bar of H_2 at 1500°C, oxidation is hardly possible if hydrogen, of course, was not contaminated with traces of oxygen. The authors of the original work[55] speculated that the ($\sqrt{3}\times\sqrt{3}$)R30° order is already induced during this preparation, ultimately forming a well-ordered silicate layer upon subsequent exposure to air. Later it was shown that the ordered silicate films could as well be prepared under UHV-based conditions using oxidation of the Si-rich 6H–SiC(000-1)-(2×2) surface at 500°C at relatively high O_2 exposures (ca. 10^5 L).[63] The long-range order remained by further oxidation at high temperatures, up to 1000°C. Furthermore, ordered films were also obtained on 4H–SiC(0001) surface under UHV conditions by exposing the Si-rich surfaces to O_2 contaminated with water at temperatures of about 900°C.[64] It appears that surface enrichment with Si plays important role for the film ordering. Apparently, the Si-rich surfaces are prone to O_2 dissociation even at room temperature.[65] On the basis of combined STM, XPS and IRAS results, it was concluded that oxygen readily dissociates at the surface with subsequent incorporation into the silicon layer. In contrast to pure Si, this initial oxidation is likely to relax the highly strained SiC surface, suggesting that a more uniform silica film can be produced on SiC than on Si substrates.

Medium energy ion scattering (MEIS) and XPS experiments on the ordered films prepared in UHV showed that it consists of a silicate layer (Si_2O_3), which partly accompanies a Si_2O_5 structure.[63] However, with increasing the oxidation temperature and at longer oxidation time, the Si_2O_5 silicate layer starts to dominate. Moreover, monitoring isotope exchange reaction with MEIS on the film prepared with $^{16}O_2$ and then exposed to isotopically labeled oxygen ($^{18}O_2$) at 900°C, it was possible to shed some light on the oxidation kinetics. The four steps were considered, namely: (i) oxygen dissociative adsorption on the surface Si atoms; (ii) oxygen adsorption to the back-bond of the Si layer thus forming Si–O–C linkages to SiC; (iii) oxygen desorption from the silica layer; and (iv) oxygen transfer from the silica layer to the back-bond of the Si layer. After the Si_2O_3 silica layer is formed rather promptly, the process (ii) is believed to be the rate-determining step, which accompanies the processes of (iii) and (iv). The oxidation showed Arrhenius type temperature dependence, from which the apparent activation barrier of 130 $\pm$ 20 kJ/mole was determined.

The angle-resolved photoelectron diffraction (PED) measurements of the ordered films prepared on 4H–SiC(0001) in UHV well agreed with the Si_2O_5 layer model.[64] Comparative analysis of PED data for the ordered silica layer and amorphous ultrathin silica film (prepared by oxidation in 10^{-4} mbar of O_2 at 800°C) revealed some similarities (Fig. 3.9). The diffraction patterns of the Si atoms in a SiC substrate are nearly identical for both ordered and disordered films since in both samples the Si atoms of the SiC bulk are located in identical atomic environments. The same was found for the Si^+ diffraction patterns, thus indicating that the local atomic environment of the Si atoms at the interface is very similar in both systems. However, considerable differences were observed for diffraction patterns of Si^{4+} species in silica, especially at high polar angles (marked with circles in Fig. 3.9(c,g)). Further analysis showed that the only structural difference between these two films is the smaller size of ordered regions in the non-ordered film and the absence of oriented bonds of the near-interface Si atoms to oxygen atoms.

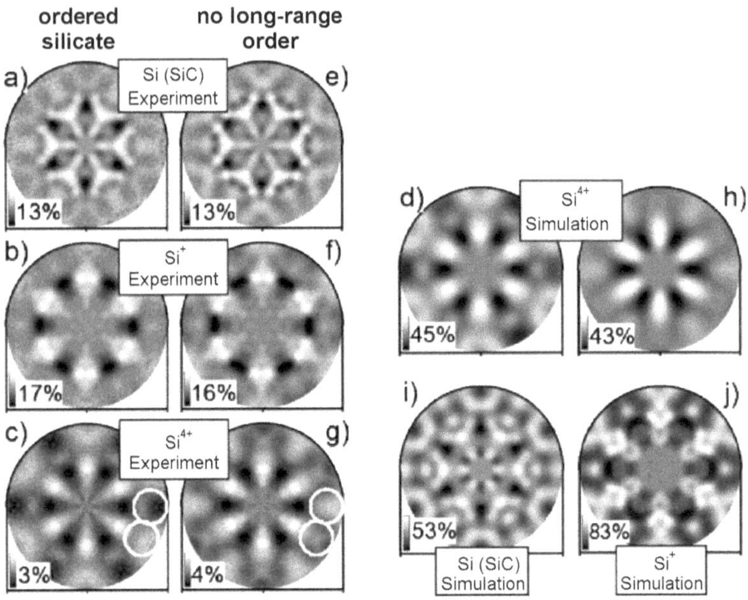

Figure 3.9 Experimental and simulated x-ray photoelectron diffraction patterns of the Si 2p emission lines in a $(\sqrt{3}\times\sqrt{3})R30^\circ$ ordered film (a–d) and an amorphous film of similar thickness (e–j). The photon energy is 180 eV. Adapted with permission from ref. 64.

It is a bit surprising, that ordered silica films on SiC have only recently been investigated with STM.[66] The films were prepared following the originally reported recipe for 4H–SiC(000-1), i.e., by etching in 1 bar of H_2 at 1400°C. Then the through-air-transferred samples were annealed in UHV at 655°C to remove hydrocarbon contaminations. Although the films displayed a sharp $(\sqrt{3}\times\sqrt{3})R30^\circ$ diffraction pattern, STM images revealed, however, a surface covered by small particles of several nanometers in size. The films became better ordered after thermal flash to 970°C such that flat terraces exposing a honeycomb-like structure were observed in agreement with the proposed models shown in Fig. 3.7. In addition, structural defects were imaged by STM tentatively assigned to missing SiO_n units. Tunneling spectroscopy measurements revealed the band gap (5.5 ± 0.5 eV) which is much smaller than 8.9 eV for bulk SiO_2. This reduced band gap at the surface could be reproduced theoretically[67]

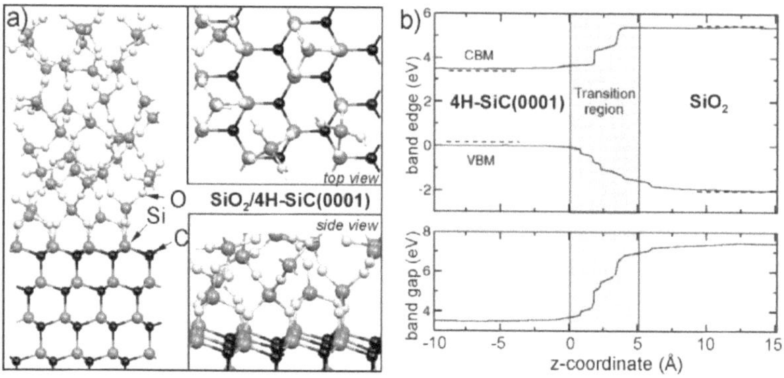

Figure 3.10 (a) Side view of the SiO_2/4H–SiC(0001) interface. The transition region is zoomed in insets. (b) Evolution of the conduction band minimum (CBM), valence band maximum (VBM), and a band gap across the interface. Dashed lines correspond to VBM and CBM obtained from bulk calculations of SiC and SiO_2 and aligned at the interface through the average electrostatic potential. Adapted with permission from ref. 68.

only for the Si_2O_3 layer, whereas the calculated band gap of the Si_2O_5 layer formed on the Si-terminated surface is about 9 eV, i.e., very close to the experimental value of bulk SiO_2. It appears that the difference is largely due to Si–C bonds in the Si_2O_3 layer which are missing in the Si_2O_5 layer only having Si–O bonds in the film.

Inspired by experimental observations of ordered silica layers on SiC surfaces, atomically abrupt interfaces have been generated in theoretical modeling of the SiO_2/SiC interfaces in thicker silica films, although the films themselves were amorphous. One such model is shown in Fig. 3.10a presenting the transition region between SiC and SiO_2 without any coordination defect.[68] Analysis of the electronic structure showed the full band gap of silica at a distance about 5 Å away from the interface (Fig. 3.10b).

As mentioned above, well-ordered silica layers on SiC were produced by annealing at 1300–1500°C in 1 bar of H_2 and subsequent exposure to the air. It has turned out that switching H_2 to N_2 and further annealing in nitrogen for 2–30 min prior to the exposure to air results in the N-containing film showing the same $(\sqrt{3}\times\sqrt{3})R30^\circ$ diffraction pattern. I/V LEED analysis (R-factor 0.14) suggested the atomic structure shown in Fig. 3.11a,

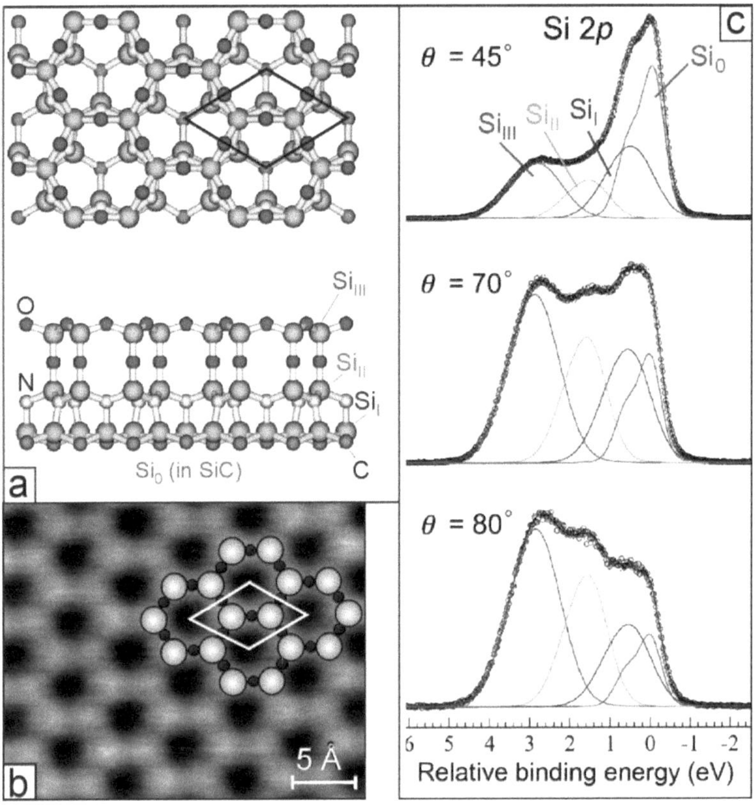

Figure 3.11 (a) DFT-optimized structure of a SiO(N) film on 4H–SiC(0001) in top and side views. (b) STM image superimposed with the topmost silicate layer. The $(\sqrt{3}\times\sqrt{3})R30°$ unit cell is indicated. (c) Photoelectron Si 2p spectra measured with photon energy 700 eV at the emission angle of 45°, 70°, and 80° with respect to the surface normal. The spectra deconvoluted into four components corresponding to different Si coordination to surrounding atoms (Si_0, Si_I, Si_{II}, and Si_{III}) marked in model (a). Adapted with permission from ref. 58.

which was referred to as a silicon oxynitride layer, in which a silica layer sits on top of a silicon nitride layer through the Si–O–Si linkages.[69] The LEED derived model was corroborated by DFT calculations resulting in the $Si_4O_5N_3$ stoichiometry in the film.[67,70] High-resolution STM images (see Fig. 3.11b) showing a honeycomb-like surface structure fit the proposed model.

Photoemission spectra of the O 1s and N 1s core levels showed a single component, in full agreement with the structural model suggesting only one type of the N atoms and, in essence, one type of the O atoms all involved in the Si–O–Si bonds, albeit having different bond angles. However, at least three different Si species exist beyond the Si atoms in the SiC substrate, which differ by Si coordination to the surrounding atoms, i.e., $Si–O_4$, $Si–ON_3$, and $Si–NC_3$, marked as Si_{III}, Si_{II}, Si_{I}, respectively, in Fig. 3.11a. All species are resolved in photoelectron spectra presented in Fig. 3.11c. Si_{III} species dominate spectra at grazing emission as they belong to the outermost surface layers, whereas the Si_{I} component attenuates because the corresponding atoms are closer to the SiC support, in full agreement with the model.

In addition, tunneling spectra (current vs. bias) revealed the band gap about 8.7 eV which is very close to that of bulk silica. Again, this finding suggests that the band gap is primarily associated with the presence of solely Si–O bonds in the topmost layer. Another interesting characteristic to compare between ordered pure silica and silicon oxynitride layers is the interface, which possesses one dangling Si bond per unit cell in the case of pure silica (see Fig. 3.7c,d) thus giving rise to interfacial states within the band gap. The presence of the N layer allows to achieve a bond density reduction at the interface (Fig. 3.11a). There are no more dangling Si bonds and hence no need to invoke bond saturation by hydrogen.

Experimentally, the band structure profile for a SiON film grown on 6H–SiC(0001) has been addressed in elegant way using soft x-ray absorption spectroscopy (XAS) and x-ray emission spectroscopy (XES).[71] Since XAS and XES probe electronic transitions from a core level to conduction states and from valence states to the core hole, respectively, element-specific conduction- and valence band partial densities of states could be obtained. The K-edge absorption lines of O (O 1s→2p) and N (N 1s→2p), and the O KLL and N KLL Auger transitions were used for analysis. Accordingly, N 2p→1s and O 2p→1s electronic transitions were measured by XES. The obtained XAS and XES data allowed to draw the energy-band profiles across the surface layers. The band gaps were found to be of 6.3 ± 0.6 eV at the silicon nitride layer and 8.3 ± 0.8 eV at the silicon oxide layer as shown in Fig. 3.12. The energy-band profile of the SiON layer bears

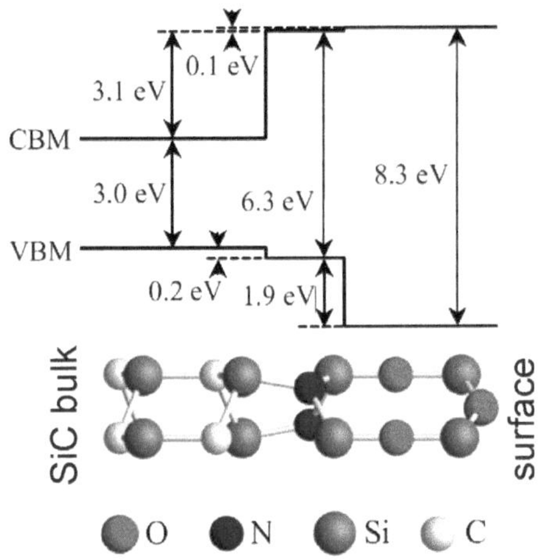

Figure 3.12 Experimentally obtained energy-band profiles of SiON/SiC layer. The bandgap of the SiC substrate is set to 3.0 eV as in the bulk. Reproduced with permission from ref. 73.

close similarities to the one theoretically predicted and shown in Fig. 3.10b for a SiO_2 layer. Indeed, further ab initio calculations[72] found only small deviation between these two systems.

Thanks to tremendous development of computational techniques in the recent years, more comprehensive DFT studies were possible to carry out to elucidate oxidation mechanisms resulting in silica layers on SiC. The reaction starts with oxygen adsorption (dissociation). As oxygen coverage considerably increases, the oxygen atom incorporates into the sub-surface region that, in turn, triggers the formation of a surface oxide.[74] Figure 3.13 depicts several DFT-optimized structures obtained at increasing oxygen coverage up to 2 ML (one monolayer (ML) corresponds to one O atom per surface unit cell).

This initial oxidation is relatively fast process, whereas oxide growth at thicknesses above ~2 nm is supposed to be limited by reactions at the interface. These latter exhibit the apparent activation energy of 3.8 ± 0.1 eV as measured on 4H–SiC(0001) using XPS and grazing incident x-ray reflectivity.[75] The interface reaction rate was proposed to be limited by the elementary step of CO escape from the interface into the gas phase according to

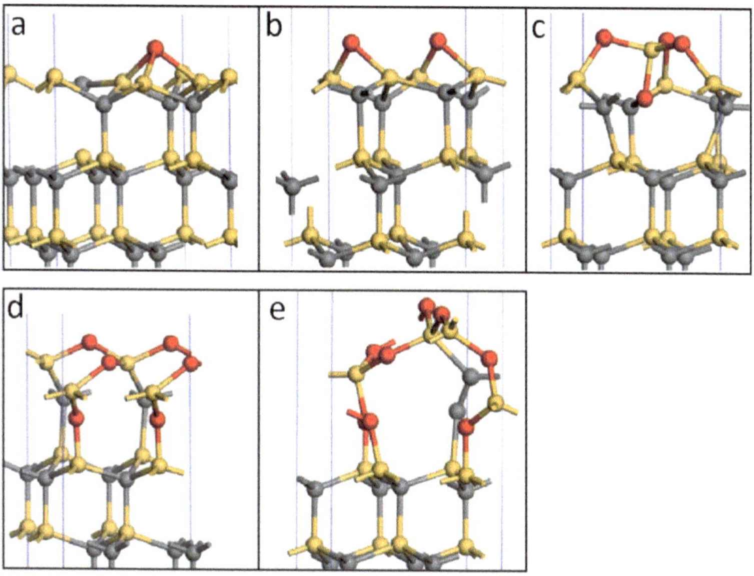

Figure 3.13 DFT-optimized structures formed upon oxygen adsorption on 4H–SiC(0001) at 0.25 ML (a), 0.5 ML (b), 1 ML (c), 1.5 ML (d), and 2 ML (e) oxygen coverage. Adapted with permission from ref. 74.

the reaction $SiC + 3/2O_2 -> SiO_2 + CO$. Accordingly, the growth rate change in sub-nanometer regime was explained by the change of CO escaping pathway.

On the other hand, CO release may compete with carbon dimer formation.[76] In a more recent DFT study,[77] the reaction pathways for O_2 reached the SiO_2/4H–SiC interface showed the energy barriers of CO formation much lower than those of C–C bond formation. This implies that the reaction forming CO molecules preferentially occurs during the interfacial reaction processes of SiC oxidation. Based on the calculated energies, the reaction is more facile on the C-face than on the Si-face, in qualitative agreement with experimental observations. The activation barriers also suggested that the formation of CO molecule at the SiO/SiO_2 interface depending on the orientation could be the rate-limiting step.

Finally, ab initio study[78] pointed to an important role of oxidizing species (molecular O_2 vs. atomic O) depending on the bonding requirement at the interface. In principle, three scenarios are possible: direct molecular oxygen mechanism when the required density of bonds is low, atomic oxygen mechanism when the required density of bonds is high, and defect-assisted (i.e., defect can crack O_2 molecule) molecular mechanism when the required density of bonds is very high.

3.3 Chemical Deposition of Thin Silica Films on Si Substrates

Silicon is known to form various volatile compounds which can be used as precursors for chemical vapor deposition (CVD). In principle, CVD growth can take place on any solid substrate heated to the temperatures high enough to decompose precursor molecules. With respect to thin silica films, CVD-based approach has received a certain interest in attempts to prepare silica layers at considerably lower temperatures than usually required for thermal oxidation of Si and SiC surfaces discussed in previous sections.

Plasma-mediated preparations often suffered from surface damage effects caused by highly reactive plasma species, ultimately leading to films with high density of defects. Atomic layer deposition (ALD) was invoked for fabrication of uniform ultrathin silica films at low temperatures. The technique is based on sequential and self-limiting surface reactions, thus allowing an accurate thickness control.

In early studies, precise determination of adsorbed species formed upon exposure to Si-containing gas molecules like silane (SiH_4) and their derivatives was quite difficult.[79] Silane exhibits very low sticking coefficient (i.e., the ratio of the number of adsorbed molecules to the total number of molecules impinging the surface during the same time) in contrast to di-silane (respectively, 10^{-5} and 0.5) on both Si(001) and Si(111) surfaces, as the Si–H bond is much stronger than the Si–Si bond in di-silane. Depending on temperature and exposure, adsorption of silanes is accompanied by sequential dehydrogenation resulting in SiH_x species and hydrogen desorption. The film growth was found to be a rather slow process that allowed controlling the film thickness with a high precision.

Growth of thin ($\leq$10 nm) SiO_2 films on Si(001) (that had a native oxide layer, though) by thermal pyrolysis of tetra-ethoxysilane at 900 K[80] showed kinetics that was fit well by the mechanism where by-products inhibit the film deposition. Adding the oxidizing agent, such as O_2, to silane to accelerate oxidation may, however, additionally lead to reactions in the gas phase.[81] In particular, surface species containing OH groups were found in such films. The prepared films showed low-quality electrical performance such as high leakage currents and low-field breakdown. It was, therefore, concluded that low-pressure, low-temperature CVD reaction of silane and O_2 is not well suited, at least for gate-oxide applications.

Another route has been considered for silica films that uses chlorinated Si precursors of general formula SiH_xCl_{4-x} with oxidizing agent like H_2O. However, HCl as the reaction product causes undesirable etching of a substrate. To reduce the reaction temperatures and large reactant exposures, it was suggested to separate the reaction $SiCl_4 + 2H_2O \rightarrow SiO_2 + 4HCl$ into so-called $SiCl_4$ and H_2O "half-reactions", and each half-reaction was performed in alternating sequence and catalyzed with pyridine or ammonia.[82–84] The catalyst allows to lower the deposition temperature down to 300 K as well as the reactant flux required for complete reactions. The obtained growth rate was about 2 Å (in nominal thickness) per cycle. The fabricated films were stoichiometric SiO_2 and exposed relatively smooth surface, with a roughness nearly identical to the initial substrate surface. Also, the films demonstrated dielectric breakdown strengths similar to thermally deposited SiO_2 films.

A variety of Si precursors was examined for ALD processes avoiding the use of halogen precursors. For example, tri-(*tert*-butoxy)silanol rapidly formed a self-limited 1 nm-thick SiO_2 film on a clean reconstructed Si(100)-(2×1) surface at 300 K as studied by XPS and IRAS.[85] One of the most striking features observed in this study is the low-temperature loss of about 95% of the hydrocarbon present in the precursor at 300 K, i.e., in contrast to previously studied tetraethylorthosilicate or diethyldiethoxysilane precursors, where the majority of hydrocarbon loss does not occur below 600 K. The difference was associated with the silanol functional group which react quickly with the silicon "dimers" present on the Si(100)-

(2×1) surface (see Fig. 3.1 for the surface structure). Apparently, the previously studied precursors first undergo substantial degradation via Si–O bond cleavage to react with the Si(100) surface.

Although ALD provides precise thickness control, its application suffers from low deposition rates, typically about one monolayer per cycle. In so-called "rapid" ALD process, the deposition rate was about 100 times higher (~12 nm per cycle) while still maintaining the self-limiting behavior of "conventional" ALD.[86] This was achieved by alternating exposure to vapors of trimethylaluminum (TMA, i.e., Al-containing) as a catalyst and tris(tert-butoxy)silanol (i.e., Si-precursor) in the temperature range of 225–250°C. Although the prepared films were highly conformal and followed the topography of a substrate, they were not pure SiO_2. Based on cross-sectional SEM images, the film structure was described as a nanolaminate in which very thin layers of alumina are interspersed by thicker layers of silica. Later, it was proposed that rapid ALD resulted from the growth of siloxane polymer chains at the Al catalytic sites with the subsequent cross-linking of these polymer chains to form a dense SiO_2 film.[87] The possible scenario is schematically shown in Fig. 3.14. First, TMA reacts with available hydroxylated sites. Subsequently, the silanols (in this case, tris(*tert*-pentoxy)silanols) react at the Al center to release CH_4. Additional silanol precursors insert at the Al catalytic center and release *tert*-pentanol. The polymerization reaction is believed to occur as long as the Si precursors can diffuse to the Al catalyst. After the polymerization reaction stops, TMA is redeposited on the surface to repeat the reaction sequence. In general, the SiO_2 film thickness deposited during one silanol exposure was dependent on the substrate temperature, silanol pressure, and silanol exposure time. The experiments were usually performed on fused quartz or commercial Si wafers as substrates.

Many silane derivatives as precursors were examined in order to elucidate the effects of steric hindrance and activation barriers on the silicon reactivity using water as an oxidant.[88] Note, however, that in these experiments the silica deposition was carried out on ZrO_2 particles in order to image the process by TEM in situ. Among the precursors studied, tris(dimethylamino)silane showed

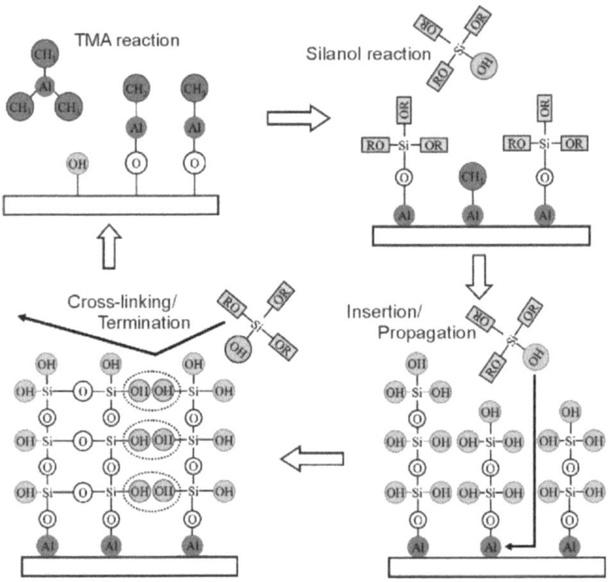

Figure 3.14 Main steps in the growth mechanism of rapid SiO_2 ALD. Adapted with permission from ref. 87.

the highest reactivity. However, remaining SiH surface species were difficult to remove by water and could only be removed by hydrogen peroxide (H_2O_2) at temperatures above 450°C. The film growth correlated with the removal of the SiH surface species. The latter can be enhanced by plasma treatment, thus providing a method for performing all ALD steps even at room temperature. Recent infrared spectroscopy studies showed that the O_2/Ar plasma treatment removes CH_x groups but also transforms SiH surface species into SiOH that was difficult to achieve by H_2O and H_2O_2 during thermal ALD at low temperatures.[89]

Another, yet unexplored deposition method recently proposed for the fabrication of thin silica films employs a sol-gel process, which is routinely used for the preparation of various silica-based materials. Prior to gelation, the sol is to be deposited as a thin film on a substrate, for instance, by dipping, spinning, or spraying. In one of the first such preparations, tetraethylorthosilicate dissolved in ethanol was spin-coated onto a Si wafer covered with a thin, vapor-deposited Ag film in order to employ IR spectroscopy in reflection mode.[90] The films were then heated in vacuum at 200°C for 2 h to remove residual solvent, water and other volatile organic species. (Note that in these experiments, the annealing temperature was limited because of Ag migration into the Si substrate.) By adjusting the concentration of the precursor solutions, planar silica glass films of thicknesses between 3 and 12 nm were obtained.

Following this approach and aimed at the preparation of ultrathin silica films, in particular, on metal supports, Pemberton and co-workers[91] made use of precursor sol very dilute in the silica component. However, the hydrophobic nature of metal (Ag and Au) substrates prevented significant wetting of silica to the metal surface. To improve the film adhesion, a self-assembled monolayer of (3-mercaptopropyl)-trimethoxysilane was used as a buffer layer. Once assembled, the trimethoxysilane functionality can be polymerized through hydrolysis and condensation reactions to form a homogeneous surface uniformly covered by silanols acting as anchors to the silica film. Therefore, one end of the buffer molecule binds covalently to the metal substrate, while the other end binds covalently to the silica. The prepared silica films on Ag and Au, about 20 nm in thickness showed good characteristics for nanoelectronics applications. Importantly, the preparation did not require high-temperature annealing.

The approach was further modified in that good adhesion of the silica films to a gold substrate was achieved through a gold oxide as the intermediate linker (see inset in Fig. 3.15).[92] A gold substrate is oxidized either electrochemically or using UV photo-oxidation in an ozone atmosphere. As examined by AFM and voltammetry, the resulting thin silica films were dense, uniform, of high purity, and exhibited surface chemistry identical to that of bulk silica. The fabrication procedure was highly reproducible and did not require sophisticated equipment. The key parameter for thickness control of the resulting silica films was the Si/H_2O ratio, typically between 1:1000 and 1:100. Figure 3.15 shows IRAS spectrum of a 5.5 nm-thick silica film (measured by ellipsometry) which exhibits all vibrational characteristics of silica: the ν_{as}(Si–O–Si) modes at 1235

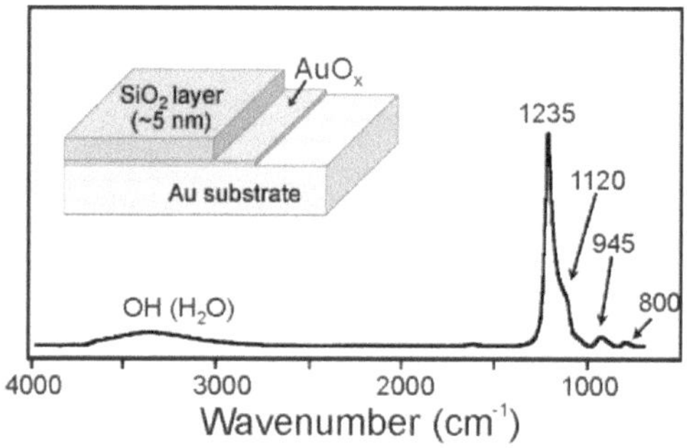

Figure 3.15 Schematic structure and IRAS spectrum of a 5.5 nm-thick silica film deposited by a sol-gel method on a gold substrate. Adapted with permission from ref. 92.

and 1120 cm^{-1}, the ν_S(Si—O—Si) mode at 800 cm^{-1}. In addition, the film is hydroxylated as manifested by the ν_S(Si—OH) mode at 945 cm^{-1}. The broad ν(OH) band centered at 3400 cm^{-1} whose intensity increases linearly with film thickness is indicative for some amounts of trapped water in the films and/or H-bonded silanols within the film matrices. The absence of ν(C—H) modes in the region around 2900 cm^{-1} suggests complete hydrolysis and condensation of the precursor methoxy groups.

Recently, the epitaxial growth of nanostructured polycrystalline quartz films with tunable textures on Si(100) was reported using "soft-chemistry-based route."[93] This approach is based on the solution deposition and gelation of amorphous silica thin films, followed by thermal treatment. Key to the process is the combined use of either a Sr^{2+} or Ba^{2+} catalyst with an amphiphilic molecular template, which is responsible for the homogeneous crystallization of a low α-quartz film. Cross-sectional TEM images together with XRD analysis revealed very good epitaxial relationship despite the relatively large (~9%) lattice mismatch between α-quartz and Si(001), and no trace of amorphous silica was observed at the interface. Although the prepared films were more than 100 nanometers in the thickness, the results indicated that film preparations using "wet chemistry" at mild temperatures may even be successful for preparation of crystalline, ultrathin silica films.

References

1. Chabal, Y. J., *Fundamental Aspects of Silicon Oxidation*. 2012: Springer Berlin Heidelberg.
2. Wagner, C., *Beitrag zur Theorie des Anlaufvorgangs*, in *Zeitschrift für Physikalische Chemie*. 1933. p. 25.
3. Cabrera, N., and N. F. Mott, Theory of oxidation of metals. *Reports on Progress in Physics*, 1948. **12**: pp. 163–184.
4. Fehlner, F. P., Formation of ultrathin oxide films on silicon. *Journal of The Electrochemical Society*. 1972. **119**(12): pp. 1723–1727.
5. Deal, B. E., and A. S. Grove, General relationship for the thermal oxidation of silicon. *Journal of Applied Physics*. 1965. **36**(12): pp. 3770–3778.
6. Irene, E. A., Silicon oxidation studies: Some aspects of the initial oxidation regime. *Journal of The Electrochemical Society*. 1978. **125**(10): pp. 1708–1714.
7. Adams, A. C., T. E. Smith, and C. C. Chang, The growth and characterization of very thin silicon dioxide films. *Journal of The Electrochemical Society*, 1980. **127**(8): pp. 1787–1794.
8. Raider, S. I., and L. E. Forget, Reversal of relative oxidation rates of <111> and <100> oriented silicon substrates at low oxygen partial pressures. *Journal of The Electrochemical Society*, 1980. **127**(8): pp. 1783–1787.
9. Ngau, J. L., P. B. Griffin, and J. D. Plummer, Silicon orientation effects in the initial regime of wet oxidation. *Journal of The Electrochemical Society*, 2002. **149**(8): pp. F98–F101.
10. Murali, V., and S. P. Murarka, Kinetics of ultrathin SiO_2 growth. *Journal of Applied Physics*, 1986. **60**(6): pp. 2106–2114.
11. Rochet, F., et al., The thermal oxidation of silicon: the special case of the growth of very thin films. *Advances in Physics*, 1986. **35**(3): pp. 237–274.
12. Stoneham, A. M., C. R. M. Grovenor, and A. Cerezo, Oxidation and the structure of the silicon/oxide interface. *Philosophical Magazine Part B*, 1987. **55**(2): pp. 201–210.
13. Mott, N. F., et al., Oxidation of silicon. *Philosophical Magazine Part B*, 1989. **60**(2): pp. 189–212.
14. Gusev, E. P., et al., Growth mechanism of thin silicon oxide films on Si(100) studied by medium-energy ion scattering. *Physical Review B*, 1995. **52**(3): pp. 1759–1775.

15. Finster, J., et al., Interpretation of XPS core level shifts and structure of thin silicon oxide layers. *Surface Science*, 1985. **152–153**(PART 2): pp. 1063–1070.

16. Grunthaner, P. J., et al., The localization and crystallographic dependence of Si suboxide species at the SiO_2/Si interface. *Journal of Applied Physics*, 1987. **61**(2): pp. 629–638.

17. Derrie, J., and M. Commandré, SiO_2 ultra thin film growth kinetics as investigated by surface techniques. *Surface Science*, 1982. **118**(1): pp. 32–46.

18. Schaefer, J. A., and W. Göpel, Initial stages of oxidation of Si{100}(2×1): A combined vibrational (EELS) and electron binding energy (XPS) study. *Surface Science*, 1985. **155**(2–3): pp. 535–552.

19. Himpsel, F. J., et al., Microscopic structure of the SiO_2/Si interface. *Physical Review B*, 1988. **38**(9): pp. 6084–6096.

20. Hirose, K., et al., Photoelectron spectroscopy studies of SiO_2/Si interfaces. *Progress in Surface Science*, 2007. **82**(1): pp. 3–54.

21. Feltz, A., U. Memmert, and R. J. Behm, In situ STM imaging of high temperature oxygen etching of Si(111)(7×7) surfaces. *Chemical Physics Letters*, 1992. **192**(2): pp. 271–276.

22. Feltz, A., U. Memmert, and R. J. Behm, High temperature scanning tunneling microscopy studies on the interaction of O_2 with Si(111)-(7×7) surfaces. *Surface Science*, 1994. **314**(1): pp. 34–56.

23. Ono, Y., M. Tabe, and H. Kageshima, Scanning-tunneling-microscopy observation of thermal oxide growth on Si(111)-(7x7) surfaces. *Physical Review B*, 1993. **48**(19): pp. 14291–14300.

24. Kageshima, H., and K. Shiraishi, First-principles study of oxide growth on Si(100) surfaces and at SiO_2/Si(100) Interfaces. *Physical Review Letters*, 1998. **81**(26): pp. 5936–5939.

25. Tsai, V., et al., Conformal oxides on Si surfaces. *Applied Physics Letters*, 1997. **71**(11): pp. 1495–1497.

26. Ourmazd, A., et al., Si ->SiO_2 transformation: Interfacial structure and mechanism. *Physical Review Letters*, 1987. **59**(2): pp. 213–216.

27. Rochet, F., et al., Growth of epitaxial silica on vicinal Si(001) surfaces during thermal oxidation in O_2. *Philosophical Magazine Part B*, 1989. **59**(3): pp. 339–363.

28. Takahashi, I., T. Shimura, and J. Harada, X-ray diffraction evidence for epitaxial microcrystallinity in thermally oxidized SiO_2 thin films on the Si(001) surface. *Journal of Physics: Condensed Matter*, 1993. **5**(36): p. 6525.

29. Munkholm, A., et al., Observation of a distributed epitaxial oxide in thermally grown SiO_2 on Si(001). *Physical Review Letters*, 1995. **75**(23): pp. 4254–4257.
30. Herbots, N., et al., The formation of ordered, ultrathin SiO_2/Si(100) interfaces grown on (1×1) Si(100). *Materials Science and Engineering: B*, 2001. **87**(3): pp. 303–316.
31. Shaw, J. M., et al., Atomic displacement free interfaces and atomic registry in SiO_2/(1×1) Si(100). *Journal of Applied Physics*, 2006. **100**(10): p. 104109-4.
32. Tanaka, N., et al., First observation of SiO_2/Si(100) interfaces by spherical aberration-corrected high-resolution transmission electron microscopy. *Journal of Electron Microscopy*, 2003. **52**(1): pp. 69–73.
33. Kimoto, K., et al., Metastable ultrathin crystal in thermally grown SiO_2 film on Si substrate. *AIP Advances*, 2012. **2**(4): p. 042144.
34. Queeney, K. T., et al., Silicon oxidation and ultra-thin oxide formation on silicon studied by infrared absorption spectroscopy. *Physica Status Solidi (A)*, 1999. **175**(1): pp. 77–88.
35. Queeney, K. T., et al., Infrared spectroscopic analysis of the Si/SiO_2 interface structure of thermally oxidized silicon. *Journal of Applied Physics*, 2000. **87**(3): pp. 1322–1330.
36. Herbots, N., et al., The formation of ordered, ultrathin SiO_2/Si(100) interfaces grown on (1×1) Si(100). *Materials Science and Engineering B: Solid-State Materials for Advanced Technology*, 2001. **87**(3): pp. 303–316.
37. Devine, R. A. B., Structural nature of the Si/SiO_2 interface through infrared spectroscopy. *Applied Physics Letters*, 1996. **68**(22): pp. 3108–3110.
38. Martinet, C., and R. A. B. Devine, Analysis of the vibrational mode spectra of amorphous SiO_2 films. *Journal of Applied Physics*, 1995. **77**(9): pp. 4343–4348.
39. Chowdhuri, A. R., et al., Strain and substoichiometry at the Si(100)/SiO_2 interface. *Physical Review B*, 2003. **67**(24): p. 245305.
40. Giustino, F., and A. Pasquarello, Infrared spectra at surfaces and interfaces from first principles: Evolution of the spectra across the Si(100)–SiO_2 interface. *Physical Review Letters*, 2005. **95**(18): p. 187402.
41. Giustino, F., A. Bongiorno, and A. Pasquarello, Atomistic models of the Si(100)–SiO_2 interface: Structural, electronic and dielectric properties. *Journal of Physics: Condensed Matter*, 2005. **17**(21): p. S2065.

42. Kovačević, G., and B. Pivac, Modeling the interface between crystalline silicon and silicon oxide polymorphs. Physica Status Solidi (A), 2013. **210**(4): pp. 717–722.

43. Woo-Byoung, K., M. Taketoshi, and K. Hikaru, Ultrathin SiO_2 layer with a low leakage current density formed with ~100% nitric acid vapor. *Nanotechnology*, 2010. **21**(11): p. 115202.

44. Reiner, J. W., et al., Crystalline oxides on silicon. *Advanced Materials*, 2010. **22**(26–27): pp. 2919–2938.

45. Weldon, M. K., et al., The surface science of semiconductor processing: Gate oxides in the ever-shrinking transistor. *Surface Science*, 2002. **500**(1–3): pp. 859–878.

46. Kang, J., et al., Direct and defect-assisted electron tunneling through ultrathin SiO_2 layers from first principles. *Physical Review B*, 2008. **77**(19): p. 195321.

47. Tomoya, O., et al., First-principles study of the electronic structures and dielectric properties of the Si/SiO_2 interface. *Journal of Physics: Condensed Matter*, 2007. **19**(36): p. 365202.

48. Cooper, J. J. A., Advances in SiC MOS Technology. Physica Status Solidi (A), 1997. **162**(1): pp. 305–320.

49. Raynaud, C., Silica films on silicon carbide: A review of electrical properties and device applications. *Journal of Non-Crystalline Solids*, 2001. **280**(1–3): pp. 1–31.

50. Liu, G., B. R. Tuttle, and S. Dhar, Silicon carbide: A unique platform for metal-oxide-semiconductor physics. *Applied Physics Reviews*, 2015. **2**(2): p. 021307.

51. Buczko, R., S. J. Pennycook, and S. T. Pantelides, Bonding arrangements at the $Si–SiO_2$ and $SiC–SiO_2$ interfaces and a possible origin of their contrasting properties. *Physical Review Letters*, 2000. **84**(5): pp. 943–946.

52. Presser, V., and K. G. Nickel, Silica on silicon carbide. *Critical Reviews in Solid State and Materials Sciences*, 2008. **33**(1): pp. 1–99.

53. Hornetz, B., H. J. Michel, and J. Halbritter, AR XPS studies of SiO_2–SiC Interfaces and oxidation of 6H SiC single-crystal Si-(001) and C-(00-1) surfaces. *Journal of Materials Research*, 1994. **9**(12): pp. 3088–3094.

54. Guinel, M. J.-F., and M. G. Norton, Oxidation of silicon carbide and the formation of silica polymorphs. *Journal of Materials Research*, 2006. **21**(10): pp. 2550–2563.

55. Bernhardt, J., et al., Epitaxially ideal oxide–semiconductor interfaces: Silicate adlayers on hexagonal (0001) and (000-1) SiC surfaces. *Applied Physics Letters*, 1999. **74**(8): pp. 1084–1086.
56. Starke, U., et al., Reconstructed oxide structures stable in air: Silicate monolayers on hexagonal SiC surfaces. *Journal of Vacuum Science & Technology A*, 1999. **17**(4): pp. 1688–1692.
57. Lu, W., P. Krüger, and J. Pollmann, Atomic and electronic structure of silicate adlayers on polar hexagonal SiC surfaces. *Physical Review B*, 2000. **61**(20): pp. 13737–13744.
58. Tochihara, H., and T. Shirasawa, The epitaxial crystalline silicon-oxynitride layer on SiC(0001): Formation of an ideal SiC–insulator interface. *Progress in Surface Science*, 2011. **86**(11–12): pp. 295–327.
59. Hollering, M., et al., Electronic states of an ordered oxide on C-terminated 6H–SiC. *Surface Science*, 1999. **442**(3): pp. 531–542.
60. Amy, F., et al., Si-rich 6H- and 4H–SiC(0001)-(3×3) surface oxidation and initial SiO_2/SiC interface formation from 25 to 650°C. *Physical Review B*, 2002. **65**(16): p. 165323.
61. Simon, L., et al., X-ray spectroscopy of the oxidation of 6H–SiC(0001). *Physical Review B*, 1999. **60**(8): pp. 5673–5678.
62. Virojanadara, C., and L. I. Johansson, Oxidation studies of 4H–SiC(0001) and (0001). *Surface Science*, 2002. **505**: pp. 358–366.
63. Hoshino, Y., R. Fukuyama, and Y. Kido, Oxidized surface structure and oxidation kinetics of the C-terminated 6H–SiC(000-1)-(2×2) surface. *Physical Review B*, 2004. **70**(16): p. 165303.
64. Schürmann, M., et al., Structure of the interface between ultrathin SiO_2 films and 4H–SiC(0001). *Physical Review B*, 2006. **74**(3): p. 035309.
65. Amy, F., et al., Atomic scale oxidation of a complex system: O_2/alpha-SiC(0001)-(3×3). *Physical Review Letters*, 2001. **86**(19): pp. 4342–4345.
66. Tochihara, H., et al., Scanning tunneling microscopic and spectroscopic studies on a crystalline silica monolayer epitaxially formed on hexagonal SiC(000-1) surfaces. *Applied Physics Letters*, 2014. **104**(5): p. 051601.
67. Krüger, P., B. Baumeier, and J. Pollmann, First-principles investigation of an epitaxial silicon oxynitride layer on a 6H-SiC(0001) surface. *Physical Review B*, 2008. **77**(8): p. 085329.
68. Devynck, F., et al., Structural and electronic properties of an abrupt 4H-SiC(0001)/SiO_2 interface model: Classical molecular dynamics

simulations and density functional calculations. *Physical Review B*, 2007. **76**(7): p. 075351.

69. Shirasawa, T., et al., Epitaxial silicon oxynitride layer on a 6H–SiC(0001)surface. *Physical Review Letters*, 2007. **98**(13): p. 136105.
70. Devynck, F., Ž. Šljivančanin, and A. Pasquarello, Electronic properties of an epitaxial silicon oxynitride layer on a 6H–SiC(0001) surface: A first-principles investigation. *Applied Physics Letters*, 2007. **91**(6): p. 061930.
71. Shirasawa, T., et al., Atomic-layer-resolved bandgap structure of an ultrathin oxynitride-silicon film epitaxially grown on 6H-SiC(0001). *Physical Review B*, 2009. **79**(24): p. 241301.
72. Devynck, F., and A. Pasquarello, Band gap opening at the 6H–SiC(0001) surface passivated by an epitaxial silicon oxynitride layer: A first-principles investigation. *Surface Science*, 2008. **602**(18): pp. 2989–2993.
73. Shirasawa, T., et al., Atomic-layer-resolved bandgap structure of an ultrathin oxynitride-silicon film epitaxially grown on 6H–SiC(0001). *Physical Review B*, 2009. **79**(24): p. 241301.
74. Li, W., et al., Insight into the initial oxidation of 4H–SiC from first-principles thermodynamics. *Physical Review B*, 2013. **87**(8): p. 085320.
75. Kikuchi, R. H., and K. Kita, Interface-reaction-limited growth of thermal oxides on 4H–SiC (0001) in nanometer-thick region. *Applied Physics Letters*, 2014. **104**(5): p. 052106.
76. Gavrikov, A., et al., First-principles-based investigation of kinetic mechanism of SiC(0001) dry oxidation including defect generation and passivation. *Journal of Applied Physics*, 2008. **104**(9): p. 093508.
77. Akiyama, T., et al., First-principles investigations for oxidation reaction processes at 4H–SiC/SiO_2 interface and its orientation dependence. *Surface Science*, 2015. **641**: pp. 174–179.
78. Shen, X., B. R. Tuttle, and S. T. Pantelides, Competing atomic and molecular mechanisms of thermal oxidation—SiC versus Si. *Journal of Applied Physics*, 2013. **114**(3): p. 033522.
79. Neergaard Waltenburg, H., and J. T. Yates, Surface chemistry of silicon. *Chemical Reviews*, 1995. **95**(5): pp. 1589–1673.
80. Ong, T. P., P. Tobin, and T. Mele, Deposition of ultrathin silicon dioxide film from pyrolysis of tetraethoxysilane. *Journal of Applied Physics*, 1995. **77**(11): pp. 6055–6057.
81. Liehr, M., and S. A. Cohen, Low pressure chemical vapor deposition of oxide from SiH_4/O_2: Chemistry and effects on electrical properties. *Applied Physics Letters*, 1992. **60**(2): pp. 198–200.

82. Klaus, J. W., O. Sneh, and S. M. George, Growth of SiO_2 at room temperature with the use of catalyzed sequential half-reactions. *Science*, 1997. **278**(5345): pp. 1934–1936.

83. Klaus, J. W., et al., Atomic layer deposition of SiO_2 using catalyzed and uncatalyzed self-limiting surface reactions. *Surface Review and Letters*, 1999. **6**(3–4): pp. 435–448.

84. Klaus, J. W., and S. M. George, SiO_2 Chemical vapor deposition at room temperature using $SiCl_4$ and H_2O with an NH_3 catalyst. *Journal of The Electrochemical Society*, 2000. **147**(7): pp. 2658–2664.

85. Miller, K. A., et al., Self-limiting chemical vapor deposition of an ultra-thin silicon oxide film using tri-(tert-butoxy)silanol. *Thin Solid Films*, 2001. **397**(1): pp. 78–82.

86. Hausmann, D., et al., Rapid vapor deposition of highly conformal silica nanolaminates. *Science*, 2002. **298**(5592): pp. 402–406.

87. Burton, B. B., et al., Rapid SiO_2 atomic layer deposition using tris(tert-pentoxy)silanol. *Chemistry of Materials*, 2008. **20**(22): pp. 7031–7043.

88. Burton, B. B., et al., SiO_2 atomic layer deposition using tris(dimethy-lamino)silane and hydrogen peroxide studied by in situ transmission FTIR spectroscopy. *The Journal of Physical Chemistry C*, 2009. **113**(19): pp. 8249–8257.

89. Lu, Y., et al., Chemical reactions during plasma-enhanced atomic layer deposition of SiO_2 films employing aminosilane and O_2/Ar plasma at 50°C. *Japanese Journal of Applied Physics*, 2014. **53**(1): p. 010305.

90. Allara, D. L., Z. Wang, and C. G. Pantano, Characterization of organic adsorbates on model glass surfaces. *Journal of Non-Crystalline Solids*, 1990. **120**(1–3): pp. 93–101.

91. Robertson, J. W., M. Cai, and J. E. Pemberton, Insulating ultrathin silica films formed by a room-temperature sol–gel process. *Advanced Materials*, 2001. **13**(9): pp. 662–667.

92. Macech, P., and J. E. Pemberton, Ultrathin silica films immobilized on gold supports: Fabrication, characterization, and modification. *Langmuir*, 2007. **23**(19): pp. 9816–9822.

93. Carretero-Genevrier, A., et al., Soft-chemistry–based routes to epitaxial α-quartz thin films with tunable textures. *Science*, 2013. **340**(6134): pp. 827–831.

Chapter 4

Ultrathin Silica Films on Metals

In this chapter, we address the preparation of well-ordered thin silica films on metallic substrates. It may be interesting, that the first "surface science" studies of silica layers on metal surfaces were carried out in attempts to get rid of Si impurities usually present in metal single crystals and foils.[1,2] In order to understand the possible role of adventitious Si on surface structures and adsorption properties of catalytically active metals like Pt and Pd, the surfaces were deliberately contaminated by Si using vapor deposition following by annealing in vacuum and in oxygen. The results suggested that Si even deeply diffused in the metal bulk may affect oxygen adsorption eventually resulting in silica clusters rather than forming "platinum oxide" layer as originally thought. The formation of Si stabilized surface oxide was also observed on the Si-contaminated Pd surfaces, albeit the situation was not so obvious due to the concomitant formation of PdO_x layers.[3] The adsorption studies of Pt foils doped with Si revealed (not surprisingly, though) silica deposits only blocking adsorption of ambient gases on the Pt surface.[4] There were no other findings observed which would be indicative of a more complex Si-induced effect on reactivity of Si-containing metal surfaces.

Introduction to Ultrathin Silica Films: Silicatene and Others
Shamil Shaikhutdinov

ISBN 978-981-4877-28-2 (Hardcover), 978-1-003-08862-2 (eBook)
www.jennystanford.com

To the best of our knowledge, the studies really targeted on the controlled preparation of well-defined thin silica films on metal substrates were first initiated by Goodman and co-workers in the early 1990s.[5–7] This group aimed at the fabrication of *thin* and *pure* silica films at *low* temperatures. Such films were considered as good model planar supports for studying catalytic reactions on silica supported metal catalysts. As a substrate, single crystal surfaces of molybdenum have been chosen at first, because it is relatively easy to clean the Mo surface in ultra-high vacuum (UHV) simply by thermal flash to very high temperatures as high as 1800 K. To date, many other metal substrates and preparation conditions have been tested, which will be discussed below in detail. In some places, the discussion is accompanied with a historical digression in order to better understand the reasons and ideas that have been put forward later.

4.1 Molybdenum

First silica overlayers on the Mo(110) surface were prepared by Si evaporation in O_2 (typically, 10^{-6}–10^{-5} mbar) onto the sample kept at 300 K.[5,6] Actually, pure Si deposition from a quasi-melt Si rod used in evaporators only takes place at relatively lower oxygen pressures. Since hot Si surfaces in the oxygen atmosphere undergo gasification (etching) (see, for example, Fig. 3.3 in Chapter 3), it is rather SiO molecules than Si atoms that impinge at a substrate surface at increasing O_2 pressures. Nonetheless, full oxidation to SiO_2 only occurs on the metal surface. Oxygen readily dissociates on Mo even at room temperature. Therefore, during deposition the Mo surface is pre-covered by O ad-atoms which react with Si precursor sticking to the surface. There are no systematic studies available in the literature about the initial stages of the Si (SiO_x) layer formation on metal substrates. In case of Mo, even the atomic structure of the O-covered surfaces, the Si atoms are reacting with, is not fully understood.

The "as deposited" overlayer on Mo(110) showed no long-range order when inspected by low energy electron diffraction (LEED). Actually, this holds true for all metal substrates studied thus far because of relatively low temperatures used for Si deposition. Stepwise UHV annealing at elevated temperatures certainly improved ordering, yet the films showed no diffraction. However, the principal band in the infrared reflection absorption spectra (IRAS) became considerably narrower, and the peak gradually shifted to low frequencies as shown in Fig. 4.1. Since only the vibrations causing dipole moment change in the direction normal to the surface are IRAS active (see Chapter 2), the two bands, at 1252 cm^{-1} and (as a shoulder) 1160 cm^{-1}, are similar to those observed for the thermally grown SiO_2 films on Si and correspond to the longitudinal optical (LO) vibrations (see Chapter 2, and also Fig. 3.5 in Chapter 3).

Later, the same group reported the preparation of ultrathin silica films on a Mo(100) substrate.[7] The prepared films were additionally characterized by x-ray photoelectron spectroscopy (XPS), which

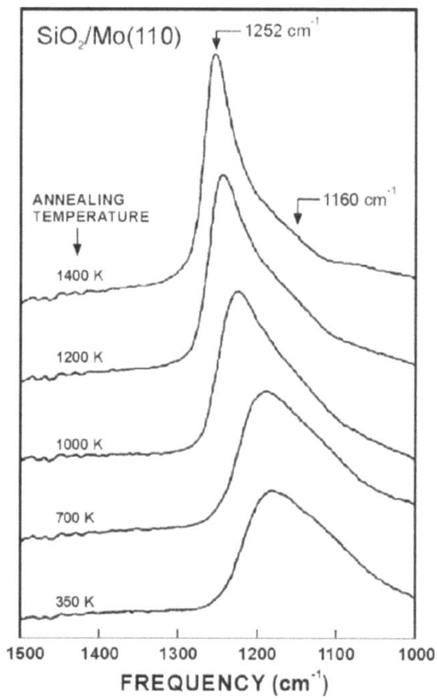

Figure 4.1 IRAS spectra of a thin silica film deposited on Mo(110) at 350 K and subsequently annealed to the indicated temperatures. Adapted with permission from ref. 6.

is more sensitive with respect to the chemical states of Si than previously used Auger electron spectroscopy (AES). It was found that the films as deposited at 350 K in 10^{-5} Torr of O_2 consisted of SiO_2 with a small fraction of sub-oxides. However, annealing to 1300 K resulted in a fully stoichiometric SiO_2 film: only Si^{4+} states were detected by XPS.

Albeit promising, these studies did not inspire other research groups to work on the preparation of metal-supported thin silica films. To some extent, this can be explained by the lack of long-range ordering in the prepared films and their close similarities, both structurally and electronically, to the widely studied amorphous silica films thermally grown on Si surfaces (see Chapter 3). Also, the preparation of stoichiometric SiO_2 films still required high-temperature annealing, and as such it was not beneficial for technological applications, either. Therefore, further studies were focused primarily on finding recipes for preparation of *crystalline* silica films.

The surface unit cells of the Mo(110) and Mo(100) surfaces do not match lattice parameters of silica polymorphs and, therefore, hardly favor the formation of crystalline films requiring good epitaxial relationships at the interface. One may note, however, that both cristobalite and tridymite possess hexagonal layers in their structures (see Fig. 1.2 in Chapter 1). Solely on the symmetry grounds, it appears more reasonable to use the hexagonal and close-packed (111)-oriented surfaces, instead. In the case of Mo which has a body centered cubic structure, the Mo(111) surface does not exhibit a close-packed structure. Moreover, the surface readily undergoes adsorbate induced faceting at relatively low temperatures, in particular into (112) planes in case of oxygen.[8] Therefore, in the next studies, the Mo(112) surface has been used, first by Freund and co-workers in the late 1990s. The preparation consisted of deposition-oxidation cycles, i.e., vapor deposition of sub-monolayer amounts of Si onto clean Mo(112) kept in UHV at room temperature and subsequent oxidation at 800 K in 5×10^{-5} mbar O_2.[9–12] After the films were oxidized in 5×10^{-6} mbar O_2 at elevated temperatures, up to 1150 K, the films showed a c(2×2)-Mo(112) LEED pattern (Fig. 4.2a), thus indicating high degree of crystallinity and good epitaxial relationships between the silica

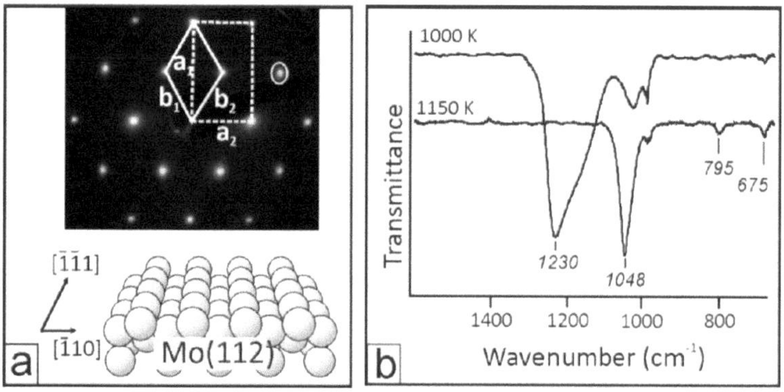

Figure 4.2 (a) Typical LEED pattern of the crystalline silica films grown on Mo(112). A perspective view of the Mo(112) surface is shown in the lower part of (a). The unit cells are indicated. Slightly elongated shape of the silica-related diffraction spots is highlighted by the oval. (b) IRAS spectra of the "as deposited" silica film upon annealing to the indicated temperatures.

overlayer and the metal surface. In addition, XPS of the annealed films revealed Si only in the 4+ state, whereas two oxygen species were distinguished in the O 1s region, thereafter assigned to O atoms in the silica and O at the interface. Finally, IRAS spectra shown in Fig. 4.2b revealed a broad band at 1230 cm^{-1} in the "as deposited" films, which disappears upon high-temperature annealing giving rise to a strong band at 1048 cm^{-1}. This latter falls in the range of frequencies corresponding to the TO band (1030–1060 cm^{-1}) observed on SiO_2/Si films (see Fig. 3.5 in Chapter 3).

The atomic structure of the clean Mo(112) surface is shown in Fig. 4.2a. It exposes atomic rows along the Mo(-1-11) direction, which are separated by troughs in the (-110) direction. A rectangular surface unit cell exhibits two vectors, a_1 = 2.73 Å and a_2 = 4.45 Å. The c(2×2) LEED pattern observed for the silica overlayer corresponds to unit vectors, b_1 and b_2, both of the length 5.22 Å. The angle between them is 63°, i.e., very close to 60° in the perfect hexagonal structure. A closer look at the diffraction spots revealed broadening of the c(2×2) spots and its oval shape elongated in the a_1 direction (as highlighted in Fig. 4.2a), whereas all the fundamental spots of Mo(112) remained sharp and isotropic. Since disorder in a certain direction produces diffraction streaks perpendicular to this direction, the broadening in the direction indicates the existence of line defects running preferentially along the direction (see Chapter 2). Accordingly, spot profile analysis suggested anti-phase domain boundaries formed between two hexagonal domains as film imperfections. The defects attenuate upon vacuum annealing at higher temperatures (1250 K), henceforth used in the films preparation.

Further ultraviolet photoelectron spectroscopy (UPS) and metastable impact electron spectroscopy (MIES) studies[12–15] allowed to determine the valence band which showed close similarity to the valence band of quartz, thus suggesting corner-sharing SiO_4 tetrahedra as a building block in such silica films. Also, the band gap (8.8 eV) measured in these experiments is very close to that in the bulky silica. Again, all spectral features become much sharper after UHV annealing at high temperatures due to film ordering in agreement with LEED.

Based on LEED patterns showing pseudo-hexagonal symmetry of a silica overlayer, several bulk-derived structures of silica polymorphs exhibiting hexagonal layers, like in HP-tridymite and β-cristobalite, were first considered.[12] These two structures show a surface unit cell length of 5.05 Å, but a different stacking sequence, i.e., ABAB... and ABCABC... along the (0001) and (111) directions of HP-tridymite and β-cristobalite, respectively (see Fig. 1.2 in Chapter 1). Another candidate, the unreconstructed (0001) plane of the low-density quartz, has a unit cell length of 4.91 Å. Therefore, all these structures exhibit lattice constants substantially shorter than ~5.22 Å measured in prepared silica films. Accordingly, the silica layer must be stretched at the interface that implies a very strong interaction with the metal support. In addition, the film must be very thin to keep this stretched structure throughout the entire film.

It became clear that any structure verification needs additional information and, even more importantly, theoretical modeling. On the other hand, theoretical calculations also need input data about surface symmetry, stoichiometry and thickness of an oxide overlayer. This latter was the least known in that time. Indeed, thickness determination with electron spectroscopies such as XPS and AES is based on the precise knowledge of a mean free path of electron in a material under study, but this often suffers from large uncertainty when applied to ultrathin films where each layer counts (see Chapter 2). As a result, the reported values for Mo-supported films considerably scattered, between 3 and 8 Å, even though the films were similarly prepared.[11,14,16]

In attempt to establish the film structure with the help of density functional theory (DFT), Ricci and Pacchioni[17] used a Mo(112)-(2×2) slab having three layers of corner-sharing SiO_4 tetrahedra on top as depicted in Fig. 4.3a. Silica binds to the unreconstructed Mo(112) surface via Si–O–Mo linkages. Basically, the interface is formed by cleaving the bulk structure along the (111) direction of β-cristobalite or the (001) direction of tridymite. However, in order to accommodate a c(2×2)-Mo(112) symmetry suggested by LEED, the silica lattice had to be expanded. To saturate the surface bonds, cristobalite and tridymite surfaces exposed terminal silanols, Si–OH, whereas a film of quartz, also computed for comparison, was terminated by geminal silanols, $Si{-}(OH)_2$ (see more about water/silica surfaces in Chapter 8). Among the structures considered, β-cristobalite showed the highest adhesion energy. For hydroxyl-free surfaces only observed in experiments, the dehydroxylated silica was reconstructed via two-membered silica rings formed by edge-sharing SiO_4 tetrahedra at the surface (see Fig. 4.3a). Although quite unusual for the silica polymorphs, these surface elements showed vibrational modes at around 800 cm^{-1}, which was then proposed as the origin for the weak 795 cm^{-1} band observed by IRAS (Fig. 4.2b).

Basically, the DFT-predicted models were derived from truncation of the bulk silica polymorphs and could, in principle, be of any thickness by increasing the number of silica layers between the surface and the interface. Meanwhile, Goodman and co-workers[16] came to the conclusion that the silica film on Mo(112) is just of only one layer in thickness or a "monolayer". Studied by high-resolution electron energy loss spectroscopy (HREELS), the thicker films showed the losses at 496, 800, and 1176 cm^{-1}, which are virtually identical to those obtained on the quartz (0001) surface (see Fig. 2.9 in Chapter 2). However, at smaller film thicknesses, the

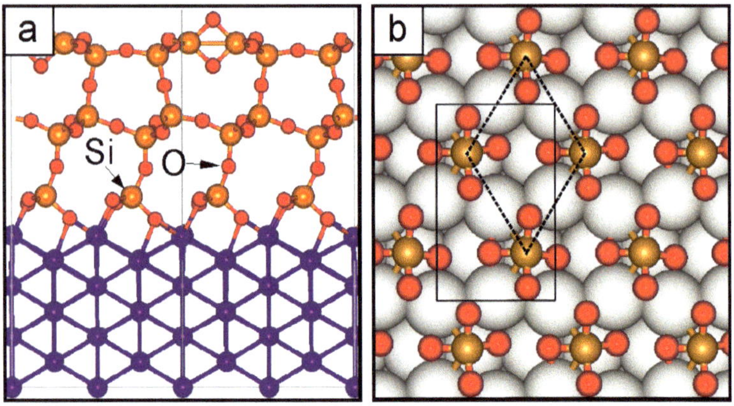

Figure 4.3 First models of a c(2×2)-ordered silica film on Mo(112). (a) Side view of a cristobalite film exposing two-membered silica rings at the surface. (b) Top view of the "cluster" model with isolated SiO_4 clusters adsorbed in the c(2×2) sites.

signals transform into the bands peaked at 672, 768 and 1048 cm^{-1}, which are very close to 675, 795, and 1048 cm^{-1} bands measured by IRAS on well-ordered films (Fig. 4.2b). Analysis of vibrational spectra of various Si–O–M (M = Al, Ti, Zr) mixed oxides reported in the literature pointed to that the principal 1048 cm^{-1} band most likely originates from the vibrations including Si–O–Mo bonds which must exist at the silica/metal interface. Among different binding geometries of a single layer of SiO_4 units to the Mo surface, the authors favored the "cluster" model, in which isolated SiO_4 tetrahedra are adsorbed on Mo(112) in c(2×2) positions as depicted in Fig. 4.3b. However, this model contradicted the XPS results[11, 18] showing two different O species in the film, which are impossible to explain within the cluster model with all oxygen atoms being in the Si–O–Mo bonds only.

In principle, none of the experimental findings obtained solely by a single, albeit state-of-the-art, technique can serve as the compelling evidence for the correctness of the proposed structure model, although each finding may reject some other models contradicting this finding. In fact, only full structural, electronic, and vibrational characterization of the *same* film in the *same* experimental setup in combination with the high-level theoretical calculations makes it possible to establish the atomic structure. Such an approach was employed in our own laboratories.

The preparation was first optimized in order to obtain well-ordered films showing very sharp LEED spots with a low background intensity. This includes: (i) exposure of the clean Mo(112) surface to 5×10^{-8} mbar O_2 at 900 K that results in a Mo(112)-(2×3)O surface reconstruction; (ii) deposition of about one atomic layer of Si at 900 K in the same oxygen pressure and temperature; (iii) final annealing in UHV at high temperatures (~1250 K) for ca 5 min. Annealing at lower temperatures resulted in nanoparticulate morphology of the film, whereas overheating causes partial film dewetting and/or sublimation into the vacuum.

The prepared well-ordered films displayed large atomically flat terraces in scanning tunnelling microscopy (STM) images (see Fig. 4.4a) and showed a strong and very sharp IRAS band at 1060 cm^{-1} never seen on the thermally grown amorphous silica films on Si (Fig. 4.4b). In addition, weaker signals are observed at 770 and 675 cm^{-1}. As all peaks shifted to the lower frequencies (by 41, 7, and 19 cm^{-1}, respectively) if measured on the films prepared with $^{18}O_2$ isotope, the vibrational bands involve solely the Si–O bonds. Further deposition of small amounts of Si on top of the crystalline film in oxygen ambient gives rise to the broad band centered at 1240 cm^{-1}, originally seen on the "as deposited" samples prior to the UHV annealing, and as such it is attributed to silica nanoparticles and aggregates (Fig. 4.4c). Subsequent UHV annealing at 1250 K, however, restores both the flat morphology and the IRAS-spectrum, indicating a "self-limited" film growth such that thicker overlayers readily sublime in UHV at higher temperatures until only the crystalline ultrathin film is left on a Mo(112) substrate.

High-resolution STM images revealed a honeycomb-like surface morphology with a ~5.4 Å periodicity (Fig. 4.5), in fair good agreement with the LEED results yielding 5.2 Å, on average, since the surface lattice is not perfectly hexagonal. Although the image contrast depends on the tunneling parameters and tip conditions, the two line defects marked by the arrows in Fig. 4.5a and zoomed in in Fig. 4.5b can readily be formed at the boundary between two domains both showing a hexagonal lattice, a half-lattice shifted with respect to each other, ultimately resulting in a line of alternating

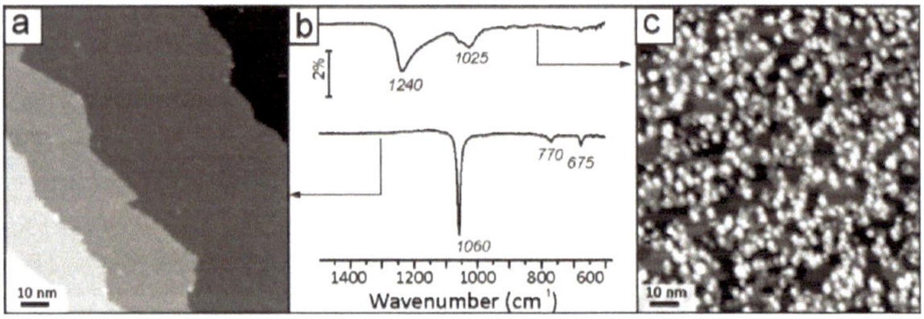

Figure 4.4 Large-scale STM images and IRAS spectra of a crystalline silica film grown on Mo(112) and of the same film with additional silica particles deposited on top.

eight- and four-membered silica rings as schematically shown in Fig. 4.5c.

These STM images provided solid basis for a model in which a silica film on Mo(112) is nothing else, but a single silicate layer, with one apical oxygen in SiO_4 tetrahedron having a bond to the Mo surface atoms. Moreover, this model nicely fits the XPS results showing two different chemical states for the O 1s core levels with binding energies (BE) at 532.5 and 531.3 eV and the intensity ratio of 3:2 determined by signal deconvolution (Fig. 4.6a). The two states are better resolved by using synchrotron radiation.[19] Comparison of the two signals ratio at normal and grazing (i.e., more surface sensitive) emission allow to assign the high BE signal to surface

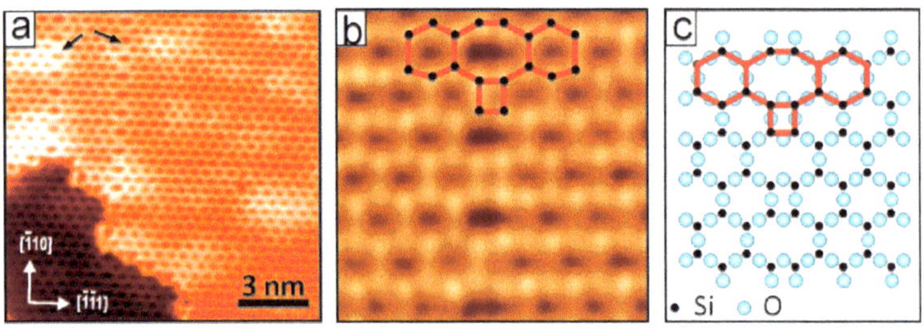

Figure 4.5 (a,b) High-resolution STM images of a crystalline silica film on Mo(112). The line defects running along the (-110) direction in the image (a) is atomically resolved in the image (b). Panel (c) displays the top-view of the monolayer of corner-sharing SiO_4 tetrahedra having a half-lattice shift between the two domains on the left and on the right.

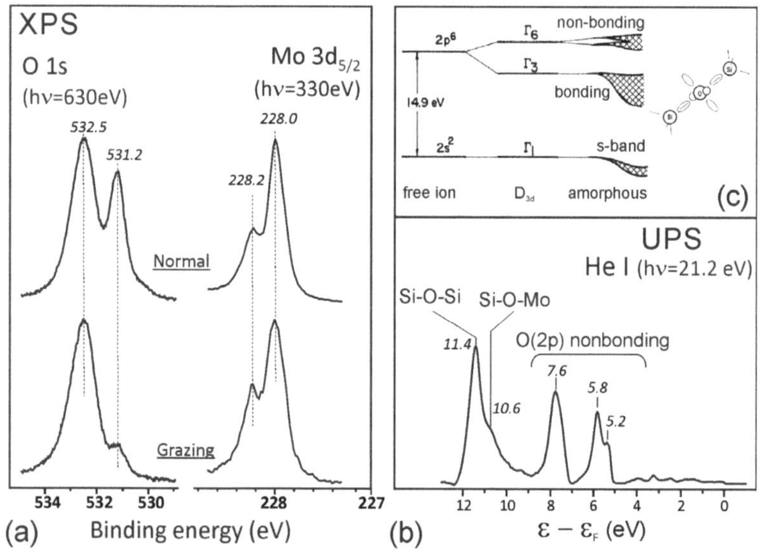

Figure 4.6 (a) Photoelectron spectra of the O 1s and Mo $3d_{5/2}$ core levels in a silica film on Mo(112) obtained with the synchrotron light at photon energies as indicated. (b) Ultraviolet (He I) photoelectron spectrum. The scheme illustrates splitting of the O 2p non-bonding states due to variations of Si–O–Si bonds angles.

O atoms, whereas the low BE state is associated with Si–O–Mo linkages in the "sub-surface" region. The Mo atoms involved in silica bonding are manifested in XPS through a weak signal in the Mo $3d_{5/2}$ core level at 228.2 eV (i.e., only 0.2 eV shifted with respect to the metallic Mo atoms in the substrate) that gains in intensity at grazing emission. On the basis of XPS results, the formation of Mo oxide phase underneath the silica film can be ruled out.

Finally, we address the valence band in the silica film measured by UPS as shown in Fig. 4.6b. All spectral features are quite sharp because of a high degree of crystallinity in the films under study. Small features at low energies (<4 eV) originate from a Mo substrate. The signals between 4 and 9 eV are commonly associated with non-bonding O 2p derived valence states, and the signals between 10 and 12 eV - with the O bonding states formed by hybridized O 2p - Si 3s,3p orbitals.[20,21] Comparative analysis of UPS spectra measured on several films of different thicknesses and degree of ordering linked the principal band at 11.4 eV to the Si–O–Si bonds, and the prominent shoulder at 10.6 eV to the Si–O–Mo bonds at the interface, where the O atoms accumulate electron density from the Mo substrate.[22] Splitting of the O 2p non-bonding states can be rationalized in terms of Si–O–Si bond angle variation as illustrated by the scheme depicted in Fig. 4.6c. In cubic SiO_2 having D_{3d} symmetry, this angle is 180° and the $2p^6$-degeneracy is split to produce two non-bonding orbitals (Γ_6) which are perpendicular to the Si–O–Si axis, and one bonding orbital (Γ_3) which links the Si and O centers. The *s* orbitals remain as the Γ_1 level. In going to other and eventually amorphous structure, the Γ_6 level will split into two non-bonding bands, corresponding to non-bonding orbitals which lie perpendicular to and parallel to a plane determined by the Si–O–Si chain.[20]

The model of a single layer silicate film, favored on the basis of experimental results presented above, and other possible models were examined theoretically by Sauer and co-workers.[19,23–25] The thermodynamic stability of several proposed structures was tested by computing the formation energy of a silica film (ΔE_{form}) from a clean Mo(112) surface, bulk α–quartz, and oxygen, following the reaction:

$$\text{Mo}\,(112) + m\,(\text{SiO}_2)_{\text{quartz}} + n\frac{1}{2}\text{O}_2 \rightarrow (\text{SiO}_2)_m\,\text{nO/Mo}\,(112),$$

where m and n are the number of SiO_2 units and oxygen excess in a surface (2×2) unit cell used in calculations, respectively. Thus, ΔE_{form} is defined as the difference between the total energy (E) of a given model and the corresponding energies of the clean Mo(112) surface, bulk α–quartz, and an O_2 molecule:

$$\Delta E_{\text{form}} = E - E_{\text{Mo(112)}} - m \times E_{\text{quartz}} - n \times \frac{1}{2} E_{\text{O}_2}$$

In addition to the previously proposed structures (Fig. 4.3), several single- (1 ML) and double-layer (2 ML) films were tested, some are depicted in Fig. 4.7. In the absence of the dispersion corrections (which turned out to be important for bilayer films as will be shown in the following sections), the calculations revealed the 2 ML film to be unstable since the formation energy is positive ($\Delta E_{\text{form}} = 0.8$ eV). Even though the introduction of "extra" oxygen (O*) atoms only adsorbed on the Mo(112) surface (Fig. 4.7b) substantially stabilizes

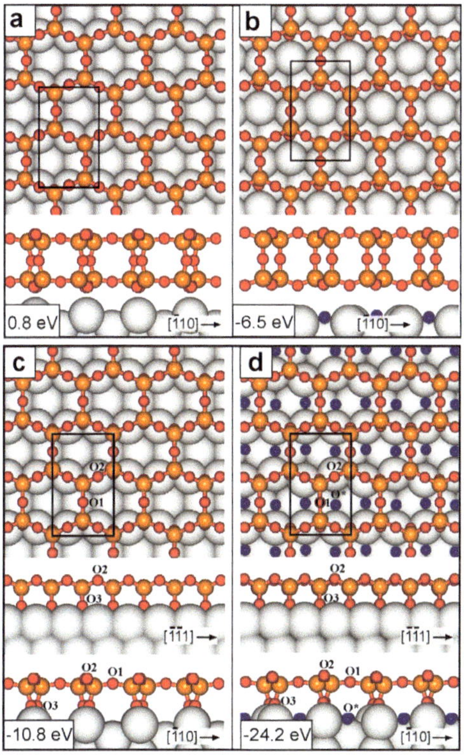

Figure 4.7 Top and side views of the single- (c,d) and double-layer (a,b) silica films on Mo(112). Structurally inequivalent oxygen species are labelled as O1–O3. Oxygen (O*) atoms only adsorbed on the Mo(112) surface are shown in blue. The formation energies ΔE_{form} are given in insets.

the structure, $\Delta E_{\text{form}} = -6.5$ eV, the monolayer silicate film is considerably more stable (-10.8 eV), and the formation energy increases even further by adding O* ad-atoms, up to -24.2 eV for the case of four O* atoms in the unit cell (Fig. 4.7d), which is by far the lowest formation energy. For comparison, ΔE_{form} computed for the β-cristobalite model shown in Fig. 4.3a, which can be referred to as a tri-layer (3 ML) film, is considerably smaller, i.e., -9.1 eV. The "cluster" model (Fig. 4.3b) yielded $\Delta E_{\text{form}} = -11.5$ eV, indicating its relatively high stability.

One has to bear in mind, however, that the models differ by the chemical composition. Therefore, the relative stability has to be compared at a given chemical potential of oxygen using the free energy change, normalized to the surface area A:

$$\Delta\gamma\,(T,\,p) = \frac{1}{A} \times (\Delta E_{\text{form}} - n\Delta\mu_{\text{O}}\,(T,\,p));$$

$$\Delta\mu_{\text{O}}\,(T,\,p) = \mu_{\text{O}} - \frac{1}{2}E_{\text{O}_2},$$

where μ_{O} is the oxygen chemical potential which is related to temperature and partial oxygen pressure by assuming that the surface is in thermodynamic equilibrium with the gas phase O_2.

If one compares $\Delta\gamma$ of considered structures as a function of chemical potential of oxygen (see Fig. 4.8), it becomes clear that a monolayer silicate film, either with or without the O* atoms adsorbed on Mo, hereafter referred to as "O-poor" and "O-rich" structures, is the most stable at the experimentally relevant conditions.

In addition to thermodynamic considerations clearly favoring the monolayer structures, the calculated vibrational properties further validate the model. Note that for comparison with experimental

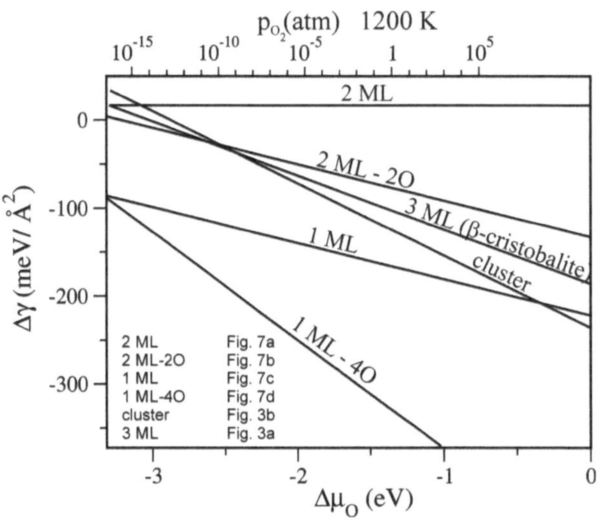

Figure 4.8 Surface-related free energy of formation $\Delta\gamma(T,p)$ for selected models presented in Figs. 4.3 and 4.7 as a function of oxygen chemical potential, which is transformed into corresponding O_2 pressure at fixed temperature (1200 K) usually used for the film preparation.

spectra obtained by IRAS which obey the surface selection rules, only vibrational modes having non-zero dynamic dipole moment normal to the surface must be considered. Three such IRAS-active modes are schematically shown in Fig. 4.9 together with the computed bands presented as bars. Not only the position, but also the relative intensities of the bands perfectly match the experimental results. The principal band at 1060 cm^{-1} originates from asymmetric stretching of the Si–O bond pointing to the Mo substrate. The second mode at 770 cm^{-1} corresponds to the Si–O–Si symmetric stretching coupled with Si–O–Si bending, and the third mode at 675 cm^{-1} is attributed to a coupling of Si–O–Si bending modes. Interestingly, the frequencies considerably shift (by about 50 cm^{-1}) depending on the orientation of the silicate layer with respect to the Mo(112) surface, which is practically impossible to determine experimentally, at least with the techniques used thus far. In addition, the presence of O* atoms adsorbed directly on Mo(112) causes small red-shift of the principal band (see Fig. 4.9), which was also observed experimentally on the "O-poor" and "O-rich" samples.

It should be mentioned here that interpretation of vibrational spectroscopy results for the atomic structure of a crystalline silica on Mo(112) revealed some confusion in using transverse optical (TO) and longitudinal optical (LO) modes terminology for the metal supported ultrathin films.[19] As originally introduced from analysis of an infinite cubic crystal with two atoms in a unit cell, TO modes correspond to a frequency ω_{TO} where the imaginary part of the complex dielectric function (ε^*) has a maximum value. Accordingly, the LO modes occur at a frequency ω_{LO} where ε has a minimum absolute value. In the infinite cubic crystal, the LO phonons induce a long-range electric field in the crystal which opposes the motion of the ions. In analogy with the infinite crystal, the TO and LO terminology has been used for thin crystalline films.[26,27] In one normal mode, the vibrations are parallel to the film surface, and the frequency is ω_{TO}. The term TO is used here because the vibrations parallel to the film surface produce no electric field, so they occur when the dielectric function has a very large absolute value. In the other mode, the vibrations are normal to the film surface and the frequency is ω_{LO}. Again, the term LO is used here

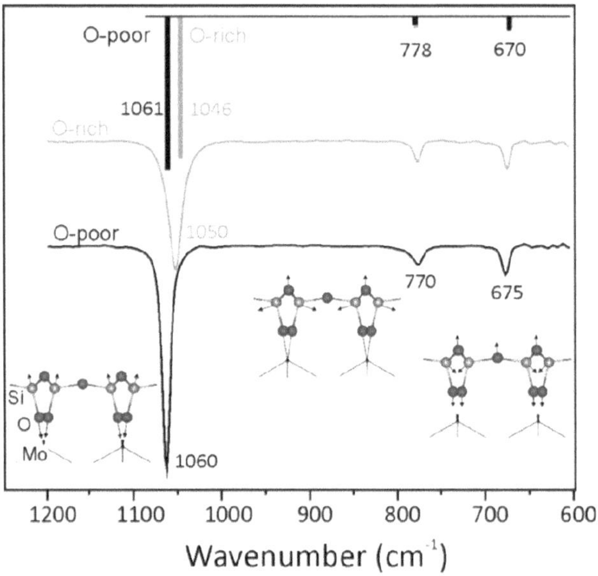

Figure 4.9 Experimental IRAS spectra obtained for the "O-poor" and "O-rich" silica films on Mo(112). Vibrational spectra calculated for proposed models, depicted in Fig. 4.7c,d, are shown as vertical bars with their height proportional to the intensity.

because the vibrations cause a surface polarization which results in an electric field normal to the surface equal to that component of polarization, and opposite in direction, so that the absolute value of ε reaches a minimum. According to the Berreman effect[26] discussed in detail in Chapter 2, the vibrations parallel to the surface (means, TO modes) are suppressed if using p-polarized light. This is because the vibrational dipole moment of the TO modes is entirely suppressed by the response of the metal substrate underneath the film.

The classification of vibrational modes in terms of LO and TO modes is not straightforward for *ultrathin* films. In our case of a silicate layer on Mo(112), all the asymmetric Si–O–Mo stretching vibrations are pure LO vibrations, since they occur normal to the surface. However, the physics of this LO vibration is completely different from that of the LO modes in "thick" bulk-like films, because there is no long-range electric field induced by this vibration that could oppose the motion of the atoms. Indeed, there is just one Si–O–Mo bond perpendicular to the surface with the metal surface below and vacuum above it, so no long-range electrostatic response is possible. Moreover, there is no TO mode associated with these vibrations since it would require Si–O–Mo bonds aligned parallel to the metal surface. Similarly, all combinations of the asymmetric Si–O–Si stretching vibrations are pure TO modes since they have no polarization perpendicular to the surface. The corresponding LO vibrations are impossible, because it would require Si–O–Si bonds perpendicular to the surface, but such oriented bonds exist in amorphous thicker films.

The silicate layer model received further support from studies using fast hydrogen and helium atom scattering at a grazing incidence.[28–31] The experimental technique is quite sophisticated and is not widely used in the "surface-science" community. The methods are based on scattering of He atoms and H_2 molecules with energies about a few thousands electronvolts under very small angles of incidence (0.2°–1.8°) from the surface. Scattering along low indexed atomic strings in the topmost surface layer results in diffraction pattern owing to diffraction effects for the elastically scattered projectiles. The diffraction patterns are then analyzed in terms of semi-classical trajectory calculations making use of interaction potentials derived from DFT. The experimental results obtained for the crystalline films on Mo(112) were in perfect agreement with the DFT-derived model shown in Fig. 4.7c. Moreover, the positions of topmost surface atoms could be determined with accuracy in the range of 0.01 Å.

Also, more recent I/V LEED measurements[32] revealed good agreement with the model (the reliability R factor $= 0.18$), and the structural parameters (bond distances and angles) were very close to those calculated by DFT. In particular, the silicate structure exhibits two Si–O–Si bonds with bond angles 133° and 163°, which, in turn, explain splitting of the O 2p non-bonding state in UPS spectra as discussed above (Fig. 4.6b).

Therefore, all these findings combined together provide compelling evidence for a single-layer silicate as the ultimate structure of a crystalline silica film on Mo(112). However, we were still curious whether the film thickness can be directly measured, for instance, by STM (or AFM). For this, we prepared samples only partially covered by silica so that a metal substrate could be used as a reference to measure the film thickness. It was thought that silica islands will be randomly distributed on a metal surface. To our surprise, STM images revealed a strip-like morphology (Fig. 4.10a).[33] If merged, the stripes form domains (Fig. 4.10b) showing the same honeycomb-like structure as observed on fully covered surfaces (see Fig. 4.5). Analysis of atomic protrusions both on stripes and on island revealed their registry, suggesting identical nature of protrusions forming the STM contrast on both stripes and islands. The spacing between adjacent stripes (13.4 Å) is equal to the periodicity observed on the reconstructed Mo(112)-p(2×3)O surface formed prior to the Si deposition.[34] Apparently, the O-covered surface acts as a template for the formation of silica first as stripes which then coalesce and ultimately form a silicate layer.

The template effect helped in finding a structure model with DFT. By implementing elegant genetic algorithm,[35] the calculations led to the structure shown in Fig. 4.10d, which additionally includes several oxygen species only bonded to the Mo atoms. Vibrational characteristics of the proposed structure agreed fairly well with the experimental results (Fig. 4.10c) and also explained the changes in comparison to the dense film. Basically, the largest difference is found for the vibrational band at 770 cm^{-1} in the film which is shifted to 875 cm^{-1} in the stripes (*cf.* Figs. 4.9 and 4.10c), because the mode coupling the Si–O–Si symmetric stretching and Si–O–Si bending in the film has less Si–O–Si bending character in the stripe. These silicate "wires" can be referred to as one-dimensional (1D) silica.

Inspired by establishing the atomic structure of a crystalline monolayer silica film on Mo(112), there were many attempts to grow "multilayer" crystalline films in a layer-by-layer mode. However, all efforts to prepare ordered silica layers on top of a pre-existing monolayer film were unsuccessful and resulted solely in amorphous silica. Annealing of thicker films at elevated temperatures in order to induce crystallinity led either to the film decomposition or its partial

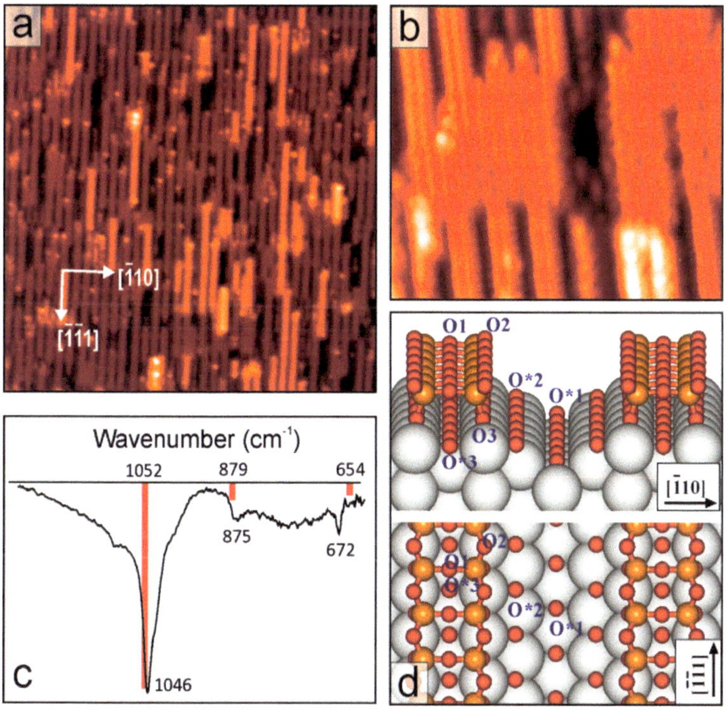

Figure 4.10 STM images (a, b) and IRAS spectrum (c) of a silica film on Mo(112) at sub-monolayer coverage. (d) Perspective and top views of the atomic structure of silica stripes predicted by DFT. Different oxygen species are labelled. The calculated frequencies for this structure are shown in (c) as vertical bars with their height proportional to the intensity.

sublimation, finally resulting in the monolayer film. It was thought that the lack of layer-by-layer growth was caused by that the silicate surface is formed by the coordinatively saturated and hence inert oxygen layer. Therefore, one needs to create "dangling" bonds for the next silica layer to bind. This would occur if the apical oxygen atom in SiO_4 tetrahedra, which is otherwise coordinated to the Mo surface atoms, were pointed to the vacuum to bind next SiO_4 units. Following this idea, the preparation has been modified so that it begins with Si deposition onto the clean Mo(112) surface, and not on the O-precovered Mo surface as previously.[36] Subsequent UHV annealing to about 1200 K resulted in a well-ordered Mo-silicide surface showing a (2×1)-Mo(112) symmetry. Then either one or a few cycles of Si deposition and oxidation at various temperatures was performed. The results showed no Mo–O species suggesting that the interface dominated by Si–Mo bonds. In addition to Si^0 (in Si–Mo) species, the films showed primarily Si^{4+} and only small amounts of Si^{2+} and Si^{+}. However, none of the films showed long-range order in LEED. On the other hand, the resulted films having nominal thicknesses equivalent to 2–3 silicate layers showed sharp phonon spectra which could only be observed on thermally grown silica films on Si at considerably higher thicknesses, thus indicating a better ordering in the metal supported films.

It should be mentioned that the atomic structure of the interface between SiO_2 and $MoSi_2$ have previously been studied by DFT with respect to the corrosion resistance of Mo-silicide attributed to a protective silica film that forms on its surface.[37] Using β-cristobalite structure for SiO_2, the best match was observed between $MoSi_2$(001) and SiO_2(100) crystal planes, with a lattice misfit of only 0.4%, while all other yield more than 10%. DFT calculations, using both the local density approximation and generalized-gradient approximation to electron exchange and correlation, predicted two types of interfacial structures giving about 5.5 J/m^2 for the adhesion energy. The difference between these two structures is that four O atoms of SiO_2(100) are bound to four different Si atoms on the $MoSi_2$(001) surface, as shown in Fig. 4.11b, whereas two O atoms share one Si atom in the structure shown in Fig. 4.11a. Electron density difference plots revealed covalent character of the interfacial Si–O bonds. It remains unclear, however, whether these models can be applied to our silica films grown on a surface silicide layer formed on Mo(112).

To sum up, a well-ordered silica film grown on Mo(112) is one of the best understood silica/metal systems studied. The atomic structure of the film consists of a single layer of corner sharing SiO_4 tetrahedra, like in layered silicates, which is adsorbed onto Mo surface. Careful reader may notice that the film is actually identical to a $SiO_{2.5}$ silicate layer formed upon oxidation of the SiC surfaces (Chapter 3). The important role of a substrate in the film formation will be discussed later in this chapter.

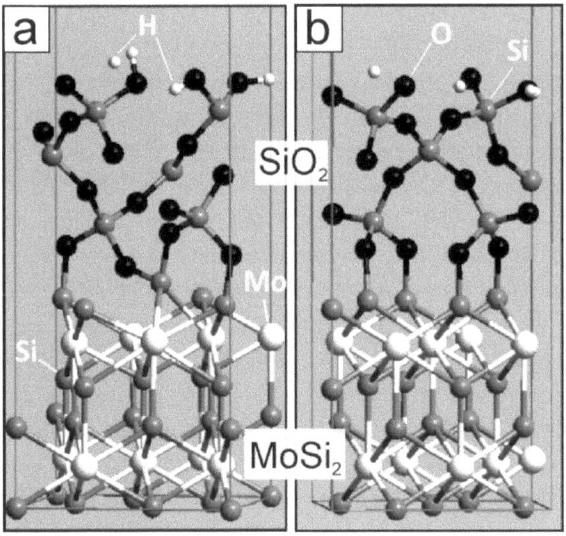

Figure 4.11 Two relaxed structures of the $SiO_2(100)/MoSi_2(001)$ interfaces. Some H atoms near the top may look isolated, but each of them is connected to an O atom in an adjacent unit cell. Adapted with permission from ref. 37).

4.2 3d-Transition Metals: Nickel, Iron, Cobalt

Thin silica films on 3d transition metals are studied to a lesser extent, although the first mention (as a private communication) of a crystalline silica film on Ni(111) can be found in the first paper reporting on silica films on Mo(112) in the year 2000,[9] but the work has only been published two years later.[38] Since no further studies have been reported on this system, we describe here the "old" preparation which considerably differs from that of used for Mo(112). The preparation began with deposition of a Si film about 3 nm in thickness onto the clean Ni(111) surface at 300 K. Then the surface is oxidized in 2×10^{-7} Torr of O_2 at 350°C for one (!) hour. Very important is, however, that oxidation was performed in the presence of *atomic hydrogen*. Although the films after this treatment showed no long-range order, without atomic hydrogen no crystalline films could be obtained after subsequent oxidation at

800°C in pure O_2 at the same pressure. The resulted surface showed a $(\sqrt{3}\times\sqrt{3})R30°$-Ni(111) symmetry in LEED.

Measured by optical ellipsometry the thickness of the crystalline silica film was about 4 nm. The preparation of thinner crystalline films was not successful, presumably due to the presence a nickel silicide phase at the interface detected by XPS. This interface between SiO_2 and Ni(111) was further inspected by in plane grazing-incidence x-ray diffraction, which allowed to propose that the prepared silica film is β-quartz in nature, since its lattice constant (5.01 Å) matches better the Ni(111) surface ($2\times\mathbf{a}_{Ni(111)}$ = 4.984 Å) than of α-quartz (4.913 Å). Unfortunately, high-resolution STM images were not obtained to judge on the flatness and surface structure of the films. In essence, the atomic structure of the crystalline films on Ni(111) remains unknown.

The crucial role of atomic hydrogen in this preparation is puzzling. Actually, the use of hydrogen for preparation of oxide films is counterintuitive since hydrogen is a reducing agent that even produces metals from their oxides. In this particular case, atomic hydrogen was invoked by the authors on the basis of previous studies showing recrystallization of amorphous silica films, deposited on quartz, upon ion implantation.[39] It was thought that implanted hydrogen atoms act as a network modifier that opens up the fully connected SiO_4 network in the amorphous film by forming non-connected tetrahedral corners (non-bridging oxygen) and hence produces a more flexible structure, thus facilitating atomic or structural rearrangements in the silica film during subsequent annealing in oxygen.

As mentioned above, deposition of Si on Ni (as well as on Mo) is accompanied by the formation of surface silicides, which are more clearly observed upon heating at elevated temperatures. It should be mentioned that the interaction of metal silicides with oxygen was widely studied in the 1980s in attempts to understand passivation of 3d-metals. It was found that initial stages of oxygen adsorption on silicides depend on the surface structure and stoichiometry. For example, O_2 exposure to the ordered $c(2\times2)$ silicide layer on Ni(001) led to the rapid formation of multiple Si–O bonds at all surface Si sites. The disordered and more Si-rich surface was more resistant towards oxidation.[40] In another AES/XPS study, the oxidation of the Ni-silicide surfaces of $NiSi_2$ and Ni_3Si stoichiometry in 5×10^{-6} Torr of O_2 is enhanced on all Ni–Si compounds studied, when compared to pure Si films under the same conditions.[41] It is interesting, however, that the strongest oxidation occurred in the Ni-rich silicide samples. Although oxygen preferentially reacted with Si surface atoms, also partial oxidation of Ni has been found at high exposures on Ni_3Si. In the same manner, AES was employed for a comparative study of the Cr, Si, and $CrSi_2$ surfaces exposed to 10^{-8} Torr of O_2 at room temperature.[42] Although the oxidation rate of Si in $CrSi_2$ was much higher than in pure Si, no oxidation of Cr in $CrSi_2$ was detected.

Apparently, the formation of silica films on 3d-metals depends on the thermodynamic stability of the respective metal silicide and metal oxide phases. Although oxidation of metal silicides remains the subject of interest in semiconductor technology and in catalysis, the above-shown study on Ni(111) seems to be the only one that was aimed at the preparation of crystalline silica films.

Recently, another approach was employed to study ultrathin silica films on metals which allowed to monitor their formation by transmission electron microscopy (TEM).[43,44] First, 3d metal like Fe and Co was vapor-deposited in situ onto the commercial TEM grids having SiO_2 coatings about 10 nm in thickness. Then the sample was annealed at 700°C in a high vacuum (10^{-7} mbar) to form relatively large (~50 nm) metal particles. After cooling the sample to 450°C, an intense high energy (200 keV) electron radiation ("beam shower") was applied to a selected spot presumably resulting in clean metal surface in this area. A few minutes later at the same temperature, a silica film covering the metal surface was observed. Figure 4.12 displays TEM images, which according to the image simulation are consistent with the formation of a monolayer silicate film on a metal particle surface. In addition, EELS spectra measured at the edge of Fe and Co particles provided solid evidence that the film bears close similarities to SiO_2.

The precise mechanism of the silica film formation in the above-mentioned experiments is not well understood. In principle, nucleation and growth of a silica overlayer can be explained either by bulk migration or by surface diffusion of silicon and oxygen atoms through (on) the metal particle. The diffusion seems to be facile

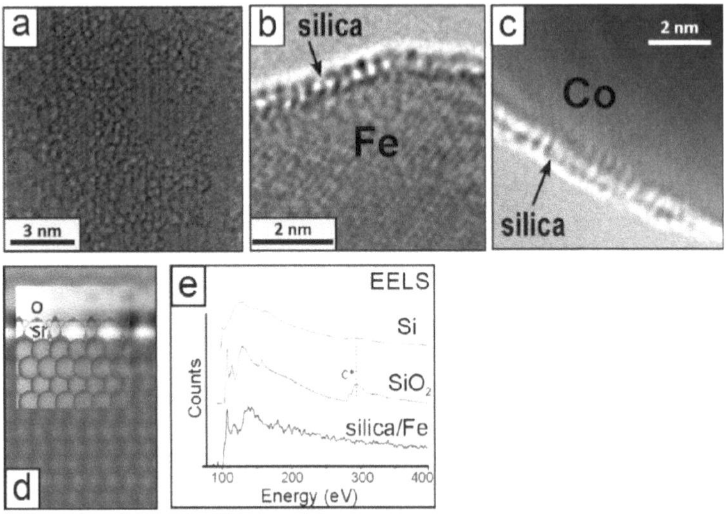

Figure 4.12 Plan-view (a) and cross-view (b,c) TEM images of a silica film grown on Fe (a,b) and Co (c) particles. (d) Simulated TEM image of a monolayer silicate film, superimposed with the model, showing close similarities to the image contrast in panel (b). (e) Electron energy loss spectrum measured at the edge of the Fe particle imaged in (b). The spectra of SiO_2 and Si are shown for comparison. The latter showed traces of adventitious carbon (C*). Adapted with permission from refs. 43, 44.

along grain boundaries. Since the interstitial formation enthalpy is relatively high (e.g., 2.5 eV calculated for Co), the solubility of Si in the 3d-metals is low, albeit in amounts sufficient to form a monolayer. In addition, the migration can be governed by kinetics rather than energetics. It may well be that metal particles dissolve small amounts of Si (and O) atoms upon first heating to 700°C, which then precipitate upon cooling as silicate layer. It is very important to have the metal surface "clean" prior to the film precipitation that was achieved by intense electron beam. However, once the film has been formed it is quite sensitive to the electron beam, and ultimately disappears under prolonged exposure.

Although the electron beam induced effects have to be taken into account in such experiments, further TEM studies[44] pointed to the possible formation of 1D-silica structures as well. These were

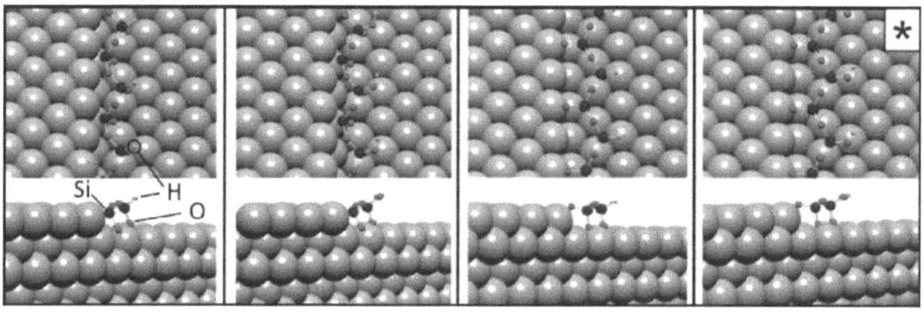

Figure 4.13 Top and side views of proposed 1D-silica structures at a step edge on the Ru(0001) surface. Among considered structures, the one marked (*) is found the most stable in the entire range of oxygen chemical potential. Adapted with permission from ref. 44.

imaged as lines, presumably on terrace step edges. On the Fe(111)-oriented facets, the lines formed a hexagonal pattern, whereas a rectangular pattern appeared on Fe(100). Also, on the Co(0001) surface, most of the lines formed a hexagonal pattern. The steps have been tentatively assigned as pinning sites for migrating Si and O atoms which react towards silica. Several hypothetical models depicted in Fig. 4.13 were examined by DFT with respect to its stability. (NB: The calculations were carried out for the Ru(0001) surface, also used in experiments. Silica films on Ru(0001) will be discussed in very detail in the next section).

The proposed 1D structures are quite different from the silica stripes observed on Mo(112) and discussed in the previous section (see Fig. 4.10). Indeed, in order to saturate the dangling bonds, all the models involved H atoms to form either Si–H or Si–OH species, which are hardly relevant to the above-described TEM experiments carried out in vacuum at high temperatures. In addition, the electron beam is known to sputter hydrogen atoms easily. Nonetheless, the DFT analysis confirmed their thermodynamic stability and suggested the 1D structures to be even more stable than the 2D films. Among the structures used in calculations, the structure shown in the right panel of Fig. 4.13 is found the most favorable.

To conclude this section, we note that silica overlayers on 3d-metal particles have not yet been characterized by surface sensitive techniques to draw more solid conclusions about their atomic structure. It is still awaiting the studies of silica films grown on single crystal surfaces of 3d metals.

4.3 Ruthenium

Although preparation of a well-ordered monolayer film on Mo(112) led us to a much better understanding of the silica/metal interface, attempts to grow thicker, "multilayer" silica films in crystalline phase were yet unsuccessful. This made us look for other metal supports better suited for such objectives. Apparently, the substrate should be more resistant towards oxidation than Mo, as the formation of the strong Si–O–Mo bonds at the interface seems to terminate the growth above one silicate layer. It is also beneficial to have the metal surface of a hexagonal or trigonal symmetry in order to better fit the symmetry of a silicate layer. It remained unclear whether the ability of a metal to form a silicide is advantageous for the film growth or not.

In this respect, the Ru(0001) surface was considered as a good candidate, and in fact it is widely used for preparation of thin oxide films. In particular, well-ordered and stoichiometric $CeO_2(111)$ films can be grown on Ru(0001) despite very large (~40%) lattice mismatch.[45] Although Ru is not as noble as, for instance, Pt and Au and may even form thin surface oxide layers, the Ru–O bond is definitely much weaker than the Mo–O bond. Oxygen atoms produced by oxygen dissociation on the Ru(0001) surface can readily react with incoming Si atoms and thus prevent Ru–Si intermixing. On the other hand, Ru is known to form surface silicides which may, in principle, improve the adhesion of the silica layers, and perhaps stabilize two-dimensional structures. On another note, Ru is not known for migration into the oxide films as observed, for example, for Ni and Mg substrates, that may be critical for using high-temperature annealing to make the films ordered. Irrespective of a substrate, one may think about preparation using Si deposition at low temperatures in order to suppress Si ad-atoms diffusivity and hence to favor the formation of a uniform overlayer.

As will be shown in detail below, well-defined ultrathin films can, indeed, be prepared on a Ru(0001) substrate, with principal film structures being dependent on the amounts of Si deposited (Si coverage).[46,47] There are slightly different preparations recipes reported thus far which remain the subject of continuous modifications in order to prepare dense, uniform and defects-free films. Nonetheless, it begins with O_2 adsorption on the clean Ru(0001) surface resulting in 3O-(2×2) ad-layer structure which can be viewed as a close-packed layer of the O atoms with one quarter of those missing in (2×2) positions. Then Si is vapor-deposited onto the O-precovered surface in O_2 ambient. Finally, the surface is annealed at about 1200 K in 10^{-6} mbar of O_2. Deposition at low temperatures (100 K) results in films of better quality as compared to the deposition at 300 K and above, because of the formation of more uniformly covered silica clusters in the "as deposited" films prior to the high-temperature annealing. Inspection of the samples after each preparation step with XPS showed that the O ad-layer prevents intermixing of Si and Ru and suppress the surface silicide formation. Simultaneously, it supplies more reactive oxygen species for oxidation of the Si deposits. However, the full oxidation only occurs at high temperatures, and it is accompanied by film ordering.

4.3.1 Monolayer Films

We first address the films prepared at relatively low Si coverages, although historically these were studied later.[47,48] IRAS spectra measured on the films prepared by deposition of about 0.5 ML of Si (here 1 ML is defined as the number of the surface atoms on Ru(0001)), showed the strong band at 1134 cm^{-1} and weaker signals at 1074, 790 and 687 cm^{-1} (Fig. 4.14a). This spectrum somewhat resembles the one obtained for a monolayer silicate film on Mo(112) showing a strong principal band (at 1060 cm^{-1}) and two weak bands at low wavenumbers (see Fig. 4.9). Also the O 1s XPS signal bears some similarity to the Mo(112)-supported films by showing at least two components separated by 1.5 eV. Solely based on these results, one can propose that the resulted film on Ru(0001) is a single silicate layer in nature.

Large-scale STM images (Fig. 4.15a) showed films following the morphology of the Ru substrate underneath, although the films on a large scale may also show holes and/or ill-defined silica

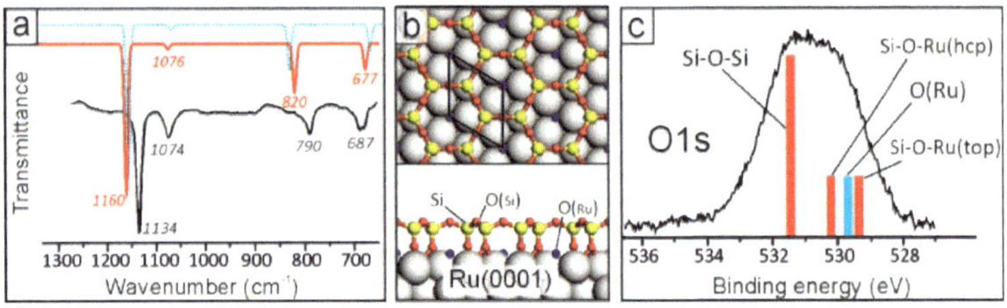

Figure 4.14 (a) IRAS spectra of a silicate monolayer film on Ru(0001) (black line). Red (solid) line shows the spectrum computed for the most stable DFT-optimized structure, which is depicted in panel (b) in top and side views. For clarity, the O atoms only bonded to the Ru surface atoms are shown in blue. For the structure where the O(Ru) atoms are missing, the calculations lead to the spectrum shown as dashed (blue) line. (c) Experimental XPS spectrum of the O 1s region. Blue vertical bars show binding energies calculated for the different O atoms in the model shown in (b). The bars height is proportional to the amount of corresponding O atoms in the unit cell.

nanoparticles. In turn, the terraces expose domains showing a honeycomb-like structure with a ~5.5 Å periodicity (Fig. 4.15b). If compared to the Mo(112)-supported films showing wide, atomically flat and defect-free terraces, the terraces on Ru(0001) show a mosaic structure, also evidenced by LEED via split (2×2) diffraction spots (Fig. 4.15c). Using the holes as a reference, the apparent film thickness measured by STM is about 1.5 Å, which is consistent with its assignment to a silicate layer.

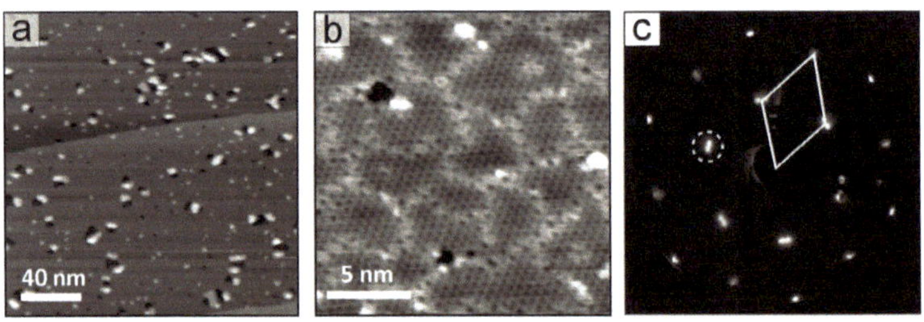

Figure 4.15 Large-scale (a) and high-resolution (b) STM images of a $SiO_{2.5}$/Ru(0001) film. Two monoatomic steps of the Ru surface underneath are running across the image, from the left to the right. (c) LEED pattern (at 70 eV) showing the (2×2)-Ru(0001) symmetry as indicated by the rhomb. Dashed circle highlights one of the six diffraction spots which are split due to the mosaic morphology of the film as imaged by STM in panel (b).

The DFT calculations fully supported the proposed structure and even found that the monolayer silicate film is stable only if it contains O atoms directly adsorbed on Ru (Fig. 4.14b). In analogy to the films supported on Mo(112), such a film may be referred to as an "O-rich" film. A more recent DFT studies performed by Krasheninnikov and coworkers, however, questioned the crucial role of the O ad-atoms on the film stability.[43] According to their calculations, the presence of O atoms at the Ru surface only slightly affects the energetics of the electronic ground state.

The harmonic IRAS spectrum simulated for the structure, shown in Fig. 4.14b, revealed the most intense IRAS active mode at 1160 cm^{-1} (see Fig. 4.14a) associated with the in-phase combination of asymmetric stretching vibrations in the Si–O–Ru linkages. The much weaker mode at 1076 cm^{-1} involves combinations of symmetric O–Si–O stretching vibrations. The bands at 820 and 677 cm^{-1} originates from the combinations of asymmetric stretching of Si–O–Ru linkages and O–Si–O bending modes. The positions and relative intensities of the calculated bands (according to the metal selection rules applied to IRAS) are in good agreement with the experimental data, even though the precise position of Si–O–Ru linkages with respect to the Ru(0001) surface in a mosaic-like film is hard to determine experimentally. Note that, in agreement with thermodynamic stability, also the simulated IRAS spectrum for an "O-rich" structure fits better than the film with the O(Ru) ad-atoms missing, both shown in Fig. 4.14a for direct comparison. Finally, the O 1s core levels simulated for the monolayer silicate film match well the experimentally measured one shown in Fig. 4.14c. A more recent high-resolution XPS study[49] using synchrotron radiation could resolve the O 1s signal and confirmed their assignment, although there were small quantitative deviations in signal intensities as compared to the DFT predictions. Based on angular resolved XPS data, the authors could map the electronic band dispersion which revealed four bands below the Fermi level. At high binding energy, two of them show a linear dispersion at their crossing K point (so called Dirac cones) and two others form semi-flat bands. While the latter two correspond to hybridized states between the silica oxide and the metal substrate, the former relate to the topmost Si-O layer not directly coupled to the substrate.

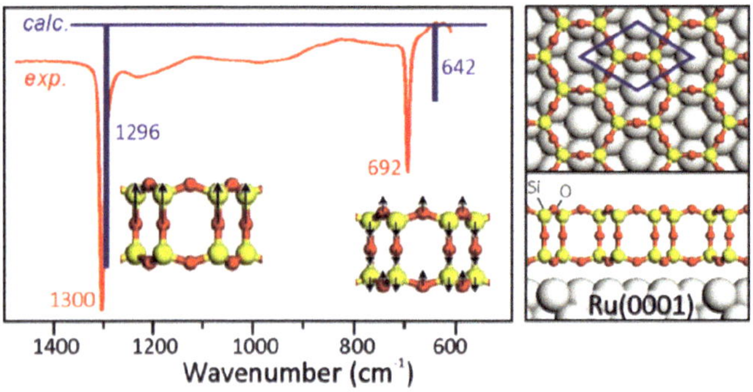

Figure 4.16 IRAS spectrum of a bilayer silicate film grown on Ru(0001) shown in the top and side views. For comparison, the simulated spectrum is shown as vertical bars with the height proportional to its intensity. The inset depicts the associated vibrational modes.

4.3.2 Bilayer Films

In contrast to a Mo(112) support, on which all films above one monolayer in thickness only showed amorphous structure spectroscopically very similar to the thermally grown films on Si, deposition of additional 0.5 ML Si on top of a crystalline monolayer film on Ru(0001) (or the deposition of totally 1 ML Si in one step) followed by oxidation resulted in a crystalline film but showing very sharp and intense IRAS bands at 1300 and 692 cm^{-1} (Fig. 4.16a) which have never been observed on silica films before. At intermediate Si coverages (between 0.5 and 1 ML) deposited in one step, the new bands coexisted with those of a monolayer film. Therefore, the new film structure is thermodynamically stable and only depends on the total Si coverage.

Solely the fact, that the amount of Si in the new films is twice of the monolayer film, suggests its bilayer structure. The principal IRAS band at 1300 cm^{-1} is substantially higher in frequency than previously observed for Si–O–Ru or Si–O–Mo interfacial bonds in monolayer films, i.e., 1135 and 1060 cm^{-1}, respectively. Applying a reduced mass analysis to an oscillator, one comes to the conclusion that the corresponding vibration involves atoms lighter than Mo and Ru, means Si–O–Si bonds which are normal to the surface to be IRAS-active. Combined together the results provided a good basis for a model of a bilayer silicate film, shown in Fig. 4.16, composed of two silicate layers linked together through bridging O atoms in the mirror plane.

In fact, the bilayer structure have previously been considered in the DFT calculations of silica films on Mo(112) (see Fig. 4.7a,b), but rejected because it showed positive formation energy. Largely, it is because the weak dispersive (van der Waals) forces between O-terminated layer and a metal surface have been neglected in the calculations. Therefore, further DFT studies of a bilayer silicate film on Ru(0001) included semi-empirical dispersion (+D) correction. Within the so-called DFT+D approach, the adhesion energy computed for the model depicted in Fig. 4.16 becomes slightly negative, i.e. −3.1 kJ mol^{-1} $Å^{-2}$, with the main contribution to the total energy coming from the dispersion term, -3.8 kJ mol^{-1} $Å^{-2}$.[46] The simulated vibrational spectrum revealed two IRAS active modes as in experiment. The most intense mode at 1296 cm^{-1}, which almost coincides with the experimental value 1300 cm^{-1}, originates from an in-phase combination of asymmetric Si–O–Si stretching vibrations of the Si–O–Si linkage between two layers. The second mode at 642 cm^{-1} reminiscent of the experimental 692 cm^{-1} band results from a combination of symmetric stretching vibrations of the Si–O–Si bonds parallel to the surface. Thus, the DFT and IRAS results provided solid evidence for the bilayer model.

Following previously introduced models of "O-poor" and "O-rich" structures, also bilayer films can accommodate additional O* atoms only adsorbed onto the Ru(0001) surface (Fig. 4.17).[50] Not surprisingly, the presence of O* ad-atoms lowers the adhesion energy due to their repulsive interaction with the bottom O-layer in the film (see Table 4.1). As a result, the distance between the silicate bilayer and Ru support considerably increases. However, the O* atoms show almost no effect on the vibrational bands which only shift by a few wavenumbers. However, they are clearly detected by XPS as a small O 1s signal (at ~530 eV) well separated from the main signal at ~532 eV associated with all kinds of the Si–O bonds.

To get rid of the O* atoms and transform the film into the "O-poor" state, the samples are annealed in UHV at elevated temperatures. Annealing does decrease the O* coverage as measured by XPS, but also causes gradual shift of the O 1s and Si 2p signals

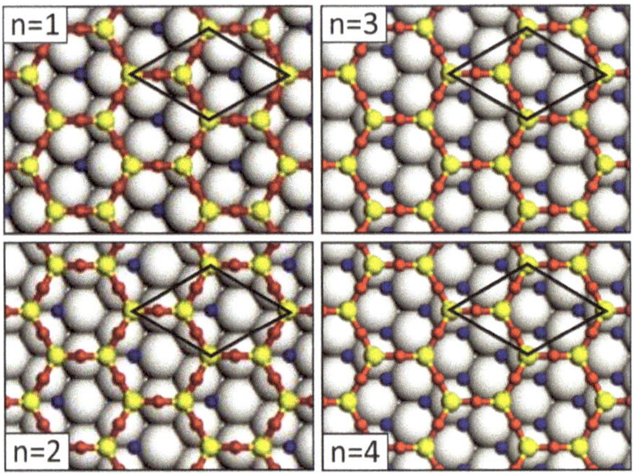

Figure 4.17 The top views of the "O-rich" bilayer silica structures on Ru(0001). The number of O* atoms directly adsorbed on Ru per (2×2) unit cell (n) are indicated and shown as blue balls (NB: some are screened by the Si atoms above).

Table 4.1 Distance (d) between the bottom O-layer in the silicate film and the Ru surface layer, and adhesion energy ($\boldsymbol{E}_{adh}$, including zero-point energy) as a function of the number (n) of O* atoms per Ru(0001)-(2×2) unit cell

n	0	1	2	3	4
$\boldsymbol{d}$ (Å)	2.75	3.08	3.68	3.81	3.85
$\boldsymbol{E}_{adh}$ (kJ mol^{-1} Å^{-2})	−3.12	−2.39	−1.80	−1.82	−1.85

Source: From ref. 50.

accociated with a silica film to higher BEs by about 0.8 eV, whereas the O* signal only attenuates (Fig. 4.18a,b). The transition between the O-poor and O-rich structures is fully reversible: Re-oxidation of the annealed samples fully restores the initial spectrum. Also the valence band responds in a similar way: Silica-related features all shift by about the same value as the Si 2p and O 1s core level signals.

Theoretical calculations[50] of the energy separation between the O 1s states in SiO_2 and of the O* atoms performed for the "O-rich" structures shown in Fig. 4.17 yielded 2.5, 2.2, 2.0, and 1.8 eV for n = 1, 2, 3, and 4, respectively. Therefore, the principal O 1s peak will shift to high BEs at decreasing O* coverage by 0.7 eV (= 2.5 −1.8), in full agreement with the experimentally observed shift. One may argue that the O* atoms also alter the distance at the interface (Table 4.1) and hence may affect the ground electronic state. However, additional calculations performed by artificially varying the distance revealed this effect negligible.

The observed simultaneous shift of both the Si and O core levels in silica caused by O* adsorption/desorption from the metal surface highlighted its relation to the band alignment at the oxide/metal junction, which is driven by the change in the work function (WF) of a metal substrate the silicate layer weakly adsorbed on. Indeed, WF decreases from 6.45 to 5.50 eV upon UHV annealing, that nicely correlates with the BE shifts of the core levels (Fig. 4.18e). Recent DFT study[51] of the charge redistribution in O*/Ru and SiO_2/nO*/Ru systems revealed the charge transfer from Ru(0001) to the O* atoms, resulting in negative surface dipole moment and hence increased WF as compared to the clean surface. On the other hand, the O atoms in the bottom silicate layer lose electrons, predominantly from p_z orbitals, while the surface Ru atoms gain electrons into d_{xz} and d_{yz} orbitals. This interfacial charge transfer from the SiO_2 film to the Ru substrate leads to the positive interface dipole moment. The net effect depends on the number of O* atoms in the cell. Therefore, surface and interface dipole moments caused by charge transfers play a role in the energy level alignment across the silica/metal interface.

Thus far, we have not discussed long-range ordering in the bilayer films. Based on the IRAS spectra showing strong and extremely narrow bands about 10 cm^{-1} in width, one would reasonably suggest the films to be crystalline. Also theoretical calculations employing periodic (means, crystalline) structures adequately describe electronic and vibrational properties of the films. Unfortunately, conventional LEED on these systems cannot accurately examine degree of ordering in the bilayer films, because the (2×2) diffraction spots are weak and can also be attributed to the O ad-layer on Ru(0001) underneath the film and/or adventitious monolayer domains, all showing the (2×2) symmetry. Actually, the bilayer film has two degrees of ordering. While the entire film is bilayer and hence ordered in the direction *across* the film, long-

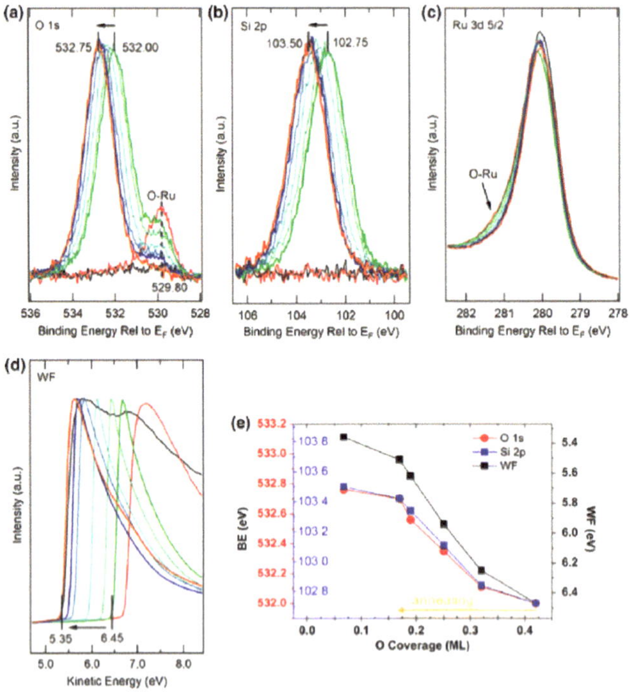

Figure 4.18 The evolution of the O 1s (a), Si 2p (b), Ru $3d_{5/2}$ (c) levels and of the work function (WF), measured via secondary electron cut-off, on the SiO_2/Ru(0001) film, after consecutive heating in UHV from 950 to 1200 K. The black and red spectra, obtained on the Ru(0001) and 3O(2×2)-Ru(0001) surfaces, respectively are used as references. Panel (e) shows the BE and WF values as a function of the O* coverage. Reproduced with permission from ref. 51.

range order *along* the film surface was found to be dependent of the preparation conditions, such as the Si deposition rate, deposition temperature, annealing time, and even cooling rate, that is at variance to the monolayer films where such small details are not critical, and the films always showed high degree of crystallinity if annealed accordingly.

To find out ordering in the bilayer films we address STM studies. Not surprisingly, to achieve STM images of bilayer films with atomic resolution is more difficult as compared to monolayer films, because these latter are strongly bonded to the metal support that makes electron transport more facile, whereas the bilayer film is weakly bonded through the van der Waals forces. Typical large-scale images of the films (of very good quality as judged by IRAS) are displayed in Fig. 4.19(a,b). The films uniformly cover the support, although some "holes" may exist which primarily expose the monolayer structure. The bare Ru surface, if present, can easily be titrated by CO since the silica does not chemisorb CO. (In fact, it is always recommended to inspect integrity of the films by CO (and H_2) adsorption experiments using, for example, thermal desorption spectroscopy since gas adsorption on many metal surfaces is well-documented in the literature).

High-resolution STM image in Fig. 4.19c shows crystalline domains with honeycomb-like structure as expected for the model depicted in Fig. 4.16. The domains are separated by less defined structures, which may even dominate the surface, as shown in Fig. 4.19e, where only small areas display a hexagonal array of depressions, and the rest of the surface exhibits both larger and smaller depressions, thus suggesting relatively broad distribution of silica rings. The lack of long-range order in the films is also reflected in corresponding LEED pattern shown as inset in Fig. 4.19c. In addition to the sharp (1×1) diffraction spots of Ru(0001) and more diffuse (2×2) spots, the pattern exhibits the ring crossing the (2×2) spots, which is characteristic for randomly oriented, "polycrystalline" structure. Depending on the preparation, the relative intensity of the (2×2) ring may vary. The defectness of the films and related issues will be discussed in details in Chapter 5.

4.3.3 "Zigzag" Films

In the course of film preparations on Ru(0001), new structures were observed under certain preparation conditions which could not be referred to either strongly bound monolayer or weakly bound bilayer films, but exhibit rather intermediate characteristics. The structure received the name "zigzag"[52] due to its appearance in STM images as parallel zigzag rows (see Fig. 4.20). This structure appears by oxidation at temperatures slightly below than used for the formation of the bilayer films, that is why it often coexists with the bilayer phase.

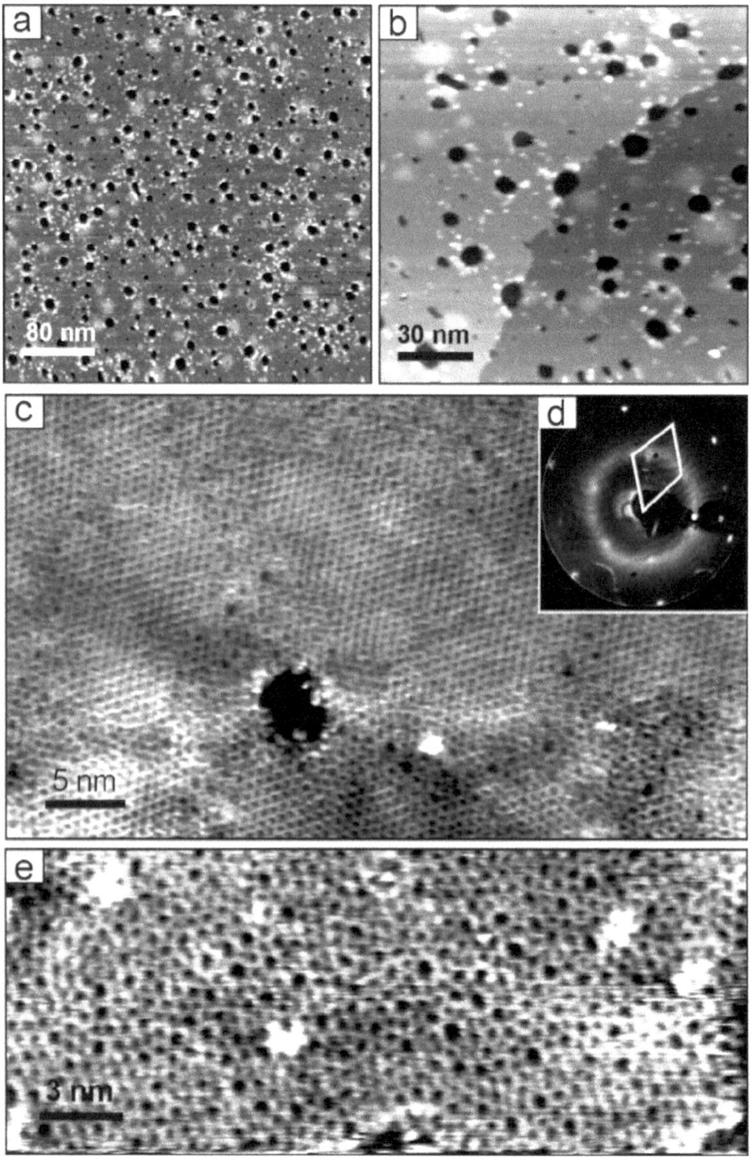

Figure 4.19 (a,b) Typical large-scale STM images of the bilayer films. (c,e) High-resolution STM image of a silicate bilayer film showing both periodic and disordered surfaces. (d) Corresponding LEED pattern (at 70 eV) of the imaged surface; the (2×2) unit cell is indicated.

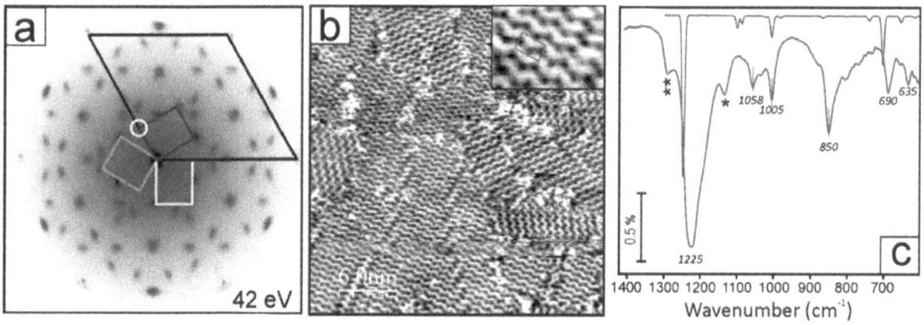

Figure 4.20 LEED pattern (a), high-resolution STM image (b), and IRAS spectrum (c) of a "zigzag" film formed on Ru(0001). (a) The rectangular unit cells are indicated for three rotational domains. Rhomb indicates the unit cell of the Ru(0001) surface. White circle highlights the missing spot in the unit cell due to a glide plane symmetry in the film. (b) Three rotational domains each showing parallel zigzag lines are clearly seen. The unit cell in real space is marked in inset. (c) Marked with ** and * are the bands originated from small regions with bilayer and monolayer structures, respectively. Blue curve shows simulated IRAS spectrum calculated for the structure model shown in Fig. 4.21.

Figure 4.20a shows typical LEED pattern obtained on these films. It is quite complex and very different from primitive (2×2) type of patterns obtained both on mono- and bi-layer films. Rectangular unit cells of the silica overlayer are shown for three rotational domains. An incommensurate structure is defined in matrix notation as $\begin{bmatrix} 4 & 2 \\ 0 & 2.8 \end{bmatrix}$, that results in the surface lattice constants 9.4 and 7.6 Å. In addition, two other features are observed by LEED. Firstly, all silica related spots are elongated in one direction independently of the diffraction order, thus pointing to a limited size of the ordered domains (see Chapter 2). The estimated size is about 15 nm × 5 nm, on average, where the smaller value refers to the size in the direction of the shorter unit cell vector (i.e., 7.6 Å). Secondly, one of the four spots in the rectangular cell is missing (highlighted by a white circle in Fig. 4.20a). Such missing spots are characteristic for surface structures exhibiting a glide plane symmetry (that is the combination of a reflection in a plane and a translation in that plane). High-resolution STM images, shown in Fig. 4.20b, revealed three rotational domains each showing parallel zigzag lines. The

STM derived rectangular unit cell possesses a glide plane symmetry in agreement with LEED.

Vibrational fingerprints are also very different from monolayer and bilayer silicate structures, which, however, may be present in the samples as minority species (Fig. 4.20c). By using Si 2p and O 1s XPS signals from mono- and bi-layer films as references, one can get an estimate for compositional stoichiometry in the "zigzag" films ($SiO_{2.3}$) that is between monolayer ($SiO_{2.5}$) and bilayer (SiO_2).

The experimental results provided solid basis for structural models which have been scrutinized using DFT calculations.[52] Among various models tested, the most stable one is shown in Fig. 4.21(a,b) in two different representations. It consists of a complex arrangement of silica tetrahedra. The parallel zigzag lines are formed by slightly distorted vertically arranged four-membered rings. These are bound together with bridging SiO_4 units in the top layer of the system, thereby forming distorted non-planar eight-membered rings in the top view (shaded in Fig. 4.21c) and vertically arranged three-membered Si rings in the side view. In addition, the unit cell contains two Si atoms connected to two O atoms each bonded to the Ru surface atoms. Therefore, the stoichiometry for the silica "zigzag" phase is $SiO_{2.17}$ in fairly good agreement with XPS-based estimation.

The model nicely explains the observed STM contrast by assigning the protrusions to the Si atoms in the top layer (Fig. 4.21c). Also the simulated IRAS spectrum for the model fits fairly well the experimental one, both shown in Fig. 4.20c, except the 850 cm^{-1} band which turned out to be much weaker in calculations than in experiment. The most prominent peak at 1225 cm^{-1} originates from antisymmetric stretching of the vertical Si–O–Si bonds that goes in-phase. The mode at 1060 cm^{-1} is antiphase counterpart. For this mode, the vertical Si–O–Si vibrations in the three-membered rings are more pronounced than those in the four-membered rings. The band at 1000 cm^{-1} is assigned to combination of antisymmetric Si–O–Si stretch vibrations in three-membered rings, horizontal antisymmetric Si–O–Si of four-membered rings, and also the antisymmetric stretch of the bridging Si–O–Si that goes further in the direction of O–Ru. Other low-frequency modes correspond

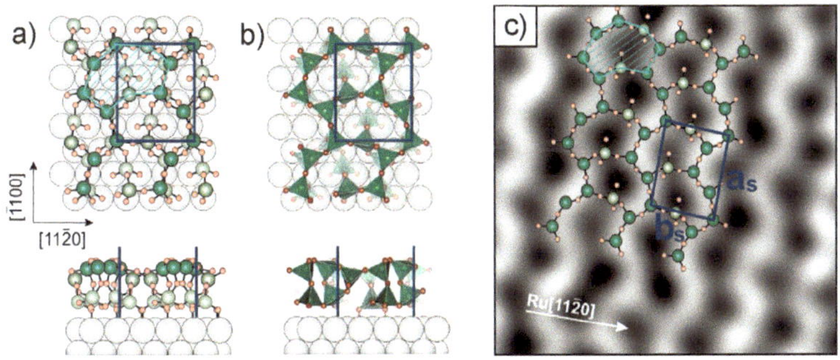

Figure 4.21 (a,b) Top and side views of the silica "zigzag" film in two different representations. The topmost Si atoms are dark green, other Si atoms are light green, O atoms are orange, and Ru atoms are gray. (c) The structure of the topmost layer superimposed with an STM image. The rectangular unit cell is indicated. Hatched area highlights eight-membered silica ring in the structure. Adapted with permission from ref. 52.

to symmetric Si–O–Si stretching, which, in addition of the inter-ring coupling, are also coupled to the O–Ru stretches. Certainly, the complex structure of the film may additionally generate vibrational modes of mixed character.

The "zigzag" film readily transforms into a bilayer film upon further oxidation at higher temperatures. However, once the bilayer film has been formed, it cannot be transformed back to the "zigzag" structure by any further treatments. It appears that the zigzag structure is an intermediate phase that forms in the course of bilayer film formation.

4.4 Platinum

Platinum, in particularly the Pt(111) surface, is widely used as a support for preparation of transition metal oxide films.[53–55] Compared to Ru(0001), the Pt(111) surface has a slightly larger lattice constant, i.e., 2.71 Å and 2.77 Å, respectively. In addition, Pt is more noble metal than Ru: The Pt–O bond is much weaker than Ru–O, and Pt does not form stable oxides. Accordingly, Pt(111) accommodates less O atoms than Ru (0001) under similar exposure conditions, which desorb upon heating to a much lower temperature.

Well-defined silica films on Pt(111) were prepared using, in essence, the same recipe as used for Ru(0001), i.e., by Si deposition onto an O-precovered metal surface at low temperatures followed by oxidation at high temperatures.[56] Interestingly, the Pt–Si bonds like in silicides were observed by XPS after Si deposition. For comparison, no Ru–Si bonds were detected in case of Ru(0001). It is because a dense 3O-(2×2) ad-layer forms on Ru that readily reacts with incoming Si atoms and thus prevents intermixing. Nonetheless, the Pt–Si features disappear upon oxidation. In fact, the Si 2p and O 1s spectra look very similar to those measured on bilayer films on Ru(0001) (Fig. 4.18(a,b)) except that less or almost no intensity in the low binding energy O 1s signal at 530 eV, which is a fingerprint of the O* atoms only adsorbed on the metal surface. As in case of Ru-supported films, the Si 2p and silica related O 1s signals gradually shift to higher binding energies upon UHV heating due to the desorption of the O* ad-atoms. However, the BE shift (about 0.25 eV) is much smaller than 0.8 eV observed on SiO_2/Ru(0001). The difference can readily be explained by that Pt(111) adsorbs smaller amounts of the O* ad-atoms.

IRAS spectra of the silica films grown at different Si coverages only show the bands at 1294 and 690 cm^{-1} (Fig. 4.22) like in bilayer, but no bands in the 1000–1100 cm^{-1} region anticipated for monolayer. The intensity of IRAS (and also XPS) signals scales with the film coverage determined by STM, suggesting the growth

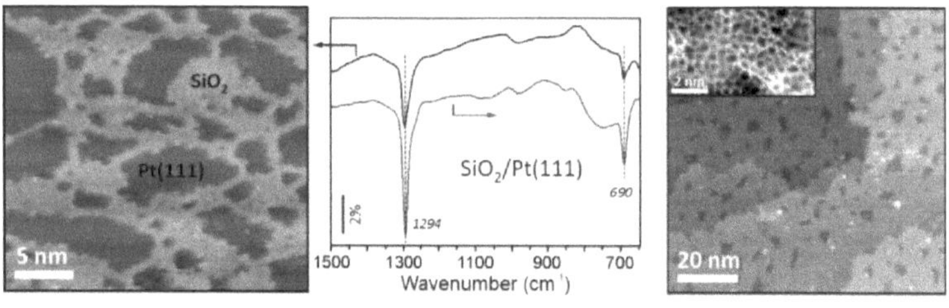

Figure 4.22 IRAS spectra and corresponding STM images of SiO_2 films grown on Pt(111). The inset shows high-resolution image indicating an amorphous bilayer silicate film.

of a bilayer silicate film from the onset. In agreement with LEED only showing the diffuse (2×2) diffraction ring and hence the absence of crystallinity, high-resolution STM images (see inset in Fig. 4.22) revealed amorphous structure like in bilayer silicate films on Ru(0001).

In principle, STM images shown in Fig. 4.22 allow to measure the film thickness by measuring the height of silica islands. To recall, this was difficult to do on Mo(112)-supported films due to the 1D-silica formation on top of heavily reconstructed O/Mo surface (see Fig. 4.10). Also on the SiO_2/Ru(0001) bilayer films, the coexistence of monolayer and bilayer structures made such measurements uncertain. Here, the area surrounding the silica islands is either pure Pt(111) or, most likely, CO-covered Pt(111) since Pt readily adsorbs CO from the UHV background. When measured at typical tunneling conditions (i.e., bias 1 V and current 0.1 nA) line profiles revealed 2 Å for the island height, whereas the geometrical height of a bilayer silicate film (i.e., the distance between the surface metal layer and the topmost O layer) determined by DFT is about 5 Å, i.e., substantially larger. On the other hand, the measured value is definitely larger than 1.4 Å obtained by STM on the monolayer films on Ru(0001) under similar tunneling conditions. The deviation between STM derived and geometrical film thicknesses can reasonably be assigned to the electronic effects contributing to the STM contrast which are frequently observed for oxide/metal interfaces (see, for example, Fig. 2.13 in Chapter 2).

Interestingly, at low coverages, the silica islands are bridged by stripes about 10 Å in width which are oriented along the three principal crystallographic directions of the Pt(111) surface. To some extent, they resemble 1D-silica structures observed on Mo(112) (see Fig. 4.10), although the atomic structure has not yet been identified.

4.5 Palladium

It is worth mentioning that very first studies on preparation of thin silica films on Pd substrates were reported even before the formation of bilayer silicate films on Ru(0001) and Pt(111) has been established.[57] The films were prepared by Si vapor deposition onto Pd(100) at 500 K in 10^{-5} mbar O_2 and subsequent cooling in oxygen ambient, i.e., without high-temperature annealing step typically required for oxide films to be well-ordered. On the basis of the Pd 3d XPS signal attenuation, the film thicknesses varied between 0.4 and 6.5 nm. More detailed structural characterization only performed on the 2.8 nm-thick films revealed substantial amounts of silicon sub-oxides, probably because of relatively low temperatures used in the preparation. Note also that no Pd-silicide phase formed under these conditions. On the other hand, vacuum annealing of the films at temperatures up to 1050 K caused Pd migration to the film surface and/or film dewetting monitored by ion surface scattering experiments. What was surprising are STM images showing atomically flat terraces exhibiting a square lattice of protrusions with a 3.6 Å periodicity. Since it is close to 1.4 (or $\sqrt{2}$) times that of Pd(100), a c(2×2) symmetry has been proposed for the film. STM results were taken as evidence for a good crystallinity of the prepared silica films, although the measured lattice constant (3.6 Å) is substantially smaller than any of silica polymorphs and of silicate sheets. In addition, HREELS spectrum measured in specular geometry revealed the principal loss at 1106 cm^{-1} (Fig. 4.23a), which is far below all previously reported frequencies measured on thin silica films and quartz. On the other hand, the band falls in the range of frequencies obtained for silicate monolayers on Mo(112) ($=1048$ cm^{-1}) and Ru(0001) ($=1134$ cm^{-1}) and associated with vibrations of Si–O-Metal bonds at the interface. Accordingly, the observed 1106 cm^{-1} band could be associated with the Si–O–Pd stretch vibrations in a single silicate layer. However, this conclusion contradicts the film thickness (2.8 nm) measured by XPS, although one may question the accuracy of XPS derived thicknesses in the case of ultrathin films.

In this respect, it is worth mentioning HREELS spectra (Fig. 4.23b) of silica overlayers formed on Pd(111) upon chemical vapor deposition of silane (SiH_4) as Si precursor.[58] Note, however, that this study was not aimed at the silica film preparation, but rather at understanding of the Pd/silica interaction. In these experiments, clean Pd(111) was first dosed with O_2 at 300 K, then with SiH_4 at 150 K, and subsequently annealed in UHV. As the

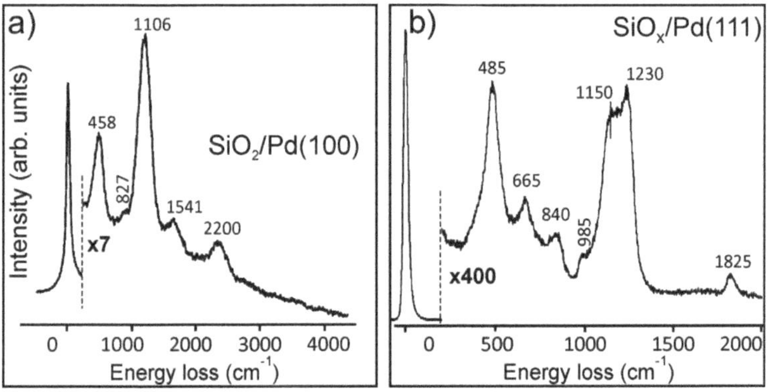

Figure 4.23 (a) HREELS spectrum of a silica film of 2.8 nm in thickness grown on Pd(100) at 500 K. Adapted with permission from ref. 57. (b) HREELS spectrum of SiO_x layer on Pd(111) formed by SiH_4 exposure to an O-precovered surface and then annealed at 950 K. Adapted with permission from ref. 58.

hydrogen atoms released from the decomposition of silane also react with the surface oxygen atoms to form water (indeed, observed by thermal desorption measurements), the preparation hardly results in fully oxidized silica. Solely with AES employed in this work, it was difficult to determine precisely the oxidation states of Si. In addition, based on CO titration experiments, the prepared silica film did not cover the entire Pd surface. In principle, the loss at 1230 cm^{-1} that develops upon high-temperature annealing is indicative of a long-range silica network structure. Tentatively, the band at 1150 cm^{-1} has been assigned to Si–O–Si asymmetric stretching, whereas the weak band at 985 cm^{-1} to Si–O–Pd asymmetric stretching. The follow-up DFT calculations of vibrational bands performed only for the case of individual SiO_x entities and their aggregates could not fit the experimental results.[59] (NB: The same group has later performed similar experiments with Pt(111) as a substrate.[60] Again, no additional oxygen treatments were applied, and as such the resulted films were most likely oxygen deficient.) It should be noticed again that the band positions observed in these HREELS studies are very different from those known for silica films, rendering the resulted structures yet poorly defined.

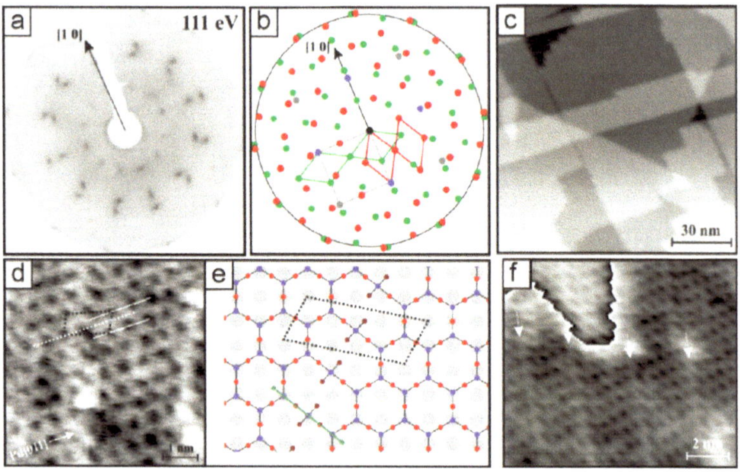

Figure 4.24 (a) High-resolution STM image of a silica film grown on Pd(100). The white lines highlight the lattice shift between the domains resulting in an antiphase domain boundary (ADB). (b) Structural model of the ADB. Gray circles represent Pd, red circles O, and blue circles Si; the darker red circles highlight O atoms at a lower depth. The dotted parallelogram connects the centers of the six-membered rings on either side of the ADB. Reprinted with permission from ref. 61.

A few years later, Altman and co-workers revisited preparation of thin silica films on Pd(100).[61] Silicon deposition in 10^{-6} mbar of O_2 onto the crystal surface kept at 550–600 K in order to mimic the previously reported preparation[57] did not show any evidence for film ordering. Only faint diffraction spots of a Pd(100) substrate were detected by LEED, therefore rejecting previous statement on the formation of ordered silica films on Pd(100) at such low temperatures. To induce long-range ordering, the surface was oxidized at temperatures above 975 K. The degree of crystallinity was found to depend on the initial film thickness, annealing time and temperature. High temperatures and insufficient oxygen may even lead to silica decomposition and migration of Si into the bulk Pd crystal.

Fairly complex LEED pattern presented in Fig. 4.24a was rationalized in terms of coincidence structures formed by a hexagonal mesh placed on top of the Pd(001) metal surface having a square lattice.

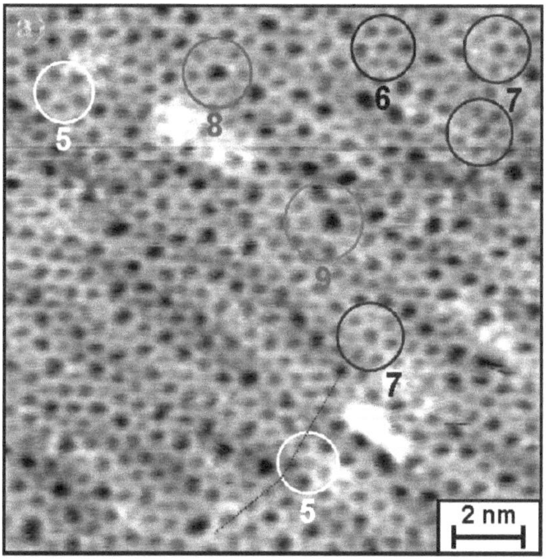

Figure 4.25 STM image of the silica film grown on Pd(100). The circles and adjacent numbers highlight *N*-membered rings seen in the center of each circle. Adapted from ref. 62.

Careful analysis revealed a periodicity twice that of the substrate and with one of the overlayer close-packed directions paralleling Pd(011). High-resolution STM images showed crystalline domains which were limited to about five unit cells along two of the three main crystallographic directions but about tens of repeat units along the third one (Fig. 4.24c,d,f). The model is depicted in Fig. 4.24e under assumption of the bilayer silicate structure.

Apparently, the formation of regular domain boundaries helps relieve stress caused by the lattice mismatch. Complementary DFT calculations, although performed on free-standing bilayers, to examine the strain effect on the film structure showed that relaxation orthogonal to the commensurate direction does reduce the strain energy. As a consequence, the square substrate promotes epitaxial growth of crystalline SiO_2 by providing an incommensurate direction along which the film can relax.

Unfortunately, these ordered silica films were not studied by vibrational spectroscopy which unambiguously differentiates mono- or bilayer structures. However, the follow-up STM study[62] reported images, one is reprinted in Fig. 4.25, which bear close similarity to those obtained on amorphous bilayer films on Ru(0001) and Pt(111) (see Figs. 4.19 and 4.22). Note, however, that the films were prepared by depositing considerably larger amount of Si (about 5 ML) in 10^{-6} Torr of O_2 at 575 K. Prolonged annealing of "thick" films at high temperatures in 10^{-7} Torr O_2 eventually led to the appearance of crystalline structures previously observed (Fig. 4.24) for the initially thinner (2 ML) films. Since high-temperature annealing may cause uncontrollable silica loss, via either silica sublimation and/or Si migration into the Pd bulk, one cannot exclude that the crystalline films reported in the previous study were, in fact, a monolayer rather than a bilayer silicate as originally thought.

Since the crystalline silicate sheet possesses the hexagonal symmetry, the Pd(111) surface appears to be better suited than Pd(001) to form commensurate "hex-on-hex" rather than "hex-on-square" structures. Indeed, first reports[63] on silica films on Pd(111) showed additional silica-related diffraction spots appearing after oxidation in 10^{-6} Torr O_2 at temperatures above 850 K (Fig. 4.26a). An inner diffraction ring consisting of twelve diffuse spots implies the presence of a relatively well-ordered film, in agreement with STM images (Fig. 4.26b) showing a ~6 Å periodicity. In addition, six satellite spots around (1/2, 1) and (1, 1/2) main spots are characteristic for Moiré-like superstructures. The latter can be reproduced by rotating the hexagonal SiO_2 layer 30° over the Pd(111) lattice.

However, a long-range film ordering suggested by LEED was not observed in our more recent study.[64] Note, however, that the preparation is slightly different: Si was vapor-deposited at 100 K onto an oxygen pre-covered Pd(111) surface and then oxidized in 3×10^{-6} mbar O_2 at elevated temperatures, typically 1000–1100 K. Figure 4.27 collects a series of IRAS spectra of the films at three different film coverages, which was, in turn, determined by CO titration of Pd. The spectra showing two intense bands at 1304 and 698 cm^{-1} look virtually identical to those obtained on Ru(0001)- (1302 and 692 cm^{-1}) and Pt(111)- (1294 and 690 cm^{-1}) supported bilayer films. The intensity is linearly proportional to the silica coverage, although the bands slightly shift to higher wavenumbers with increasing coverage. Obviously, the silica film on Pd(111) grows

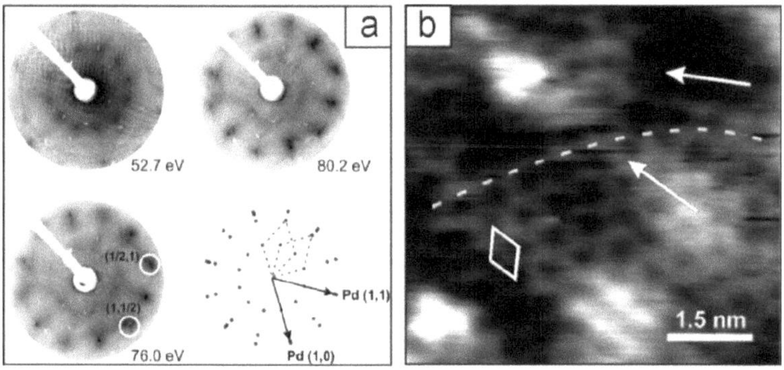

Figure 4.26 (a) LEED patterns for a SiO_2 film prepared on Pd(111) recorded at energies as indicated. Satellite spots around the (1, 1/2) and (1/2, 1) spots are highlighted within circles. A schematic of the observed spots (satellites are excluded) showing two hexagonal silica domains rotated by 30° with respect to each other. (b) High-resolution STM image showing hexagonal ring structures with two different orientations. The unit cell is indicated by white rhomb. Domain orientations are guided by two arrows, and a domain boundary is highlighted by a dotted line. Adapted with permission from ref. 63.

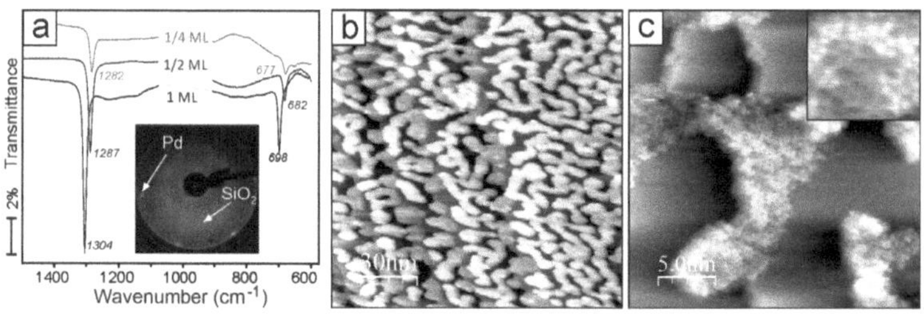

Figure 4.27 (a) IRAS spectra of silica films on Pd(111) as a function of coverage as indicated (1 ML = dense film). The inset displays a typical LEED pattern showing a faint (2×2) diffraction ring with respect to the Pd(111) spots. (b,c) STM images of a 1/2 ML silica film. The inset zoomed in the surface of silica islands. Adapted with permission from ref. 64.

as a bilayer silicate from the onset, i.e., in the same manner as on Pt(111).

Note again that the films showed no long-range order. Beyond the sharp diffraction spots coming from Pd(111), a very diffuse (2×2) ring appeared (see inset in Fig. 4.27a) thus indicating the formation of an amorphous silica film. The lack of ordering is also confirmed by high-resolution STM images (Fig. 4.27c) which only displayed very small domains showing a ~6 Å periodicity, but no really extended patches with a honeycomb-like morphology expected for the crystalline silicate films. Therefore, combined with the IRAS both the STM and LEED results point to the preferential growth of the amorphous bilayer silicate film on Pd(111) from the onset. In principle, close similarity of the film growth on Pd(111) and Pt(111) is not surprising since both metals are noble and have almost identical surface lattice constants. Perhaps, crystallinity of silicate films grown on such metals critically depends on preparation conditions. We have to recall that bilayer silicate films on Ru(0001) form crystalline and amorphous phases, often coexisting. Perhaps, crystalline silica films can as well be prepared on Pd(111) and Pt(111) under some conditions. Further studies remain to be done in this direction.

4.5.1 Metal Substrate Effects: General Trend

After having reviewed preparation of ultrathin silica films on metals, we may now speculate about the role of a metal substrate in making a silica film with a certain atomic structure. The principal structure of a film can be determined by vibrational spectroscopy since each structure shows unique phonon bands which can, therefore, be used as individual fingerprints. It is clear that the structural motif of well-ordered ultrathin silica films on metals is a silicate layer formed by corner-sharing SiO_4 tetrahedra like in sheet silicates. A monolayer silicate layer is chemically bonded to a metal substrate and exhibits an intense, narrow, and symmetric IRAS band at around 1100 cm^{-1} which originates from the stretch vibrations of the Si–O-Metal linkages at the interface. A weakly bound silicate bilayer film manifests itself by even a more intense band at around 1300 cm^{-1} coming from the Si–O–Si bonds oriented normal to the surface. In addition, both mono- and bi-layer structures show other band(s) at low frequencies (600–700 cm^{-1}) associated with bending vibrations of siloxane Si–O–Si bonds oriented almost parallel to the film surface. Apparently, additional complex structures may be formed which exhibit intermediate characteristics between mono- and bilayer films as far as coupling to a metal substrate and arrangement of silica tetrahedra are concerned. A so-called “zigzag” film is only the first example. It shows several phonon bands, with the most intense being at around 1225 cm^{-1} and 850 cm^{-1}. In contrast, a single broad and asymmetric IRAS signal peaked at 1255 cm^{-1} is always observed on films thicker than bilayer, vibrational characteristics of which are virtually identical to those in amorphous silica films grown by thermal oxidation of the Si surfaces. Basically, all thin silica films grown on metal substrates studied thus far showed either of the above-mentioned vibrational features.

Another important characteristic of the silica film, its crystallinity, cannot be determined by vibrational spectroscopy precisely. It has turned out that the differences in the band position and their width for ordered and disordered bilayer films are very small, if any, and comparable with a spectral resolution of conventional spectrometers.[65] Apparently, the long-range order in the films can only be established by diffraction methods like LEED or by direct imaging using high-resolution STM.

Monolayer silicate films on both Mo(112) and Ru(0001) are well-ordered since the silicate layer is strongly bound to the metal surface and therefore follows the topography of a metal substrate underneath. Note, however, that a monolayer film on Ru(0001) exhibits a mosaic (domain) structure at the terraces, while the films on Mo(112) show large terraces almost perfectly ordered. A bilayer silicate film on Ru(0001) exists in crystalline as well as amorphous state. Note that the amorphousness is solely referred to ordering in the lateral direction, since in the normal direction the films are uniformly bilayer. Only amorphous bilayer films were thus far formed on Pt(111). The results for Pd substrates are a bit confusing. If it was really proven (either by IRAS or STM) that the film is bilayer, then it showed no long-range ordering.[62,64]. If the films did show long-range order in LEED,[61,63] then no compelling evidence was provided that the films were really bilayer. Moreover, the comparison of preparation conditions used in these studies indicates that truly bilayer, but amorphous films were prepared at relatively high Si coverage, thus suggesting that ordered structures only seen at lower Si coverage might, in fact, belong to monolayer films. If so, then this finding would further support the conclusion on the preferential formation of crystalline films, if monolayer in thickness.

Apparently, noble metal supports like Pt favor the formation of *amorphous bilayer* films, whereas the Mo surfaces, easily reacting with oxygen, stabilize *crystalline monolayer* films. Note that ultrathin silica films obtained by segregation on the Fe and Co surfaces, i.e., on metals readily dissociating oxygen, are also monolayers in nature. Ru (and possibly Pd) occupies an intermediate position as both mono- and bilayer structures can be formed depending on the amount of Si deposited. Since both structures expose siloxane (Si–O–Si)-terminated surfaces like in silicates, their surface energies must be very close if not identical. Following general considerations for the film growth (see Chapter 2), the preferential formation of a dense monolayer film rather than bilayer islands on Ru(0001) points to the higher adhesion energy for monolayer as compared to bilayer. The interaction is governed by interfacial SiO-Metal bonds which are strong chemical bonds in case of a monolayer, and weak dispersive bonds in case of a bilayer. Therefore, the clear difference in film structures on noble and transition metals can be linked to the bond strength in Si–O-Metal linkages. The SiO–Mo bond is certainly much stronger than the SiO–Pt bond. In a crude approximation, the O-Metal bond strength correlates with the heats of dissociative adsorption of oxygen on corresponding metal surfaces. The calculated values are −544, −534, −478, −220, −170, and −133 kJ/mol for Mo, Fe, Co, Ru, Pd, and Pt, respectively.[66] Therefore, once the bond has been formed, monolayer films on Mo, Fe, and Co substrates are expected to be very stable. The SiO–Ru bond is strong enough to form a monolayer film at low Si coverage. However, upon additional Si deposition and oxidation, the monolayer transforms into the bilayer via breaking of SiO–Ru bonds and creating the Si–O–Si bonds between the layers. Such a process seems impossible on the Mo(112) support due to a very strong SiO–Mo bonding. To date, the formation of a bilayer film on Mo substrates has not been reported. In case of Pt(111), the interfacial SiO–Pt bonds are definitely weaker than on Ru(0001), that provides some flexibility for silica tetrahedra to find best coordination as if no support were present. The bilayer silicate, which is terminated by oxygen layers on either side, represents such a stable structure.

Another factor which may also influence a film growth is the lattice mismatch between the film and the metal surface. Since free-standing single silicate layer does not exist, we assume that its lattice constant is identical to the one in bilayer silicate. DFT calculations performed on unsupported bilayer films reported several different values between 5.24 and 5.32 Å, with 5.3 Å being mostly reported and hence reliable.[24,61,67] If one places the silicate film on top of Ru(0001) (surface lattice constant = 2.706 Å), Pd(111) (= 2.75 Å), and Pt(111) (= 2.775 Å), the most natural way for a film to accommodate the substrate is to form (2×2) structure resulting in a periodicity of 5.41, 5.5, and 5.55 Å, respectively. These latter are considerably larger than the lattice constant of a free-standing film (5.3 Å). Therefore, the film on these metal surfaces has to be stretched that costs considerable energy. From this point of view, Ru(0001) is better suited than Pd(111) and Pt(111), because of the smallest lattice mismatch. To some extent, this explains the formation of crystalline monolayer and (at least, partially) crystalline bilayer films on Ru(0001), and not on Pt(111) and Pd(111), where only amorphous bilayer films were formed. On the other hand, a crystalline single silicate layer readily forms on the Mo(112) surface having a rectangular unit cell (5.46 Å$\times$8.92 Å) very different from the silicate lattice. Assuming that a single silicate layer has the same lattice constant as the bilayer counterpart, in order to accommodate the observed c(2×2)-Mo(112) symmetry of the film, the silicate layer has to be expanded along the Mo(-1-11) direction up to 5.46 Å and shortened along the Mo(-311) direction to 5.2 Å. Despite this distortion, the monolayer silica film on Mo(112) is perfectly ordered, while the monolayer film on Ru(0001) shows domain structure with a high density of domain boundaries to relieve the strain. It is the strong SiO–Mo bond that forces the monolayer silicate to be in registry with the metal surface despite lattice distortion which can be energetically costly. Therefore, stabilization of a single silicate layer on the metal surface is primarily driven by strong interfacial bonds.

To rationalize the formation of a crystalline silicate film on Pd(100), i.e., of square symmetry with surface lattice constant 2.751 Å, we address DFT calculations of energy cost accompanying the lattice distortions in unsupported bilayer silicate film.[61] Two scenarios were considered, namely: (i) uniform strain along the main lattice vectors on the (111) surface; (ii) non-uniform strain along the one of the two principal directions on the (100) surface, while for another one the silica lattice was allowed to relax. Surprisingly, the results showed that allowing relaxation in the orthogonal direction may substantially reduce the overall strain energy. In particular, for a Pd(100) substrate, all the Si–O bond lengths change by no more than 1% as the film is stretched by up to 4.5%. In addition, the SiO_4 tetrahedra tilt slightly, by no more than 5°, when the film is strained in one direction, and the "vertical" Si–O–Si bond connecting the two silicate layers slightly deviates from 180°. Actually, the surface with a square symmetry may even aid the formation of the crystalline bilayer film by providing an incommensurate direction along which stress can be relieved by contraction rather than formation of antiphase domain boundaries usually observed for (111) substrates (see more in Chapter 5). According to the calculations, it seems possible to form crystalline silicate bilayers on Pt(100) or Pt(110) rather than on Pt(111).

An elegant way to experimentally examine the lattice mismatch effect has been proposed by Altman and co-workers by using metal alloy solid solutions with continuously tunable lattice constants.[68] More specifically, the Ni_xPd_{1-x} alloys with bulk lattice constants varied between 3.61 and 3.89 Å were prepared as ~100 nm-thick films on top of the hexagonal (0001) basal plane of α-Al_2O_3 films (in turn grown on Cr_2O_3/α-Al_2O_3(0001)). Assuming the lattice constant of the unstrained bilayer silicate of 5.30 Å, a (111) substrate lattice constant of 2.65 Å (i.e., a half of 5.30 Å) was targeted, which was achieved on $Ni_{0.48}Pd_{0.52}$ alloy. LEED inspection of the silica film grown on this substrate showed an anticipated (2×2) film symmetry pointing to the growth of a crystalline film that was further confirmed by STM, which revealed a honeycomb-like morphology characteristic for the crystalline films. These results could be (and, in fact, were) interpreted as a strong evidence for the lattice mismatch playing the decisive role in the film structure. However, in the absence of vibrational spectroscopy data, the resulted films on PdNi surfaces were erroneously assigned to pure silica bilayer. As it will be shown in Chapter 6 for silicate films modified with 3d-transition metals, the results should be explained in terms of the formation of $NiO_x(111)$ "buffer" layer underneath the silicate monolayer which affects the film crystallinity as was, indeed, shown in the follow-up study.[69]

Although the above-presented analysis may suffer from the somewhat limited database available, the principal structure of an ultrathin silicate film on a metal seems to result from a balance between the energy costs and gains caused by silica lattice distortions and bonding to a substrate. In this respect, we should also mention ab initio study of a bilayer silica film on graphene.[70] The DFT calculations showed that adding dispersive interactions between silicate and graphene energetically stabilizes the bilayer silica film over bulk SiO_2. Covalent bonding to unstrained graphene is a higher-energy state than the van der Waals bonded structures, likely because covalent bonding would result in a large (~7%) strain and disruption of the conjugated π-system in graphene.

4.5.2 Towards a Layer-by-Layer Growth Mode

Having found the conditions for preparation of monolayer and bilayer silicate films, one may raise the question whether well-defined silicate films can further be grown in a "layer-by-layer" mode. To date, there is no yet solid evidence of that it is feasible. To illustrate, Fig. 4.28a displays a large-scale STM image and an IRAS spectrum from a film prepared by two cycles of 1 ML Si deposition and oxidation, i.e., 2 ML in total, on Ru(0001) which is sufficient for the growth of the film two silicate bilayers in thickness. Small depressed areas seen in STM images correlate with the bilayer film remained after the second deposition-oxidation step as evidenced by strongly attenuated IRAS bands at 1300 and 694 cm^{-1}. The film is dominated by amorphous, albeit very smooth, silica showing the IRAS band at 1257 cm^{-1} having a prominent shoulder at 1160 cm^{-1}, typical for the thermally grown SiO_2 films on Si and discussed in the previous Chapter 3. Basically, the same structure is observed for "thick" films grown on Mo(112) as illustrated in Fig. 4.28b.

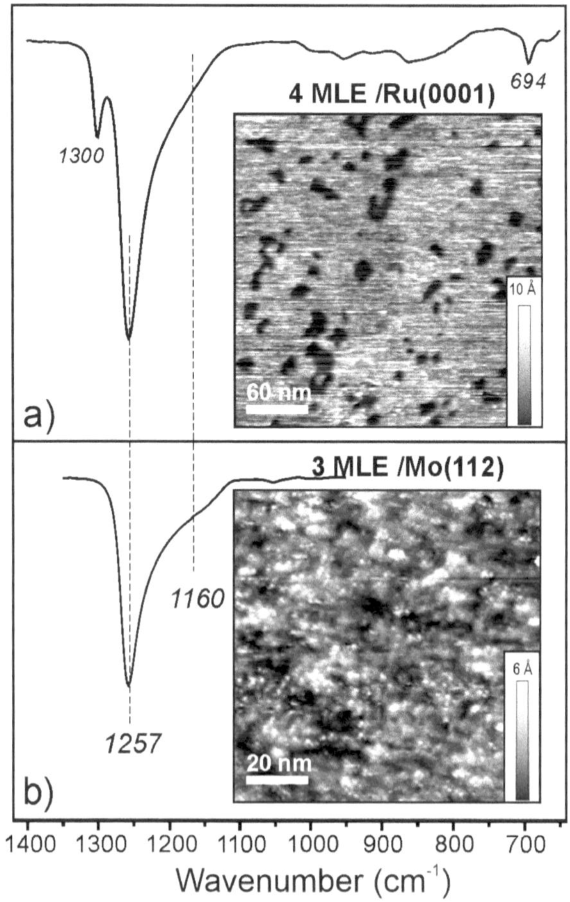

Figure 4.28 IRAS spectra and STM images of "thick" silica films prepared on (a) Ru(0001) and (b) Mo(112).

It, therefore, appears that the silica films thicker than one or two layers (depending on a metal substrate) exhibit a three-dimensional network of SiO_4 tetrahedra rather than a sheet-like structure of silicates. Accordingly, this points out to the independence of the thin silica film structures on the metal surface the films are grown on. However, it is still tempting to find preparation conditions allowing to controllably tune the thickness of *silicate* films by increasing the number of bilayer stacks.

References

1. Niehus, H., and G. Comsa, Bulk dissolved Si as a cause of the "oxide" formation on Pt(111) surfaces. *Surface Science Letters*, 1981. **102**(1): pp. L14–L20.
2. Bonzel, H. P., A. M. Franken, and G. Pirug, The segregation and oxidation of silicon on Pt(111), OR: The question of the "platinum oxide". *Surface Science*, 1981. **104**(2–3): pp. 625–642.
3. Bader, S. D., L. Richter, and T. W. Orent, Auger-UPS study of intrinsic and Si-stabilized oxides of palladium. *Surface Science*, 1982. **115**(3): pp. 501–512.
4. Ko, C. S., and R. J. Gorte, Characterization of oxide impurities on Pt and their effect on the adsorption of CO and H_2. *Surface Science*, 1985. **155**(1): pp. 296–312.
5. Xu, X., and D. W. Goodman, New approach to the preparation of ultrathin silicon dioxide films at low temperatures. *Applied Physics Letters*, 1992. **61**(7): pp. 774–776.
6. Xu, X., and D. W. Goodman, The preparation and characterization of ultra-thin silicon dioxide films on a Mo(110) surface. *Surface Science*, 1993. **282**(3): pp. 323–332.
7. He, J. W., et al., X-ray photoelectron spectroscopic characterization of ultra-thin silicon oxide films on a Mo(100) surface. *Surface Science*, 1992. **279**(1–2): pp. 119–126.
8. Zhang, C., et al., The adsorbate induced facetting of the Mo(111) surface. *Materials Research Bulletin*, 1985. **20**(10): pp. 1129–1135.
9. Schroeder, T., et al., Epitaxial growth of SiO_2 on Mo(112). *Surface Review and Letters*, 2000. **7**(1–2): pp. 7–14.
10. Schroeder, T., et al., Growth of well-ordered silicon dioxide films on Mo(112). *Microelectronics Reliability*, 2000. **40**(4–5): pp. 841–844.
11. Schroeder, T., et al., Single crystalline silicon dioxide films on Mo(112). *Solid-State Electronics*, 2001. **45**(8): pp. 1471–1478.
12. Schroeder, T., et al., Morphological and electronic properties of ultrathin crystalline silica epilayers on a Mo(112) substrate. *Physical Review B*, 2002. **66**(16): p. 165422.
13. Kim, Y. D., T. Wei, and D. W. Goodman, Identification of defect sites on SiO_2 thin films grown on Mo(112). *Langmuir*, 2002. **19**(2): pp. 354–357.
14. Wendt, S., Y. D. Kim, and D. W. Goodman, Identification of defect sites on oxide surfaces by metastable impact electron spectroscopy. *Progress in Surface Science*, 2003. **74**(1–8): pp. 141–159.

15. Kim, Y. D., et al., Surface characterization using metastable impact electron spectroscopy of adsorbed xenon. *The Journal of Physical Chemistry B*, 2002. **107**(2): pp. 592–596.
16. Chen, M. S., A. K. Santra, and D. W. Goodman, Structure of thin SiO_2 films grown on Mo(112). *Physical Review B*, 2004. **69**(15): p. 155404.
17. Ricci, D., and G. Pacchioni, Structure of ultrathin crystalline SiO_2 films on Mo(112). *Physical Review B*, 2004. **69**(16): p. 161307.
18. Min, B. K., W. T. Wallace, and D. W. Goodman, Synthesis of a sinter-resistant, mixed-oxide support for Au nanoclusterst. *The Journal of Physical Chemistry B*, 2004. **108**(38): pp. 14609–14615.
19. Kaya, S., et al., On the geometrical and electronic structure of an ultra-thin crystalline silica film grown on Mo(112). *Surface Science*, 2007. **601**(21): pp. 4849–4861.
20. DiStefano, T. H., and D. E. Eastman, Photoemission measurements of the valence levels of amorphous SiO_2. *Physical Review Letters*, 1971. **27**(23): pp. 1560–1562.
21. Di Pomponio, A., et al., Electronic properties of crystalline and amorphous SiO_2 investigated via all-electron calculations and photoemission spectroscopy. *Solid State Communications*, 1995. **95**(5): pp. 313–317.
22. Wendt, S., et al., Electronic and vibrational properties of ultrathin SiO_2 films grown on Mo(112). *Physical Review B*, 2005. **72**(11): p. 115409.
23. Weissenrieder, J., et al., Atomic structure of a thin silica film on a Mo(112) substrate: A two-dimensional network of SiO_4 tetrahedra. *Physical Review Letters*, 2005. **95**(7): p. 076103.
24. Giordano, L., et al., Structure and vibrational spectra of crystalline SiO_2 ultra-thin films on Mo(112). *Surface Science*, 2005. **584**(2–3): pp. 225–236.
25. Todorova, T. K., et al., Atomic structure of a thin silica film on a Mo(112) substrate: A combined experimental and theoretical study. *Physical Review B*, 2006. **73**(16): p. 165414.
26. Berreman, D. W., Infrared absorption at longitudinal optic frequency in cubic crystal films. *Physical Review*, 1963. **130**(6): pp. 2193–2198.
27. Fuchs, R., and K. L. Kliewer, Optical modes of vibration in an ionic crystal slab. *Physical Review*, 1965. **140**(6A): pp. A2076–A2088.
28. Seifert, J., D. Blauth, and H. Winter, Evidence for 2D-network structure of monolayer silica film on Mo(112). *Physical Review Letters*, 2009. **103**(1): p. 017601.
29. Seifert, J., et al., Structure of ultrathin silica films on Mo(112) studied via classical and quantum mechanical rainbow scattering of fast atoms. *Surface and Interface Analysis*, 2010. **42**(10–11): pp. 1575–1580.

30. Seifert, J., et al., Diffraction of fast atoms during grazing scattering from the surface of an ultrathin silica film on Mo(112). *Physical Review B*, 2010. **82**(3): p. 035436.
31. Seifert, J., and H. Winter, Structure of monolayer silica on Mo(112) investigated by rainbow scattering under axial surface channeling. *Surface Science*, 2009. **603**(17): pp. L109–L112.
32. Kinoshita, T., and S. Mizuno, Surface structure determination of silica single layer on Mo(112) by LEED. *Surface Science*, 2011. **605**(13–14): pp. 1209–1213.
33. Lu, J. L., et al., Formation of one-dimensional crystalline silica on a metal substrate. *Surface Science*, 2006. **600**(13): pp. L164–L168.
34. Kaya, S., et al., Formation of one-dimensional molybdenum oxide on Mo(112). *Surface Science*, 2008. **602**(21): pp. 3338–3342.
35. Sierka, M., Synergy between theory and experiment in structure resolution of low-dimensional oxides. *Progress in Surface Science*, 2010. **85**(9–12): pp. 398–434.
36. Stacchiola, D. J., et al., Growth of stoichiometric subnanometer silica films. *Applied Physics Letters*, 2008. **92**(1): p. 011911-3.
37. Jiang, D. E., and E. A. Carter, First-principles study of the interfacial adhesion between SiO_2 and $MoSi_2$. *Physical Review B*, 2005. **72**(16): p. 165410.
38. Kundu, M., and Y. Murata, Growth of single-crystal SiO_2 film on Ni(111) surface. *Applied Physics Letters*, 2002. **80**(11): pp. 1921–1923.
39. Roccaforte, F., et al., Epitaxial crystallization of amorphous SiO_2 films deposited on single-crystalline α-quartz. *Applied Physics Letters*, 1999. **75**(19): pp. 2903–2905.
40. Bermudez, V. M., Growth, structure and oxidation of ordered silicon layers on nickel (001). *Surface Science*, 1990. **230**(1–3): pp. L155–L161.
41. Valeri, S., U. Del Pennino, and P. Sassaroli, Oxidation behaviour of nickel silicides investigated by AES and XPS. *Surface Science*, 1983. **134**(3): pp. L537–L542.
42. Berning, G. L. P., The room temperature oxidation of Cr, Si and $CrSi_2$. *Applied Surface Science*, 1989. **40**(3): pp. 209–212.
43. Ben Romdhane, F., et al., In situ growth of cellular two-dimensional silicon oxide on metal substrates. *ACS Nano*, 2013. **7**(6): pp. 5175–5180.
44. Ben Romdhane, F., et al., Solid-state growth of one- and two-dimensional silica structures on metal surfaces. *The Journal of Physical Chemistry C*, 2014. **118**(36): pp. 21001–21005.

45. Mullins, D. R., P. V. Radulovic, and S. H. Overbury, Ordered cerium oxide thin films grown on Ru(0001) and Ni(111). *Surface Science*, 1999. **429**(1–3): pp. 186–198.
46. Löffler, D., et al., Growth and structure of crystalline silica sheet on Ru(0001). *Physical Review Letters*, 2010. **105**(14): p. 146104.
47. Yang, B., et al., Thin silica films on Ru(0001): Monolayer, bilayer and three-dimensional networks of [SiO_4] tetrahedra. *Physical Chemistry Chemical Physics*, 2012. **14**(32): pp. 11344–11351.
48. Yang, B., et al., Patterned defect structures predicted for graphene are observed on single-layer silica films. *Nano letters*, 2013. **13**(9): pp. 4422-7.
49. Kremer, G., et al., Electronic band structure of ultimately thin silicon oxide on Ru(0001). *ACS Nano*, 2019. **13**: pp. 4720–4730.
50. Włodarczyk, R., et al., Tuning the electronic structure of ultrathin crystalline silica films on Ru(0001). *Physical Review B*, 2012. **85**(8): p. 085403.
51. Wang, M., et al., Energy level shifts at the silica/Ru(0001) heterojunction driven by surface and interface dipoles. *Topics in Catalysis*, 2017. **60**: pp. 481–491.
52. Kuhness, D., et al., A two-dimensional 'Zigzag' silica polymorph on a metal support. *Journal of the American Chemical Society*, 2018. **140**(19): pp. 6164–6168.
53. Kuhlenbeck, H., S. Shaikhutdinov, and H.-J. Freund, Well-ordered transition metal oxide layers in model catalysis – a series of case studies. *Chemical Reviews*, 2013. **113**(6): pp. 3986–4034.
54. Weiss, W., and W. Ranke, Surface chemistry and catalysis on well-defined epitaxial iron-oxide layers. *Progress in Surface Science*, 2002. **70**(1–3): pp. 1–151.
55. Barcaro, G., et al., Structure of reduced ultrathin TiO_x polar films on Pt(111). *The Journal of Physical Chemistry C*, 2009. **113**(14): pp. 5721–5729.
56. Yu, X., et al., Support effects on the atomic structure of ultra-thin silica films on metals. *Applied Physics Letters*, 2012. **100**(15): pp. 151608-4.
57. Zhang, Z., et al., Preparation and characterization of atomically flat and ordered silica films on a Pd(100) surface. *Thin Solid Films*, 2008. **516**(12): pp. 3741–3746.
58. Kershner, D. C., and J. W. Medlin, Adsorption and reaction of SiH_4 and oxygen on Pd(111). *Surface Science*, 2008. **602**(3): pp. 786–794.

59. Kershner, D. C., M. P. Hyman, and J. W. Medlin, Density functional theory studies of submonolayer oxidized silicon structures on Pd(111) and Pt(111). *Surface Science*, 2008. **602**(23): pp. 3603–3610.

60. Kershner, D. C., W. Zhang, and J. Will Medlin, Investigation of submonolayer SiO_x species formed from oxidation of silane on Pt(111). *Surface Science*, 2008. **602**(21): pp. 3225–3231.

61. Altman, E. I., et al., Growth and characterization of crystalline silica films on Pd(100). *The Journal of Physical Chemistry C*, 2013. **117**(49): pp. 26144–26155.

62. Altman, E. I., and U. D. Schwarz, Structural and electronic heterogeneity of two dimensional amorphous silica layers. *Advanced Materials Interfaces*, 2014. **1**(7): p. n/a-n/a.

63. Jhang, J.-H., et al., Growth of two dimensional silica and aluminosilicate bilayers on Pd(111): From incommensurate to commensurate crystalline. *Physical Chemistry Chemical Physics*, 2017. **19**(21): pp. 14001–14011.

64. Tissot, H., et al., Ultrathin silica films on Pd(111): Structure and adsorption properties. *Surface Science*, 2018. **678**: pp. 118–123.

65. Richter, N. F., et al., Characterization of phonon vibrations of silica bilayer films. *The Journal of Physical Chemistry C*, 2019. **123**(12): pp. 7110–7117.

66. Rao, C. N. R., P. V. Kamath, and S. Yashonath, Molecularly adsorbed oxygen on metals: Electron spectroscopic studies. *Chemical Physics Letters*, 1982. **88**(1): pp. 13–16.

67. Freysoldt, C., P. Rinke, and M. Scheffler, Ultrathin oxides: Bulk-oxide-like model surfaces or unique films? *Physical Review Letters*, 2007. **99**(8): p. 086101.

68. Hutchings, G. S., et al., Epitaxial Ni_xPd_{1-x} (111) alloy substrates with continuously tunable lattice constants for 2D materials growth. *ACS Applied Materials & Interfaces*, 2017. **9**(12): pp. 11266–11271.

69. Zhou, C., et al., Structure of a two-dimensional silicate layer formed by reaction with an alloy substrate. *Chemistry of Materials*, 2019. **31** pp. 851–861.

70. Huang, P. Y., et al., Direct imaging of a two-dimensional silica glass on graphene. *Nano Letters*, 2012. **12**(2): pp. 1081–1086.

Chapter 5

Defects in Ultrathin Silica Films on Metals

Structural defects in silica were always the subject of particular interest since the defects play an important, if not crucial, role in the physicochemical properties of silica and ultimately in its performance in technological applications.[1] The defects are important not only for the electrical and optical characteristics of silica, which largely depend on defects in the silica bulk. Important are also the surface defects which may affect chemical reactions on silicas such as adsorption and hydration, and may also control the formation of catalytically active species in silica-supported catalysts. Certainly, the role of defects is enhanced in case of ultrathin films.

Defects on and in oxide films are commonly discussed in terms of "point" and "line" defects. The former are some sort of vacancies (i.e., missing atoms) in the crystal lattice or, in opposite, extra atoms located in the interstitial positions or ad-atoms at surface. Dislocations (e.g., edge and screw dislocations), domain and grain boundaries, and stacking faults can be referred to as line defects if they are solely at the film surface or planar defects if they propagate into the bulk. Due to limited thickness of the ultrathin films, surface and bulk entities are intimately connected, and any defect may,

Introduction to Ultrathin Silica Films: Silicatene and Others
Shamil Shaikhutdinov

ISBN 978-981-4877-28-2 (Hardcover), 978-1-003-08862-2 (eBook)
www.jennystanford.com

in principle, have a strong impact on the physical and chemical properties of the films. In addition, all kinds of silica/support interfaces can alter the surface chemistry in such systems. This situation appears most obvious for the monolayer silica films, the discussion of which we will begin with before addressing structural imperfections in the bilayer films. We omit in our discussion any sort of structural defects resulted from discontinuity of a film on a large-scale like holes and cracks.

5.1 Defects in Monolayer Films

As mentioned in Chapter 4, monolayer silicate films on Mo(112) are well-ordered and exhibit low amounts of defects. Low energy electron diffraction (LEED) studies using spot profile analysis (see, for example, Fig. 4.2a in Chapter 4) revealed antiphase domain boundaries as the main type of defects, which can be healed by annealing at high temperatures (1200 K). Defect structure in these films was addressed with ultraviolet photoelectron spectroscopy (UPS) and metastable impact electron spectroscopy (MIES).[2] Depending on the annealing temperature, additional electronic features were observed in the band gap by MIES, which were tentatively assigned to defect sites on surface, whereas UPS data only revealed minor features that could be related to defect sites. It was proposed that defect sites other than oxygen vacancies may be responsible for the appearance of the band gap states. Note, however, that the quality of the films under study was only judged by LEED.

To date, identification of defects in thin films is mostly based on scanning tunneling microscopy (STM) or atomic force microscopy (AFM) imaging. As a complementary approach, one can use decoration by noble metal atoms like gold, which are reasonably believed to nucleate exclusively at defect sites present on terraces and low coordination sites at step edges. The use of other defect-sensitive techniques such as electron spin resonance spectroscopy, which is routinely employed for studying paramagnetic defects in powder samples, remains challenging for metal-supported films.[3]

Figure 5.1 displays typical STM images of a well-ordered silicate film on Mo(112) illustrating various types of defects. Basically, the

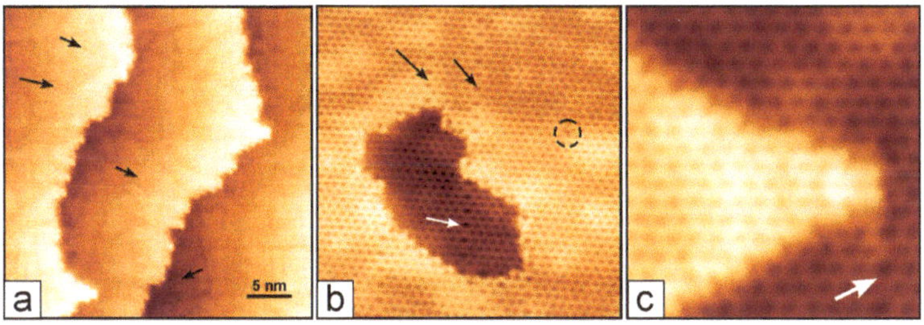

Figure 5.1 STM images illustrating various type of defects on a monolayer silicate film on Mo(112). The arrows mark line defects; circled in (b) is a point defect. Image (c) zooms in the step edge.

film follows the topography of a metal substrate. The steps about 2 Å in height between adjacent terraces corresponding to monoatomic steps on the Mo(112) surface are atomically sharp, suggesting that the film does not cover a substrate like a carpet. Indeed, the hexagonal lattice on one terrace shifts by the half of a lattice when crosses the step to another terrace, in the same manner as the metal surface lattice underneath. There should be no such shift if the silicate layer would continuously propagate across the step. In addition, this finding suggests a definite epitaxial relationship between the silicate layer and the metal surface, in nice agreement with the density functional theory (DFT) calculations,[4] showing that the vibrational frequencies are quite sensitive to the film and metal mutual orientation in plane, and only the most stable interface structure fits the experimental IRAS spectra.

The atomic structure of silica step edges has not yet been investigated in very detail. STM images in Fig. 5.1(a,b) show that the upper terrace often terminates at the step by rings visible as semi-circles. Obviously, cutting the silicate layer in any azimuthal direction will inevitably result in dangling silicon bonds. In principle, they can be stabilized by hydrogen as silanol (Si–OH) and silane (Si–H) species. Since the amount of hydroxyls in these films is negligibly small (that is not surprising for the films prepared at very high temperatures in UHV or O_2; about silica hydroxylation, see Chapter 8), the silicate layer on the upper terrace most likely forms siloxane (Si–O–Si) linkages to a silicate layer on the lower-lying terrace.

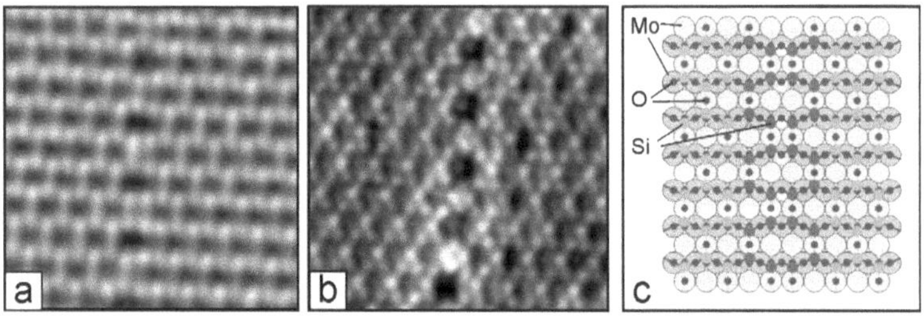

Figure 5.2 High-resolution images (a,b) and structural model (c) of a line defect consisting of alternating eight- and four-membered silica rings. The image forming protrusions are assigned to the Si atoms (in panel a) and O atoms in the topmost O-layer (in panel b).

On the terraces, the line defects dominate. They consist of alternating eight- and four-membered silica rings, running along the (-110) direction of Mo(112) as shown in Fig. 5.2. In this figure, two STM images of the same line defect were obtained at different tunneling conditions which allowed to identify either Si (in image a) or O (in image b) atoms as protrusions. The "8–4" structure can easily be explained by the lateral shift, by the half of the unit cell, between two adjacent hexagonal lattices. Such line defects represent antiphase domain boundaries predicted in early studies on the basis of LEED spot profile analysis.[5]

Only a few point defects are observed in well-ordered films. One is marked by circle in Fig. 5.1b, another example is shown in Fig. 5.3. The atomic protrusions in this image correspond to the O atoms in the topmost layer forming a hexagonal lattice. A brighter single protrusion within the defect is located above the position of a Si atom in the silicate layer, while the adjacent bridging oxygen is missing. Therefore, the defect can tentatively be assigned to upright standing Si–O or other possible species formed by breaking of a Si–O–Si bond in the surface plane. However, its electronic state is difficult to determine solely from the "topographic" STM image. Tunneling spectroscopy (i.e., tunneling current vs. bias) measurements could shed more light on its nature. Of course, at such low density it is difficult to perform statistical analysis of point defects in order to find some general trend regarding their

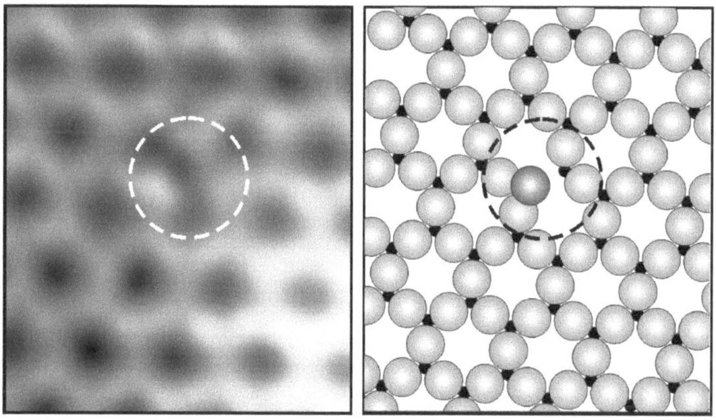

Figure 5.3 STM image and the tentative model of the point defect in silicate layer. Topmost O atoms (large gray balls) and Si atoms (small black balls) are only shown.

thermodynamic stability. As already mentioned above, making use of other techniques which would give us information about defects in the entire film remains challenging.

The atomic structure of point defects in crystalline $SiO_{2.5}$/Mo(112) films was addressed by Pacchioni and co-workers.[6] For the DFT calculations, several defect structures were generated, which are known to exist in various concentrations in chemically or mechanically activated silicas such as the non-bridging oxygen (NBO, ≡Si–O•), the Si dangling bond or so called E' center (≡Si•), the oxygen vacancy (V_O, ≡Si–Si≡), and the peroxo group (≡Si–O–O–Si≡). For comparison, the defects have been computed for a monolayer and a free-standing (unsupported) bilayer films. Table 5.1 collects the formation energies of various defects on these two systems. The corresponding DFT-optimized structures for the Mo(112)-supported silica layer are depicted below the Table.

The formation of O vacancy (V_O) is energetically the most unfavorable process for both systems studied, irrespective what oxygen atom is removed, i.e., surface O or O in the bonds normal to the silicate plane (Si–O–Mo in monolayer, Si–O–Si in bilayer). The presence of a Mo(112) substrate changes the situation since the V_O formation is accompanied by the creation of a strong Mo=O

Table 5.1 Formation energies, ΔE (in eV), of defects generated on the unsupported bilayer SiO_2 film and on $SiO_{2.5}$/Mo(112) (for the latter, the schematic models are shown below

	ΔE(NBO)	ΔE(E')	$\Delta E(V_O)^a$	ΔE(peroxo)^a
SiO_2	5.43	5.96	8.85	−1.27
	(≡Si–O$^-$; ≡Si$^+$)	(≡Si$^\bullet$; ≡Si–O$^\bullet$)	(≡Si–Si≡)	(≡Si–O–O–Si≡)
$SiO_{2.5}$/ Mo(112)	3.43	2.95	8.74, 8.94 (2.60)^b	−1.37
	(≡Si–O$^-$)	(≡Si$^\bullet$, Mo=O)	(≡Si–Si≡)	(≡Si–O–O–Si≡)

[a] The values are computed with respect to a free oxygen atom.
[b] Computed with respect to an O atom adsorbed on Mo.
Source: From ref. 6.

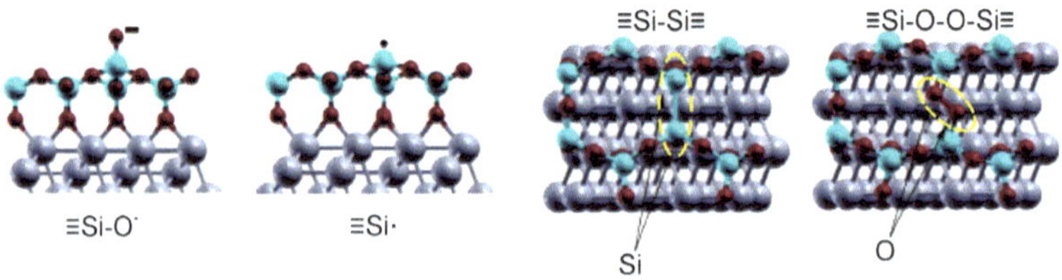

bond on the metal surface. As a result, the formation energy is drastically reduced from ~8.8 eV to 2.6 eV. Although a Mo substrate makes the formation of the dangling Si bond more facile than in the unsupported film (the energy reduces by about 3 eV), the related defect structure is metastable and readily converts into the much more stable non-defective structure, thus making this kind of defects unlikely. Surface peroxo species is the only defect having negative formation energy (−1.4 eV), calculated with respect to a free oxygen atom. However, the energy required to dissociate the O_2 molecule is very large, about 5 eV. Therefore, the formation of a peroxo group is also thermodynamically unfavorable even in excess of oxygen in the ambient. In fact, the additional oxygen atom prefers to bind at the interface with the Mo atom rather than to be included into the silica lattice. Indeed, as discussed in Chapter 4, the films prepared by annealing in oxygen readily incorporate oxygen at the Mo(112) surface forming the so-called "O-rich" structure. All in all, it appears that the silicate films only contains a limited number of oxygen vacancies as point defects.

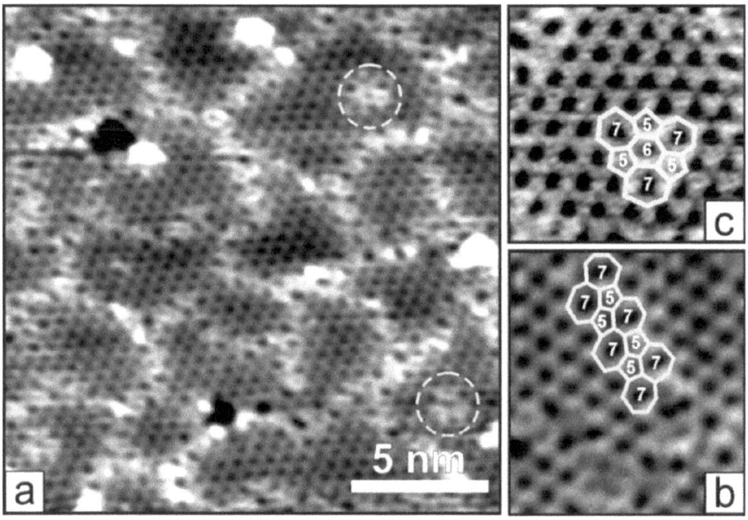

Figure 5.4 (a) STM image of a monolayer silicate film on Ru(0001). Marked by circles are "blister" defects, which are zoomed in panel (c). A domain boundary is shown in (b).

Defects in monolayer silicate films grown on Ru(0001) are structurally richer. To some extent, it is because the SiO–Ru interfacial bonds are much weaker and hence easier to break as compared to the SiO–Mo bonds. Therefore, a strain in the silicate layer caused by the lattice mismatch to the metal surface (see last sections in Chapter 4) can be relieved through the formation of domains (Fig. 5.4a). The domains about 5 nm in size are oriented in the same azimuthal direction and separated by protruding lines primarily consisting of two five-membered and two seven-membered rings in a "zig-zag" sequence, as illustrated in Fig. 5.4b. (It should be mentioned here that, even in the absence of atomic resolution within the N-members silica rings, N can be determined by counting the number of the nearest neighbor rings.) Due to threefold symmetry of the metal surface, the line defects are mostly rotated by 120° with respect to each other and form a network. However, some other azimuthal orientations and hence combinations of N-membered rings at the domain boundaries are also possible.

In addition to the line defects, a few point defects can also be found within individual domains (marked by circles in Fig. 5.4a). The close-up image in Fig. 5.4c shows that such defects are formed by a hexagon surrounded by three pentagons and three heptagons, slightly protruding with respect to surrounding hexagonal layer and hence received the name "blister" defect after Lusk and Carr, who theoretically studied defect structures in graphene.[7] Note that this type of defects differ from famous Stone–Thrower–Wales (STW) defects[8,9] observed on graphene. The latter are resulted from bonds rearrangement in four adjacent hexagons giving rise to a pair of two pentagons and two heptagons, and seem to be characteristic for all two-dimensional systems sharing a hexagonal structure.

Domain boundaries dominate the defect structure in monolayer films on Ru(0001). Its origin has been addressed by reflection high-energy electron diffraction (RHEED) measurements combined with STM.[10] Again, the films showed domains only laterally shifted with respect to each other. STM image displayed in Fig. 5.5a even resolves the O atoms residing on the Ru(0001) surface as predicted by DFT for the thermodynamically most stable structure (see Fig. 4.14 in Chapter 4). Interestingly, a long-range order in the film was found independent of the size of O ad-layer domains formed prior to the Si deposition. (To remind, the monolayer films on Ru(0001) are prepared by Si deposition onto the O-precovered surface). Despite 3O-(2×2) ad-layer domains were of several tens of nanometers, the averaged domain size in the silica films was always an order of magnitude lower. This is illustrated in Fig. 5.5 comparing STM images and corresponding RHEED patterns of the two surfaces. The diffraction streaks show a substantially stronger finite-size broadening in the silica film. The origin of the smaller domain size in the films remains poorly understood. Using energy costs estimated by DFT[11] for the lattice mismatch between Ru(0001) and a silicate film, the structure should be accommodated by the insertion of antiphase boundaries for domains of about 30 nm in size, i.e., considerably larger than 5 nm observed by STM and RHEED. It, therefore, appears that the stress induced by the lattice mismatch is not the only driving force for domain formation.

Very often, the quality of thin oxide films on metals can be improved by tuning the annealing temperature and time. The

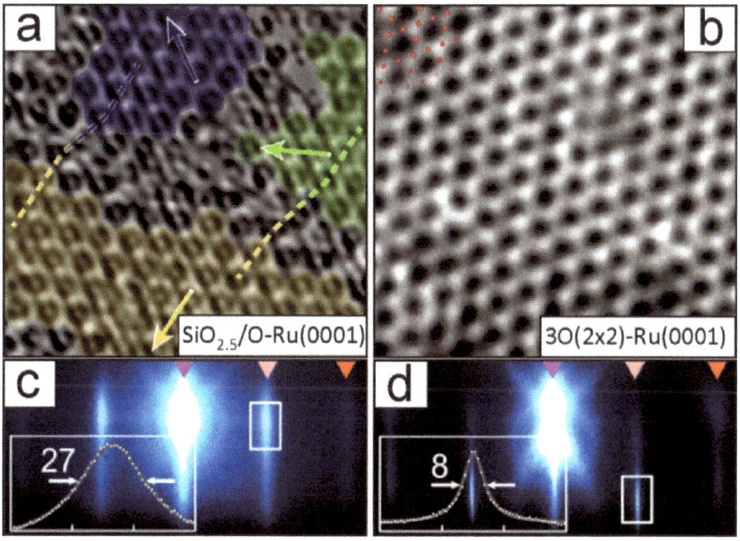

Figure 5.5 (a,b) STM images and (c,d) RHEED patterns of a silicate layer on Ru(0001) (a,c) and of the 3O(2×2)-Ru(0001) surface (b,d). The domains (colored for clarity) are laterally shifted with respect to each other (the image size 8.5 nm×6.2 nm.) Insets in (c, d) show intensity profiles averaged in the horizontal direction in the white frames; full widths at half maxima are indicated in pixel units. Adapted with permission from ref. 10.

monolayer films on Ru(0001) discussed above are usually prepared by annealing in oxygen at around 1000 K. (For comparison, perfect crystalline films on Mo(112) are formed at 1200 K and above.) In attempts to improve film ordering, the films were further annealed in UHV. In addition, this treatment could also show whether the UHV annealing can remove the O ad-atoms adsorbed on Ru which, according to the DFT results, show stabilizing effect on the film structure (see Chapter 4).

Examined by x-ray photoelectron spectroscopy (XPS), UHV annealing above 1000 K does remove the O ad-atoms: The intensity of the corresponding O 1s signal considerably decreases, whereas both the Si 2p and O 1s signals related to Si–O bonds only shift to higher binding energies (BE), by about 0.5 eV, due to the band alignment at the interface caused by the work function change (see Chapter 2). However, this procedure causes substantial structural

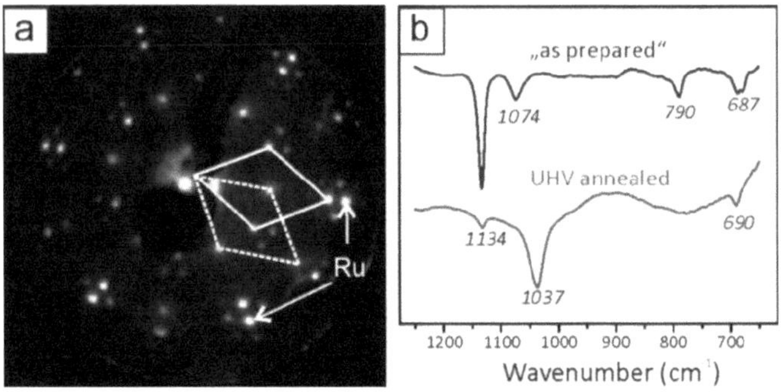

Figure 5.6 (a) LEED pattern of a monolayer silica film on Ru(0001) annealed to 1070 K in UHV. The unit cell (solid line) is rotated by 30° with respect to the original film (dotted line). Additional satellite spots are assigned to the Moiré structure. (b) IRAS spectra of the "as prepared" and UHV-annealed films.

changes, which are clearly seen in LEED pattern[12] (Figure 5.6a). A new unit cell is observed which is rotated by 30° with respect to the (2×2) unit cell in the pristine film. The lattice constant became considerably shorter, i.e., about 5.23 Å. For comparison, it is 5.42 Å in the original film. Moreover, six satellite spots found around integer spots indicate the Moiré-like "hex-on-hex" coincidence structure. In addition, structural changes manifest itself in vibrational spectra (Fig. 5.6b). The most intense band at 1134 cm^{-1} in the "as prepared" films considerably shifts to a lower frequency (1037 cm^{-1}). Accordingly, the 790 cm^{-1} band disappears, whereas the 687 cm^{-1} band is almost unaffected.

At first glance, large-scale STM images showing a contrast modulation with an approximately 25 Å periodicity (Fig. 5.7a) favor its assignment to the Moiré structure anticipated from the LEED results, and which are often formed on ultrathin films having lattice mismatch with the support (for example, see Fig. 2.17 in Chapter 2). However, high-resolution STM images showed that the annealed film no longer exhibits the hexagonal structure. Instead, the surface exposes a periodic array of blister-type defects (Fig. 5.7b). Mostly, these are blisters exhibiting a threefold symmetry and hence denoted as a T-(triangular) defect. However, rectangular (R) defects were also observed, which are formed by an octagon surrounded by

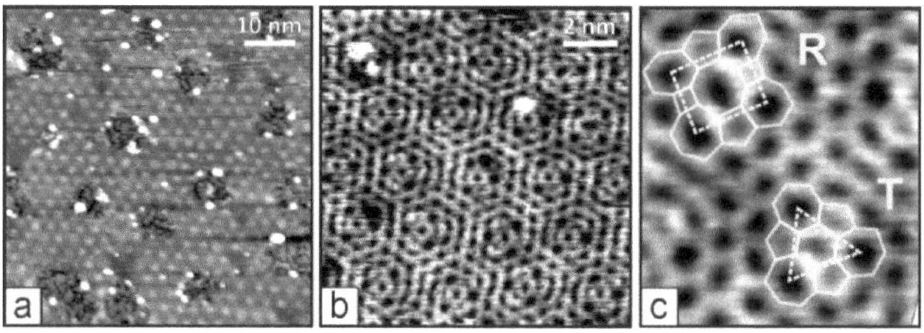

Figure 5.7 STM images of a monolayer silicate film grown on Ru(0001) and then annealed to high temperatures in vacuum. (NB: The film also contained holes exposing a metal substrate). In panel (c), the image is superimposed with the polygonal representation of triangular (T) and rectangular (R) defects.

two pentagons, two tetragons and four heptagons, thus exhibiting the twofold symmetry. In addition, both T- and R-defects have two rotational isomorphs. From the symmetry point of view, the T-defect seems to accommodate the surrounding hexagonal matrix better than the R-defect. This may also explain why the T-defects dominate the films.

Let us turn now to the vibrational characteristics. As we already know, the principal band at 1134 cm^{-1} corresponds to asymmetric stretching vibrations of the Si–O–Ru linkages oriented normal to the metal surface. In "as prepared" films showing (2×2)-Ru(0001) symmetry, the O atoms involved in these linkages equally occupy on-top and threefold hollow sites on the Ru(0001) surface according to the DFT calculations,[13] but the main contribution to the band intensity comes from atop species. A much weaker phonon band at 790 cm^{-1} corresponds to out-of-phase asymmetric stretching vibrations of the Si–O–Ru linkages in atop and hollow sites. Finally, the 690 cm^{-1} band corresponds to the "umbrella" mode which includes the vertical motion of the topmost O atoms in SiO_4 tetrahedron. If the entire film were rotated by 30° as suggested by LEED of the annealed films, a variety of different coordinations for Si–O–Ru linkages to the metal surface become possible. Since the vibration frequency depends on the O coordination to the metal surface (as previously demonstrated for the $SiO_{2.5}$/Mo(112) system, where the shift may be as high as 150 cm^{-1}),[4] the observed shift and broadening of the 1134 and 790 cm^{-1} bands in the annealed films can be explained by changes in the film orientation with respect to a metal substrate. In contrast, the azimuthal rotation of the silicate layer should not affect the "umbrella" mode which only involves the topmost oxygen atoms.

It is important to note that re-oxidation almost fully recovers the original structure of the film. Therefore, the observed transformations are thermodynamically driven. If we now recall, that a single layer silicate film on Ru(0001) is only stable (at least, according to DFT) in the presence of O ad-atoms, and that vacuum annealing is accompanied by their desorption, one may come to the conclusion that structural stability of the silicate layer on Ru(0001) is governed by the O ad-atoms. When they are present, the silicate layer follows the (2×2) registry, i.e. the same as for the O ad-atoms on Ru(0001). Accordingly, in the absence of the O ad-atoms, the silicate layer transforms into another more stable structure having periodic defects. Nonetheless, the questions remain: Why and how does the silicate layer rotate by 30°? Why upon rotation the film does not keep its crystalline structure, but creates defects, albeit periodic?

To speculate about this, let us assume that a (hypothetical) free-standing silicate layer has the same lattice constant as the one computed for an unsupported bilayer silicate, i.e., 5.30 Å. The lattice constant of a perfect silicate layer on the O/Ru(0001) surface forming a (2×2) structure is equal to 5.42 Å (=2×2.71 Å). Therefore, the layer must be stretched to accommodate the metal support. The related energy cost can be, at least partially, compensated by the formation of relatively strong Si–O–Ru bonds. In case of a crystalline layer rotated by 30° and placed on top of the O-free Ru(0001) surface, the average lattice constant measured by LEED (5.23 Å) becomes smaller than 5.30 Å in the free-standing layer, and the layer will be compressed. Therefore, both configurations suffer from the lattice mismatch. Apparently, the lattice mismatch is not the only reason for the periodic defect formation.

The periodic defect structures observed here for a silicate layer resemble those observed in layered carbon materials consisting of ordered arrangements of pentagons, hexagons, and heptagons and named haeckelites.[14] Creation of allotropes composed of patterned defects has long been considered for graphene on theoretical grounds. On the basis of DFT simulations, Lusk and Carr[7] suggested defects patterning as a tool to create artificial materials ("metacrystals") that might exhibit entirely new properties. However, any experimental evidence for such materials has not yet been reported. Surprisingly, a silicate layer on Ru(0001) can reversibly form haeckelite-type structures. Furthermore, the R-defect shown in Fig. 5.7 may be suggested as another building block or structural motif in such architectures. It seems plausible that the difference between the graphene and the single layer silicate lies in additional degree of freedom provided by a SiO_4 entity as compared to C, which constitute respective single layer materials.

5.2 Defects in Bilayer Films

To date, the bilayer films have only been grown on Ru(0001), Pt(111), Pd(100), and Pd(111) substrates. Another system that will also be discussed below refers to the films obtained in the course of chemical vapor deposition of graphene onto a copper grid,[15] henceforth referred to as "graphene-supported". Since bilayer silicate films are weakly bound to a metal support, we can reasonably assume defects observed in such films to be more intrinsic to silica and silicates in general rather than being merely the property of metal supported films.

As in case of monolayer films, defects in well-ordered bilayer films on Ru(0001) are primarily the line defects as depicted in Fig. 5.8. These are found between identically oriented domains which are shifted by a half of lattice with respect to each other, thus resulting in antiphase domain boundaries (Fig. 5.8a,b). In addition, crystalline domains can be rotated by 30° with respect to each other, thus leading to "rotational" domain boundaries (Fig. 5.8c). In essence, both types of defects can be represented by a row, often meandering, of alternating N-membered silica rings other than hexagons.

Individual (isolated) defects within crystalline domains are rarely found. An example is displayed in Fig. 5.9 for a "flower" defect which is, in fact, the smallest, only consisting of six hexagons,

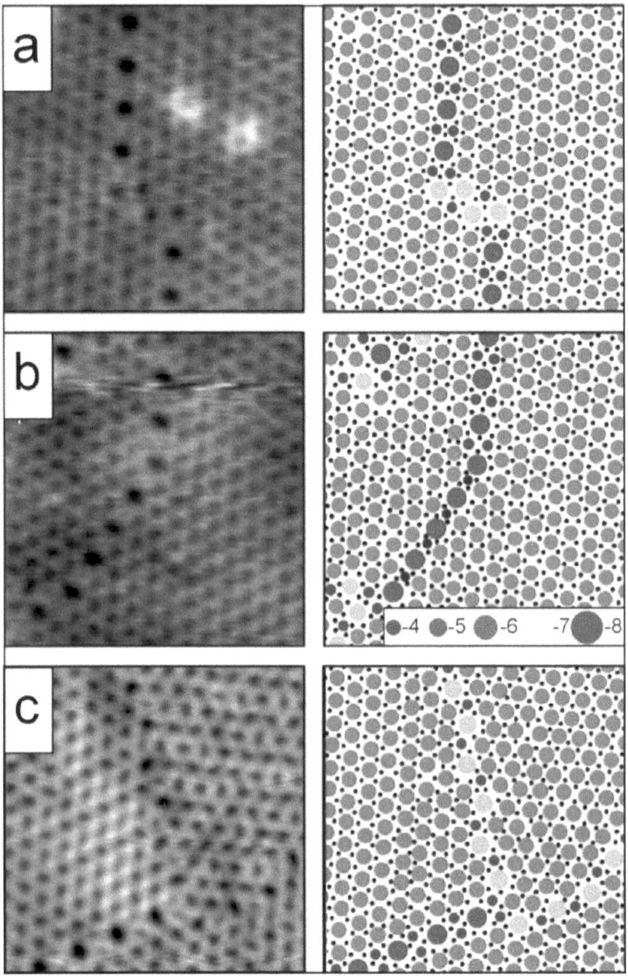

Figure 5.8 STM images (on the left) and schematic color-coded representation (on the right) of several different boundary defects observed on Ru(0001)-supported bilayer films. Black dots indicate the Si atomic positions. Adapted with permission from ref. 16.

crystalline domain rotated by 30° with respect to the surrounding lattice. The close loop boundary around the domain is formed by the five- to seven-membered rings which typically constitute the zig-zag line defect between two rotational domains (*cf.* Fig. 5.8c).

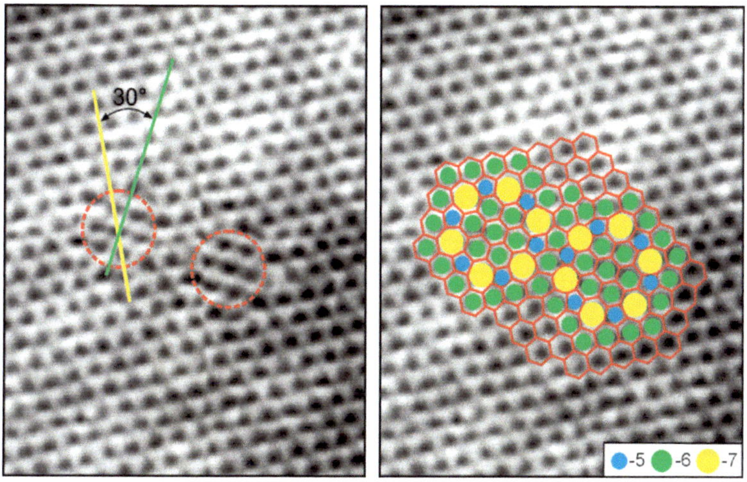

Figure 5.9 On the left: Two "flower" defects on a bilayer silicate film on Ru(0001). Those can be considered as small crystallites (outlined by red circles) rotated by 30° with respect to the surrounding lattice as indicated by yellow and green lines. On the right: The same image is superimposed with its polygonal scheme.

Basically, the same picture was observed on "graphene-supported" films (Fig. 5.10). Again, boundary defects are formed between crystalline domains either of the same azimuthal orientation or rotated by 30°. Note that the way how graphene-supported silica films have been prepared allowed to employ high-resolution transition electron microscopy (TEM), in particular to monitor structural changes in situ. In addition, with TEM it is easier than with STM to overview the entire sample and find the spot of interest.

A gallery of different isolated defects found in aberration corrected TEM images of graphene supported films is presented in Fig. 5.11. Some defects are pure topological, i.e., includes only bond rotation. Others are "non-stoichiometric" and include either vacancies or additional atoms. For example, Fig. 5.11a depicts a "classical" SWT defect which is formed by 90° rotation of one "edge" in an ideal hexagonal framework thus resulted in the 5,5-7,7 rings structure. A "flower" defect in Fig. 5.11b can be formed by six successive rotations. More complex defects involve both adding

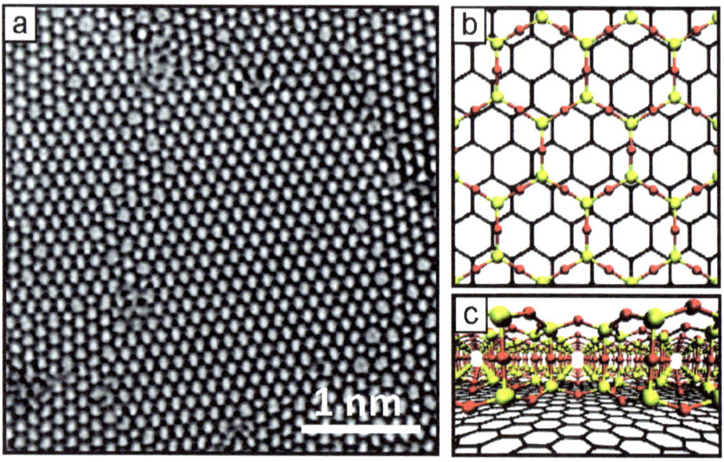

Figure 5.10 (a) TEM image of a crystalline portion of a bilayer silica film on graphene. A structural model in top (b) and perspective (c) views. Yellow balls – Si, red balls – O. Adapted with permission from ref. 15.

(removing) structural units from (to) the hexagonal lattice and subsequent rotations (Fig. 5.11c,d).

The relative probability and driving force for the defect formation in the bilayer films was thoroughly analyzed by Krasheninnikov and co-workers using the formation energy of a defect (E_f) which is defined as[17]

$$E_f = E_{def} - E_{prist} + n_{vac} \times E_{prist}/N,$$

where E_{def} and E_{prist} are, respectively, the energies of the lattice with and without the defect, n_{vac} is the number of missing units (atoms) and N is the total number of units in the system. The energies were 2.8, 1.2, and 4.8 eV, respectively, for the structures shown in Fig. 5.11(a–c). According to these numbers, the formation of a much bigger "flower" defect, which involves six rotations, costs in total less energy than creation of a SWT defect which only needs one rotation (1.2 vs. 2.8 eV, respectively). This somewhat unexpected finding is rationalized in terms of a strain field associated with rotation(s) to form defects. This process studied within a classical force field approach revealed that the strain for a flower defect is considerably smaller than for a SWT defect. On the other hand, the process goes

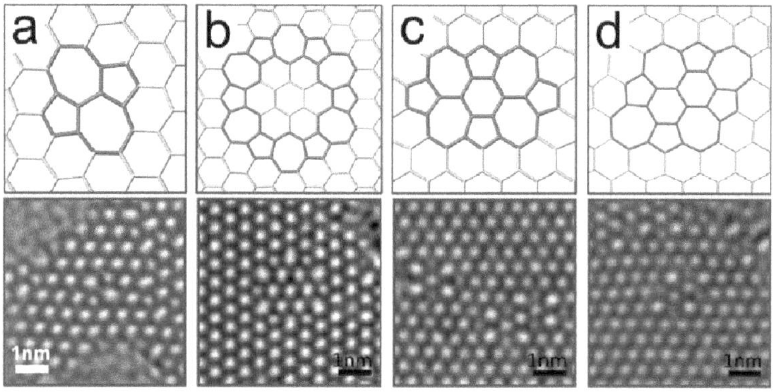

Figure 5.11 Atomic models (top panel) and aberration corrected TEM images (bottom panel) of isolated defects observed in graphene-supported bilayer silicate films. SWT and "flower" defects in (a) and (b) are purely topological defects. In contrast, double vacancy (c) and defects containing additional atoms (d) show density deficiency and excess, respectively. Adapted with permission from ref. 17.

through intermediate steps (successive rotations) whose formation energy is as higher as 8 eV.

In general, the formation energies are considerably high, in particular for the defects involving vacancies or additional atoms (about 5–10 eV), that renders creating defects by thermal activation very unlikely. However, this seems to be possible with high energy electrons used in TEM experiments[18] which first creates O rather than Si vacancies. The recorded TEM movies showed that even a relatively simple SWT transformation is probably more complex than it is thought.

5.3 Bilayer Silica Film as a Two-Dimensional Glass

As already mentioned in Chapter 4, bilayer films on Ru(0001) often show LEED patterns comprising the (2×2) spots as in crystalline samples and a diffraction ring crossing these spots. LEED studies of different bilayer films prepared at various annealing time, temperature, and oxygen pressure did not reveal a single parameter which would control the film crystallinity. Even the rate of sample cooling after high-temperature oxidation may play a role: Slow cooling somewhat favors the formation of better ordered films. This finding seems to be not surprising, as crystallization is a considerably slow process, although the films are ultrathin, and the mass transport should not be an issue. Interestingly, once the film has been prepared, it is practically impossible to alter long-range order of the film by subsequent re-oxidation or vacuum annealing, neither at slow nor fast cooling. Bilayer films grown on Pt(111) and Pd(111) only show the (2×2)-ring, not spots, suggesting all these films are amorphous.

Certainly, the term "defect" is hardly applicable to the systems lacking long-range periodicity. One can argue, however, that only ordered structures larger than ~10 nm can be detected by conventional LEED apparatus, and hence the film may still contain small crystalline areas. Indeed, diffraction rings are commonly observed in electron and x-ray diffraction measurements on powders, where such rings are associated with randomly oriented in space, small crystals (crystallites). In case of planar thin films, this implies crystalline domains randomly rotated in the surface plane, since vibrational spectra of such films only show the fingerprints of bilayer structure, irrespective of the degree of crystallinity.

The atomic structure of amorphous films has been determined by high-resolution STM. Figure 5.12a shows the first ever reported

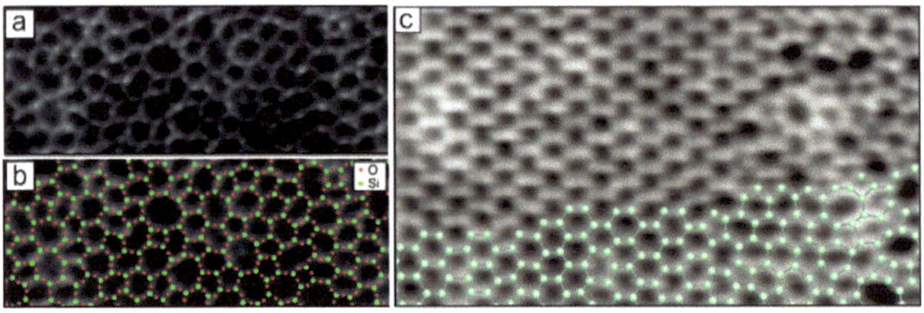

Figure 5.12 (a) High resolution STM image of an amorphous bilayer silica film on Ru(0001). (b) The same image superimposed with a model, where Si and O atoms are shown as green and red dots, respectively. Adapted with permission from ref. 19. (c) Transition region between crystalline and amorphous portions in the film. Only Si atoms are shown in the model. Adapted with permission from ref. 20.

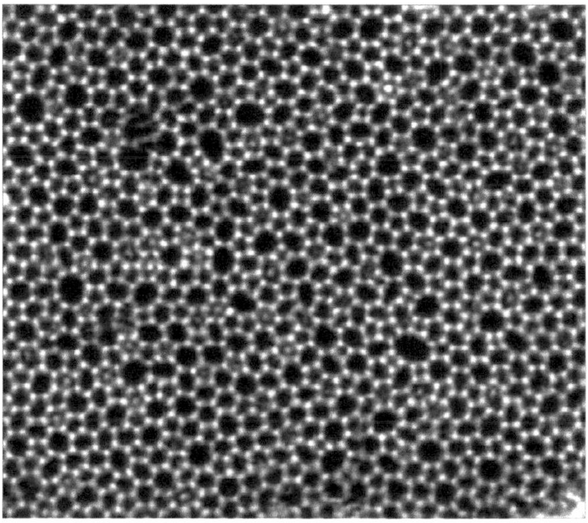

Figure 5.13 Angular dark field (ADF) scanning electron microscopy (SEM) image of an amorphous silica film on graphene. Adapted with permission from ref. 15.

atomically resolved STM image of the amorphous silica surface.[19] Each node in the imaged honeycomb-like structure is seen as a tripod-like protrusion which is assigned to three surface O atoms out of four in the SiO_4 tetrahedra. The position of a Si atom can easily be deduced for each node, and the resulted model is superimposed with the image in Fig. 5.12b. This experimentally observed network of N-membered silica rings almost coincides with the two-dimensional cartoon used by Zachariasen in his famous paper[21] to illustrate the Random Network (RN) model of silica glasses (see Chapter 1). Note also, that the (2×2) diffraction ring observed on our amorphous films is reminiscent of ring-like x-ray diffraction (XRD) patterns observed on vitreous silica and silicate glasses. The RN structure is easily recognized in the SEM images of amorphous portions in graphene-supported films (Fig. 5.13), reported[15] only two months later after STM study.[19]

Actually, the bilayer silicate films on Ru(0001) expose both ordered and disordered regions. STM images of boundaries between crystalline and amorphous phases show that the crystalline phase transforms into the amorphous one, in essence, without any defects

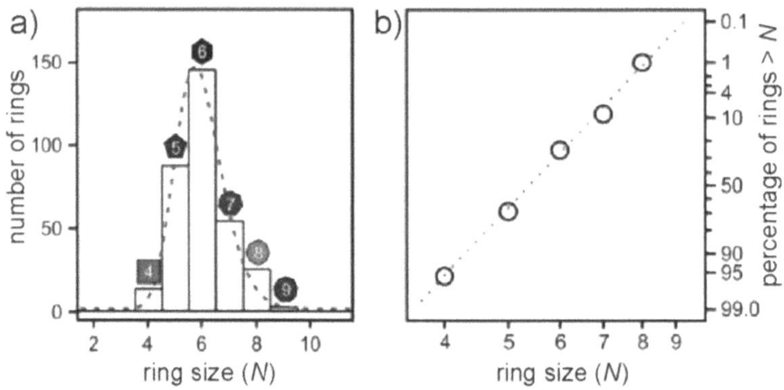

Figure 5.14 Ring size distribution obtained by STM on bilayer silicate films as a bar diagram (a) and as a test for a log-normal distribution (b). Adapted with permission from ref. 23.

in terms of unsaturated bonds or different from the corner sharing bonding arrangements, as shown in Fig. 5.12c. The transition region is relatively sharp, just about ~1.5 nm in width.

The similarity of amorphous bilayer silicate films and the RN model of vitreous silica inspired further analysis of the amorphous films which is henceforth considered as a two-dimensional analog of the silica glass. Typically, the analysis begins with ring statistics that shows a clear maximum at $N = 6$ (Fig. 5.14a), which is the only ring size present in the crystalline films. The ring distribution is asymmetric, i.e., five-membered rings are more frequently observed than the seven-membered ones, whereas the opposite relation holds true for the four- and eight-membered rings. Such a behavior has previously been pointed out by Shackelford and Brown,[22] who showed that the ring distribution in the random network follows a so-called "log-normal" distribution, i.e., the percentage of rings larger than a certain ring size (N) is a linear function of the logarithm of the ring size. If replotted (Fig. 5.14b), then the observed experimental distribution nicely matches the prediction expected for the RN model, and as such it serves as another argument for making use of the bilayer silicate film as an adequate model of glasses.

Analysis of a larger (as compared to STM) database of TEM images of graphene supported films also validated the log-normal ring distribution in the films, although the model prediction slightly deviated from the experimental results. To some extent, the deviation could be attributed to different dimensional geometric constraints. Because the above-mentioned Shackelford's model takes only geometrical factors into account, some discrepancy with experiment may be assigned to other effects not accounted for in the model, such as ring strain, and formation kinetics. Also, the observed average size of neighboring rings, which reflects medium range ordering, was compared against the prediction stemming from classical Aboav-Weaire's law empirically found for cellular structures[24,25] It turned out that the universal relationship originally established for macroscopic cell networks holds true for a network made up of covalent bonds in silica.

The experimental results for the first time visualizing silica ring distribution inspired also theoreticians working in this field to closely look at a vitreous bilayer film as a well-defined two-dimensional glass, even considered "as a paradigm system".[26] For simplicity, only a free-standing, unsupported vitreous silica bilayer was modeled using molecular dynamic (MD) procedures with two (simple harmonic, and more realistic, i.e., including electrostatic interactions and many body polarization effects) forms of inter-ion potentials. Bearing in mind that experimental microscopy images may suffer from relatively small fields of view, the MD simulations showed fairly good consistence with the model structures as far as radial distribution functions are concerned. Ring statistics obtained on computer generated samples, two are shown in Fig. 5.15, was also compared with the experimental results. The average ring size was fixed at six by topology, but the width, skewness and other moments of the distribution of ring edges were characteristics of particular samples. The experimental results slightly deviated from computer-generated bilayers. It is still unclear whether these differences are caused by the finite systems sizes or they manifest substrate effects not considered in the simulations. Certainly, the existence of a symmetry (mirror) plane in the bilayer silicate film is a very unusual situation for bulky (three-dimensional) disordered systems. However, such a symmetry is also expected in crystalline bilayers, where it is less surprising. It appears that the network takes advantage of the larger conformational space

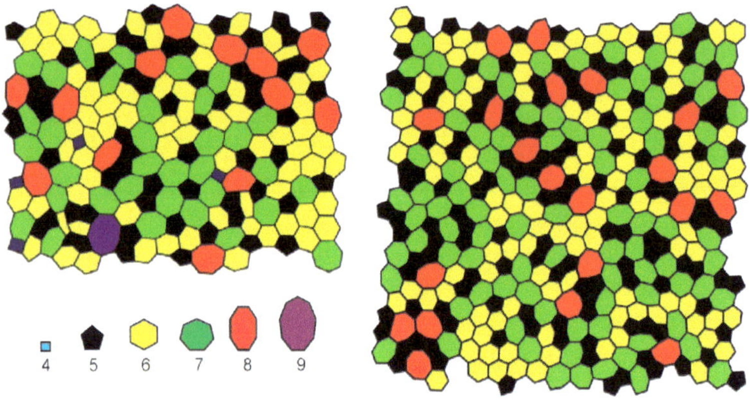

Figure 5.15 Two computer generated structures containing 432 (on the left) and 836 (on the right) Si atoms. The samples have periodic boundary conditions, and the figure here shows most of a single supercell. The color code for the ring size (from 4 to 9) is shown. Adapted with permission from ref. 27.

available when a symmetry plane is present compared to without. Although additional terms in the potential used in MD simulations, such as the Coulomb terms, may result in breaking this symmetry, this was not observed in the computer simulations at the level of accuracy used.[27]

To conclude, the atomic structure of an amorphous bilayer silicate film seems to be fairly well understood. Having amorphous bilayer silicate films representing, in essence, the two-dimensional analog of the silica glass, it is near at hand to study glass transition at atomic level. However, using these films for in situ studies of glass transitions remains challenging. To large extent, this is caused by certain difficulties of achieving high spatial resolution at elevated temperatures, in particular with STM. Although this book primarily focuses on metal-supported films, it is worth mentioning recent works on graphene supported films using the electron beam in an electron microscope both as a source for structural rearrangements and as a probe to track the material as response.[18] Electron microscopy studies revealed a complex "dance" of elastic and plastic deformations, phase transitions, and their interplay. TEM images could even identify the strain associated with individual ring rearrangements, and elucidate the role of vacancies in shear deformation.

Obviously, the bilayer silicate films weakly bound to an underlying support provide the unique possibility to study silicate glasses and the glass transition with atomic precision in real space.

References

1. *Defects in SiO_2 and Related Dielectrics: Science and Technology*. NATO Science Series II, ed. G. Pacchioni, L. Skuja, and D. Griscom. 2000: Springer.
2. Kim, Y. D., T. Wei, and D. W. Goodman, Identification of defect sites on SiO_2 thin films grown on Mo(112). *Langmuir*, 2002. **19**(2): pp. 354–357.
3. Freund, H.-J., et al., Innovative measurement techniques in surface science. *ChemPhysChem*, 2011. **12**(1): pp. 79–87.
4. Weissenrieder, J., et al., Atomic structure of a thin silica film on a Mo(112) substrate: A two-dimensional network of SiO_4 tetrahedra. *Physical Review Letters*, 2005. **95**(7): p. 076103.
5. Schroeder, T., et al., Morphological and electronic properties of ultrathin crystalline silica epilayers on a Mo(112) substrate. *Physical Review B*, 2002. **66**(16): p. 165422.
6. Martinez, U., L. Giordano, and G. Pacchioni, Nature of point defects on SiO_2/Mo(112) thin films and their interaction with Au atoms. *The Journal of Physical Chemistry B*, 2006. **110**(34): pp. 17015–17023.
7. Lusk, M. T., and L. D. Carr, Nanoengineering defect structures on graphene. *Physical Review Letters*, 2008. **100**(17): p. 175503.
8. Stone, A. J., and D. J. Wales, Theoretical studies of icosahedral C_{60} and some related species. *Chemical Physics Letters*, 1986. **128**(5–6): pp. 501–503.
9. Thrower, P. A., ed. *Chemistry and Physics of Carbon*, ed. J. P. L. Walker. vol. 5. 1969. p. 262.
10. Mathur, S., et al., Degenerate epitaxy-driven defects in monolayer silicon oxide on ruthenium. *Physical Review B*, 2015. **92**(16): p. 161410.
11. Ben Romdhane, F., et al., In situ growth of cellular two-dimensional silicon oxide on metal substrates. *ACS Nano*, 2013. **7**(6): pp. 5175–5180.
12. Yang, B., et al., Patterned defect structures predicted for graphene are observed on single-layer silica films. *Nano Letters*, 2013. **13**(9): pp. 4422–4427.

13. Yang, B., et al., Thin silica films on Ru(0001): Monolayer, bilayer and three-dimensional networks of [SiO_4] tetrahedra. *Physical Chemistry Chemical Physics*, 2012. **14**(32): pp. 11344–11351.
14. Terrones, H., et al., New metallic allotropes of planar and tubular carbon. *Physical Review Letters*, 2000. **84**(8): pp. 1716–1719.
15. Huang, P. Y., et al., Direct imaging of a two-dimensional silica glass on graphene. *Nano Letters*, 2012. **12**(2): pp. 1081–1086.
16. Burson, K.M., et al., Assessing the amorphousness and periodicity of common domain boundaries in silica bilayers on Ru(0001). *Journal of Physics: Condensed Matter*, 2017. **29**(3): p. 035002.
17. Björkman, T., et al., Defects in bilayer silica and graphene: Common trends in diverse hexagonal two-dimensional systems. *Scientific Reports*, 2013. **3**: p. 3482.
18. Huang, P. Y., et al., Imaging atomic rearrangements in two-dimensional silica glass: Watching silica's dance. *Science*, 2013. **342**(6155): pp. 224–227.
19. Lichtenstein, L., et al., The atomic structure of a metal-supported vitreous thin silica film. *Angewandte Chemie International Edition*, 2012. **51**(2): pp. 404–407.
20. Heyde, M., S. Shaikhutdinov, and H. J. Freund, Two-dimensional silica: Crystalline and vitreous. *Chemical Physics Letters*, 2012. **550**(0): pp. 1–7.
21. Zachariasen, W. H., The atomic arrangement in glass. *Journal of the American Chemical Society*, 1932. **54**(10): pp. 3841–3851.
22. Shackelford, J. F., and B. D. Brown, The lognormal distribution in the random network structure. *Journal of Non-Crystalline Solids*, 1981. **44**(2–3): pp. 379–382.
23. Büchner, C., et al., Ultrathin silica films: The atomic structure of two-dimensional crystals and glasses. *Chemistry – A European Journal*, 2014. **20**(30): pp. 9176–9183.
24. Chiu, S. N., Aboav-Weaire's and Lewis' laws—A review. *Materials Characterization*, 1995. **34**(2): pp. 149–165.
25. Aboav, D. A., The arrangement of grains in a polycrystal. *Metallography*, 1970. **3**(4): pp. 383–390.
26. Wilson, M., et al., Modeling vitreous silica bilayers. *Physical Review B*, 2013. **87**(21): p. 214108.
27. Kumar, A., et al., Ring statistics of silica bilayers. *Journal of Physics: Condensed Matter*, 2014. **26**(39): p. 395401.

Chapter 6

Ultrathin Films of (Al, Fe, Ti)-Silicates

Tetrahedrally coordinated atoms in silicates are commonly referred to as T-atoms (after "tetrahedral"). Such atoms can be substituted by other atoms. For example, the most abundant minerals in the earth's crust, the feldspars, are built of a three-dimensional network of TO_4 (T=Al, Si), where each AlO_4 tetrahedron is corner-linked to four SiO_4 tetrahedra and vice versa, thus resulting in an ordered Al/Si distribution among the available T-sites. Since the substitution of Si^{4+} cations by Al^{3+} cations obviously results in charge imbalance, the lack of positive charge in the system is compensated by alkali metal cations in the coordination sphere. As a result, the common feldspars have a general formula $X^{n+}Al_nSi_{4-n}O_8$, where X can be Na or K (for $n = 1$), or Ca (for $n = 2$).

In zeolites, one of the most important members in the large family of aluminosilicates, the TO_4 tetrahedra are arranged into regular structures forming cavities and/or channels of molecular dimensions (5–30 Å), often referred to as micropores or nanopores. The porous structure of zeolites allows to selectively sort out molecules diffusing through like in molecular sieve. In the H-form of zeolites (i.e., the X cation is proton), the charge balance is achieved by the formation of hydroxyl species bridging Si and Al cations (Si–OH–Al). This "bridged" hydroxyl shows acidic character,

Introduction to Ultrathin Silica Films: Silicatene and Others
Shamil Shaikhutdinov

ISBN 978-981-4877-28-2 (Hardcover), 978-1-003-08862-2 (eBook)
www.jennystanford.com

that is proton can be donated to another molecule. In fact, it is this property that makes zeolites be "solid acids" for many acid-base reactions in industrially important catalytic processes. It is tempting to integrate both reaction and separation characteristics of zeolites in the form of a zeolite membrane reactor. These ideas inspired studies aimed at fabrication of thin zeolite films supported by mesoporous metal or ceramic supports in order to fulfill also mechanical stability limitations often encountered under working conditions.[1–3] In addition to separation and catalysis applications, the interest in zeolite films is driven by their potential advantages in chemical sensors, non-linear optics, and batteries.[4] Zeolite thin films have even been considered as low dielectric constant insulators (so called "low-*k*" materials) for future generation computer chips, environmentally gentle corrosion-resistant coatings for aerospace alloys, and coatings for gravity-independent water separation in space stations.[5]

In the following sections, we address the preparation of thin aluminosilicate films as well as ultrathin silicate films modified by 3d-transition metals like Fe and Ti. In contrast to aluminosilicates (and other Al-containing oxides) in which Al is almost exclusively present as Al^{3+}, transition metals can, in principle, be present in several different oxidation states and thus accommodate the charge imbalance easily. Therefore, one reasonably expects to see some differences in the corresponding film structures.

6.1 Thin Zeolite Films

Three principal methods considered for the synthesis of zeolite films are depicted in Fig. 6.1. Method I includes the deposition of preformed crystals onto a substrate. Method II uses the transport of aluminosilicate (or other precursor) material to a substrate, providing favorable supersaturated conditions for nucleation and growth under hydrothermal conditions in the vicinity of the surface. Method III involves direct nucleation on the substrate surface from solution (or from a gel film), followed by growth. (Mechanistic aspects of hydrothermal zeolite synthesis can be found in the comprehensive review by Cundy and Cox).[6] For their potential

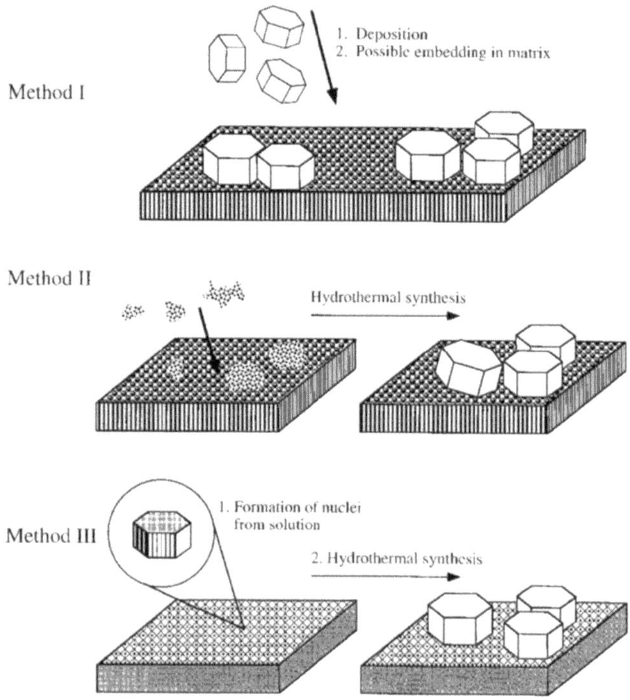

Figure 6.1 Preparation methods of zeolite layers and films. Reproduced with permission from ref. 10

applications it is often desirable to fabricate films in which the nanoscale architecture and ordering can be controlled. However, to keep intrinsic properties, such as porosity and crystallinity, of a zeolite prepared as a thin film is not a trivial task. Driven by their practical applications in separation technologies, thin films of zeolites sharing the MFI structure were mostly studied. (For zeolite structure codes and nomenclature, please see, for instance, https://asia.iza-structure.org/IZA-SC/ftc_table.php). For example, ZSM-5 films synthesized hydrothermally on fused silica glasses consisted of nanocrystals preferentially oriented with their one axis perpendicular to the substrate surface. Depending on the synthesis conditions, the individual crystals could be varied between 500 nm and 30 μm.[7] Also BEA-type zeolite films, uniformly oriented, were successfully prepared on glass substrates.[8] Thin films of LTA zeolite were reported on single-crystal Al_2O_3(0001) supports by a method including first adsorption of zeolite seeds followed by hydrothermal film crystallization.[9] Although the prepared films were referred to as "ultrathin," the thicknesses were, in fact, in the range of 200 nm and above.

As far as zeolite films on metal supports are concerned, first reports appeared in the early 1990s.[11] Actually, previous attempts using traditional methods of thin-film deposition have failed to produce thin crystalline films of zeolites on metals. The preparation of zeolite films (in this case, of zeolite Y) became successful by using modified hydrothermal alteration procedures. Although copper foils were mostly used in these experiments, zeolite coatings were as well prepared on silver, platinum, titanium, molybdenum, iron, and tin. Note, however, that the prepared films had thicknesses about one micrometer. The chemical composition and surface imperfections of metal substrates played a role for the film quality.[12] Zeolite Y crystals could grow directly on the stainless steel support only if the synthesis mixture was previously seeded.[13] The stability of the zeolite Y coating was rationalized by assuming that partly uncompleted crystals were stabilized through chemical bonding with the support. The film homogeneity was improved by an oxidic layer on the metal surface formed upon film calcination at high temperature.

As an example, Fig. 6.2a illustrates preparation used to grow thin continuous films of zeolite ZSM-5 on gold surfaces. Four steps include: (i) surface silanization, (ii) charge modification, and (iii) seeding of the surface prior to (iv) the hydrothermal crystallization of the zeolite. More specifically, the gold surface was first covered with a monolayer of S-containing silane which was hydrolyzed under acidic conditions. The surface charge is then reversed by adsorption of a cationic polymer, and a monolayer of colloidal seed crystals is adsorbed on the positively charged surface by electrostatic attraction. After decomposition of silane and cationic polymer by calcination, the seed crystals were grown into a film in a synthesis solution at hydrothermal conditions. Typical morphology of the films imaged by scanning electron microscopy (SEM) is shown in Fig. 6.2b,c. The method allowed to prepare films with a desired thickness as low as 50 nm.

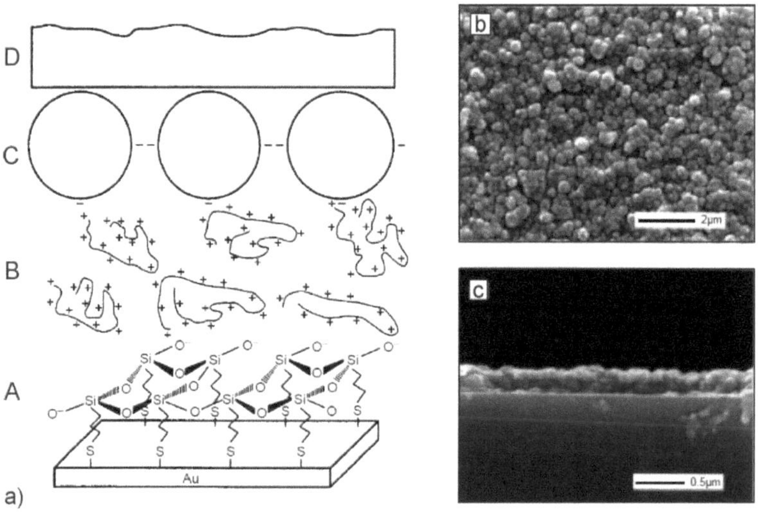

Figure 6.2 (a) Schematic representation of zeolite film preparation on a gold substrate. (A): silanization of an Au surface; (B) charge reversal of the negative siliceous interface to a positively charged interface by adsorption of a cationic polymer; (C) adsorption of negatively charged ZSM-5 crystallites; (D) the formation of a continuous zeolite film during hydrothermal treatment. (b,c) SEM images showing the top view (b) and side view (c) of the ZSM-5 zeolite film. Adapted with permission from ref. 14.

In addition to zeolite (aluminosilicate) films also mesoporous pure silicate (silicalite) were successfully grown as thin well-oriented films on mica as a support.[15] Interestingly, SEM studies revealed preferential orientation of silicate crystallites. Presumably, under acid conditions K^+ ions in mica are exchanged by protons, thereby resulting in Si–OH anchoring sites for silica precursor. The reactivity of the mica substrate probably controlled the orientation of the micellar precursor species, which in turn imposes order on the resulting solid film. The film synthesis on mica was further modified[16] using a double-headed quaternary amine surfactant. Analysis of atomic force microscopy (AFM) images suggested that films nucleate on hexagonally close packed surfactant micelles, which were then removed by calcination. The resulting mesoporous silicate films were found to be crystalline and oriented with the *c*-axis perpendicular to the plane of the film.

In the great majority of cases, only briefly surveyed above, the zeolite films had thicknesses about several hundred nanometers. Although decreasing the thickness of silica films down to the sub-nanometer range is driven primarily by miniaturization of silica-based electronic devices, the thickness seems to be not a critical issue for zeolite films, albeit for certain separation applications, decreasing thicknesses to 50 nm is desirable.[2] More important and one of the major problems in zeolite films technology is to maintain the porous and oriented structure presented in these materials. This may become even a more difficult task for nanometers-thick films. For such objectives, film preparations using spin-coating techniques were thought to be well suited, in particular for applications in microelectronics, where additional hydrothermal treatment in quite aggressive media is not desirable. Such an approach has been demonstrated on pure and gold-plated Si substrates spin-coated by a colloidal suspension of pre-synthesized microscopic zeolite particles.[17] In principle, the method allows controlled deposition of uniform films on any kind of planar substrates, although the film thickness remained in the range of hundred nanometers.

In order to reduce the film thickness and also to make the film surface smoother, an interesting approach has been suggested which is based on using a solution containing smallest zeolite crystals and their fragments ("zeosil nanoslabs")[18] which supposed to be self-assembled into a zeolite film directly on a support. To test the preparation, a precursor that has previously been identified as the intermediate formed during the synthesis of Silicalite-1 (pure silica zeolite with MFI-type structure) was deposited onto a Si(100) wafer by spin coating. The film thickness was varied by the precursor dilution factor. In this way, it was possible to grow very thin films, down to a few nanometers in thickness (see Fig. 6.3). However, analysis of electron diffraction patterns showed that the film is an amorphous arrangement of zeolite precursor particles.

Another approach to fabricate thin zeolite films is based on a stable dispersed suspension of zeolite nanosheets, for example, through exfoliation of their lamellar precursors. However, the latter process is hampered by the structure deterioration and morphological damage (e.g., fragmentation, curling, and aggregation). Tsapatsis and co-workers recently reported the synthetic route for highly

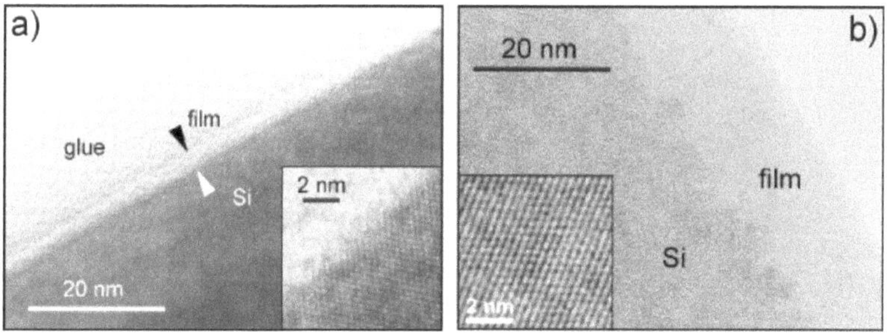

Figure 6.3 Cross sectional high-resolution TEM images of 2 nm (a) and 15 nm (b) films made of Silicalite-1 precursor on Si(001). The interface area of film (a) and the atomically resolved substrate of (b) are shown as insets. Adapted with permission from ref. 19.

crystalline nanosheets of zeolite frameworks MWW and MFI, which were then used in the fabrication of molecular sieve membranes on a porous support.[20] Stable suspensions were prepared in ethanol following acid treatment that partially removed the associated organic structure-directing agent. Exfoliated zeolite nanosheets were also deposited on non-porous Si and Au substrates.[21] Introduced at the air-water interface the nanosheets were transferred to a substrate using the Langmuir–Schaefer horizontal lifting deposition (modification of the famous Langmuir–Blodgett technique). Analysis of AFM images revealed the nanosheets about 3 nm in thickness (Fig. 6.4). Successive "layer-by-layer" depositions resulted in oriented multilayer films with control over their thickness, while secondary growth of monolayers yielded intergrown, oriented films with sub-12-nm thickness in certain cases.

Quantitative characterization of the nanosheets morphology was performed via 3D mapping of its reciprocal lattice measured with TEM, as the intensity modulation of a diffraction spot on tilting depends on the thickness, and changes in the spot shape are a measure of wrinkling. The mapping was achieved using a large-angle tilt-series of electron diffraction patterns, which showed that the exfoliated MFI nanosheet is only 1.5 unit cells ($\sim$3 nm) thick and wrinkled anisotropically with up to 0.8 nm average surface roughness.[22]

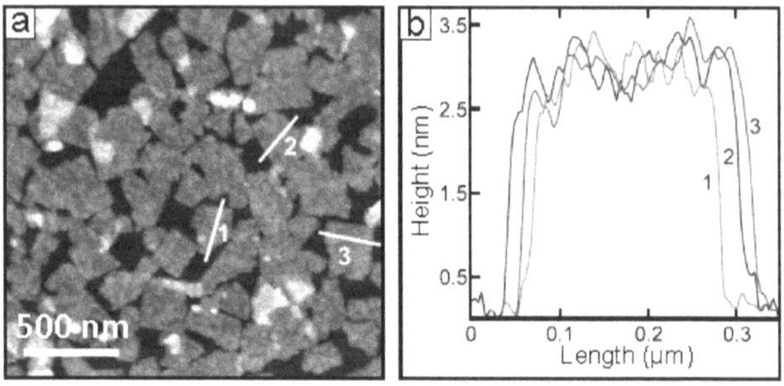

Figure 6.4 AFM image (a) and line profiles (b) of zeolite nanosheets deposited on a Si wafer using Langmuir-Schaefer technique. Adapted with permission from ref. 21.

The zeolite nanosheets were recently studied with the help of the "surface science" tools using the Au(111) surface as a substrate.[23] In order to remove carbonaceous contamination from the synthesis and exfoliation process, the samples were annealed at 775 K first in pure oxygen at atmospheric pressure and then, after transportation into the UHV chamber, in 10^{-6} mbar of O_2. Infrared reflection absorption spectroscopy (IRAS) measurements of the nanosheet samples showed two main phonon vibrations, at 1246 and 1186 cm^{-1}, which are considerably higher than 1230 and 1115 cm^{-1}, respectively, if measured on bulky samples. The difference was tentatively assigned to size effects. Interestingly, x-ray photoelectron spectroscopy (XPS) results revealed binding energy shifts upon exposure either to H_2 or O_2, which bear some similarities to effects previously observed on bilayer silicate films and attributed to the band alignment effect (see Chapter 4). Overall, the data demonstrated great potentials for the surface science to gain important information about the chemistry of zeolite materials using 2D-nanosheets as models. However, the *surface* structure of the deposited zeolite nanosheets remains unknown.

Zeolite film preparations discussed above all use liquids (e.g., solvents) at one or another stage. Preparations of aluminosilicate films not using so-called "wet" chemistry are rare. For example, alumina deposition by ion sputtering of an Al_2O_3 target onto a Si(100) substrate kept at 300°C resulted in the formation of alumina-rich films, 3–12 nm in thickness, with a relatively sharp interface.[24] The intermixed region exhibited a complex stoichiometry and crystallinity, as shown by depth-profile XPS measurements and x-ray diffraction (XRD). Post-annealing at elevated temperatures up to 900°C enhanced interdiffusion of Al and Si.

In another example, pulsed laser deposition was employed with kaoline (clay mineral with a formula $Al_2Si_2O_5(OH)_4$) pressed in carbon matrix as a target.[25] The films deposited on top of chalcogenide glass As_2Se_3 were about 200 nm in thickness, amorphous, smooth (on the μm-scale), and of chemical composition close to a parent target. As a side note, such films showed promising anticorrosion behavior.

Also chemical vapor deposition (CVD) technique was used for aluminosilicate films. Upon adsorption of various silicon alkoxides as volatile precursors onto alumina substrate sequential decomposition occurred, ultimately resulted in the silanol covered alumina surface.[26] (Actually, the study was focused on preparation of solid acid catalysts rather than zeolite films.) Instead of sequential deposition of silica on alumina (and vice versa), it appears more efficient to use a single-source precursor containing both Al and Si atoms. This was exemplified by using $[Al(OSiEt_3)_3]_2$ for film deposition on Si(001).[27] Although the resulted films about 200 nm in thickness were smooth (as observed by SEM), they were amorphous (XRD), and consisted of a mixture of alumina, silica and poorly defined aluminosilicate phases (XPS).

More recently, a combination of CVD and hydrothermal methods was applied for the preparation of zeolitic films.[28] Basically, CVD has been used for deposition of dense amorphous silica or titanium doped silica films, which were converted to a crystalline nanoporous zeolite by a hydrothermal process using an organic structure-directing agent.

6.2 Ultrathin Aluminosilicate Films

As mentioned in Chapter 1, natural and synthetic zeolites are extremely diverse in structure. On the other hand, catalytic reactions on zeolites are, in essence, governed by acidic properties of bridging (Si–OH–Al) hydroxyls. Therefore, to study solely the chemical reactions on zeolite surfaces (i.e., neglecting its nanoporous structure), a well-ordered aluminosilicate film exposing hydroxo species would be a good model system.

The first attempts to prepare planar model systems of zeolites were reported in the early 1990s by Somorjai and co-workers,[29] who prepared mixed silica-alumina thin films about 10 nm in thickness on gold foil using ion sputtering of several HY-zeolite targets. Structural characterization of the resulted films performed by XPS, SEM, and XRD revealed chemically homogeneous and amorphous films. Moreover, the films showed some activity in the test reaction (in this case, cumene cracking), whereas the films prepared from either Al_2O_3 or SiO_2 targets or a mixture of the two were inactive.

Later, Goodman and co-workers[30] used physical vapor deposition of Al onto a thin silica film grown on Mo(100). Auger electron spectroscopy (AES) and XPS results showed that Al is completely oxidized while metallic Si is formed at temperatures below 800 K. Upon annealing to higher temperatures (up to 1200 K), the Si–O–Al bonds form presumably via the migration of aluminum oxide into the bulk of the SiO_2 film with the concomitant desorption of volatile silicon monoxide. Although the results showed that the electronic structure of the prepared films was very similar to that of aluminosilicates, the films were disordered and exposed poorly defined surface. Not surprisingly, the films lack surface hydroxyls since no water was used in the preparation.

Inspired by successful preparation of a crystalline silicate film on Mo(112) (see Chapter 4) in our laboratories, further studies were directed towards preparation of the well-defined aluminosilicate films using a silicate film as a template. Although it was obvious that monolayer films cannot feature three-dimensional porous structure of zeolites, such films could represent the outer surfaces of zeolites and/or their largest cavities.

Deposition of Al onto a monolayer silicate film and subsequent annealing in oxygen usually results in quite heterogeneous films which looked in STM like alumina nanoparticles deposited onto a silica film or embedded into the silicate layer. To ease intermixing of Al and Si atoms at the surface that would, in turn, facilitate Si substitution by Al, both metals were co-deposited in oxygen ambient at low temperatures in total amount sufficient to form

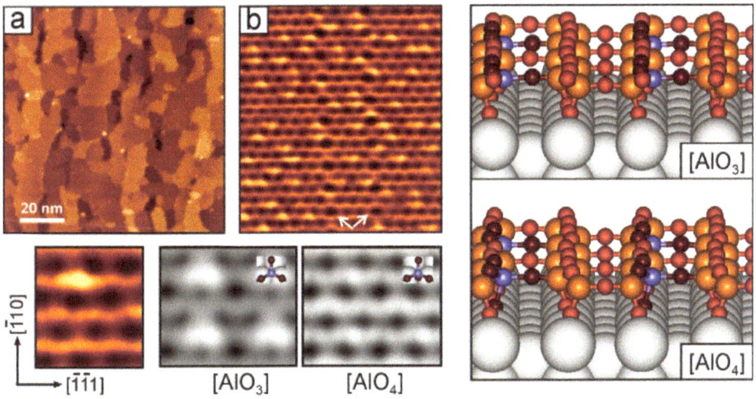

Figure 6.5 Large-scale (a) and high-resolution (b) STM images of aluminosilicate monolayer film on Mo(112). Two antiphase domain boundaries are marked by white arrows. Right panel shows two structural models in the perspective view (Al, violet; Si, orange; O, red). Simulated STM images for the two models together with the experimental one are compared.

a single silicate layer.[31] Then the films were annealed at elevated temperatures (1100 K) unless they showed ordered structure in low energy electron diffraction (LEED). Characterization of the annealed films by LEED and IRAS at relatively low Al/Si ratios (determined by XPS) showed the fingerprints of a monolayer silicate film. Also STM images showed atomically flat terraces with additional atomic features (protrusions) in a honeycomb-like structure (Fig. 6.5a,b), density of which increased with increasing the Al:Si ratio in the film. The protrusions obviously associated with Al were randomly distributed on the surface, suggesting a random distribution of Al in the film. The honeycomb-like morphology that maintains in "Al-doped" films implies that that Al substitutes Si in the silicate layer. However, one can envision two different structures. In the first one, Al replaces Si in a TO_4 tetrahedron. However, this would imply unrealistic oxidation state 4+ for Al. Therefore, the extra charge must be compensated via the charge transfer from the metal substrate. This so-called AlO_4-model was compared to the second (AlO_3-) model, in which the Al^{3+} ion is coordinated to only three O ions in the top-most layer. Both models are depicted in Fig. 6.5.

The density functional theory (DFT) calculations performed by Sauer and co-workers predicted both structures close in energy and both showing red-shift for the principal IRAS band as, indeed, experimentally observed. The only feature which allowed to discriminate the models and to ultimately favor the AlO_3 model is STM image simulation which showed a much better fit with the experimental image (see Fig. 6.5). Recall that in this configuration, both Si^{4+} and Al^{3+} cations are in their most stable oxidation states, and there is no extra charge to compensate. Accordingly, water adsorption (under UHV compatible conditions, though) results in no bridging Si–OH–Al hydroxyls.

Certainly, in monolayer oxide films having strong chemical bonds with the metal support, this latter can readily be involved in charge transfer processes if needed to stabilize the film. Synthesis of a weakly bonded, bilayer silicate films on Ru(0001) provided a good opportunity to fabricate aluminosilicate films decoupled from the support. The films were prepared by Al and Si co-deposition in oxygen ambient (in total amounts equal to the amount of Si necessary to prepare a pure bilayer silicate film) and subsequent annealing at high temperatures.[32] In contrast to the random distribution of Al ions in monolayer films on Mo(112), the aluminosilicate films grown on Ru(0001) revealed Al segregation, at least at low Al/Si ratios. Figure 6.6a displays irregularly shaped domains having small, but discernable differences in the image contrast as compared to the surrounding area which looks virtually identical to a pure silicate crystalline film. The formation of the Al-rich domains somewhat contradicts a common view, based solely on the electrostatic grounds, that the Al atoms arrange in aluminosilicates as far as possible from each other (often referred to as the Dempsey rule).[33] One of the plausible explanations for the observed segregation is that the lattice strain associated with the Al incorporation into the silicate frame is reduced if Al-containing species aggregate. This was theoretically predicted by calculations on small clusters mimicking secondary building units in zeolites.[34]

A small red-shift ($\sim$20 cm^{-1}) is observed for the principal band in the Al-modified films as compared to the pure silicate film (Fig. 6.6c). The shift depends on the Al/Si ratio. In agreement with the uniform honeycomb-like structure observed by STM, the IRAS

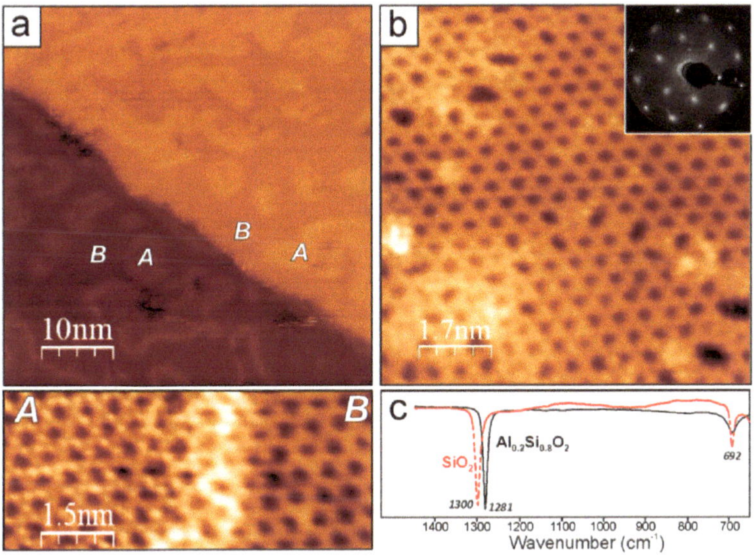

Figure 6.6 STM images of aluminosilicate films on Ru(0001) at relatively low (a) and high (b) Al:Si ratios. Inset shows a typical LEED pattern (at 70 eV). At low Al content, domains (denoted A) are formed which show slightly different image contrast as compared to surrounding areas of pure silicate (B). At high Al:Si ratio, the surface is fairly uniform. (c) Comparison of IRAS spectra obtained for pure silicate (dotted line) and Al-silicate (solid line) films.

results point to that Al substitutes Si in the silica frame rather than forms its own oxide phase. However, the two layers constituting a bilayer are not equivalent. The one that is closer to a metal substrate (a "bottom" layer) may experience some effect of proximity to the metal surface which acts as an electron reservoir for compensating any charge imbalance in the system. Therefore, at low loadings, the Al ions most likely occupy the sites in the bottom layer. The preferential population of Al in the bottom layer has been proven experimentally by measuring XPS spectra as a function of electron emission angle. As introduced in Chapter 2, the contribution from the outmost surface layers dominates in photoelectrons detected at grazing angle with respect to the surface. If Al and Si were distributed homogeneously in the film, then the intensity ratio of the Al 2s and Si 2p signals would be independent on the emission angle.

In fact, the results showed that the ratio considerably decreases at grazing emission. Therefore, the topmost layer in these films is dominated by Si, and the Al ions primarily constitute the subsurface region, i.e., the bottom layer. In addition, water adsorption did not result in bridging hydroxyls, in agreement with the absence of Al in the top silicate layer. Only at high Al:Si ratios, bridging hydroxyls have been detected by IRAS, suggesting that Al start to occupy also sites in the top silicate layer. Water adsorption and surface chemistry of hydroxyl species on silicate films will be discussed in detail in Chapter 8.

6.3 Fe-Silicate Films

Strictly speaking, the term "iron silicate" denotes minerals in the olivine group $(Mg,Fe)SiO_4$, with the Fe-end member fayalite (Fe_2SiO_4), which belong to the class of orthosilicates, in which SiO_4^{2-} ions are all isolated and do not form polymeric frames. In order to avoid confusion, henceforth we use the terms "Fe-silicate" for the framework silicates which contain iron in any form, either in or out of frame.

Zeolites (i.e., microporous aluminosilicates) and silicalites (i.e., "all Si" zeolites) containing small amounts of iron have received a great attention as efficient catalysts of industrially important chemical reactions. Also, many clay minerals contain Fe in their crystal structures, which participates in electron transfer reactions with environmental pollutants, bacteria, and biological nutrients.[35] This property is believed to be result of facile susceptibility of Fe to changes in oxidation state (redox reactions) rather than of its mere presence in the mineral structure. For example, Fe^{3+} is the more common oxidation state in smectites, but bacterial activity in soils and sediments readily reduces it to Fe^{2+}. The surface reactivity of the smectites toward pesticides, various organics, and redox-active metals is greatly increased once the structural Fe had been reduced. Also, the cation exchange and fixation capacities increase, whereas specific surface area and swelling in water decrease.

There are a few studies reported in the literature aimed at preparation of thin Fe-silicate films. Basically, they made use of preparation schemes developed for zeolite films in general (see Fig. 6.1). As an example, the films were synthesized by sol-gel method using iron triethoxide-tetraethylorthosilicate as mixed precursor.[36] The 100 nm-thick films formed upon poly-condensation reactions between Si–OH and Fe–OH groups of the hydrolyzed precursors resulting in Fe–O–Si and Si–O–Si bonds detected by infrared spectroscopy. Annealing by "rapid thermal processing," i.e., at high heating and cooling rates, yields amorphous, glassy films. Meanwhile, conventionally used heating treatments (with low heating/cooling rates) resulted in nanocrystalline films with predominantly Si–O–Si bonding in grains about 40 nm in size.

Let us now address the preparation of ultrathin Fe-silicate films following the approach previously applied for aluminosilicate films as discussed above. To date, Fe-silicate films were only prepared on a Ru(0001) substrate.[37,38] Pure Si and Fe were vapor-deposited in oxygen ambient onto a support kept at low temperatures, and the overlayer was then annealed in UHV at high temperatures. Again, as in case of Al-silicate films, the total amount of deposited Si and Fe atoms was equal to the amount of Si used to form a bilayer silicate film.

Since the principal structure of the film (e.g., monolayer, bilayer, multilayer) can immediately be established by vibrational spectroscopy, we first discuss the IRAS results in Fig. 6.7 that collects spectra as a function of the Fe/(Fe+Si) molar ratio in the films measured by XPS. Clearly, the 1300 and 692 cm^{-1} bands of the pure bilayer silicate gradually attenuate whereas a sharp and strong band at 1005 cm^{-1} together with a weaker band at 674 cm^{-1} appears and gains in intensity. Such a spectral evolution is typical for a two-phases system where the new phase (Fe-silicate) coexists with the original (pure silicate) phase. In case of the random distribution of "guest" (Fe) atoms in the silica framework, the original band(s) is expected only to shift as observed, indeed, on aluminosilicate films (Fig. 6.6c). At further increasing Fe concentration (Fe/(Fe+Si) ratio >0.5) the 1005 cm^{-1} band attenuates, shifts and ultimately disappears, thus approaching the formation solely of an Fe-oxide film which is IRAS silent in this spectral region (see the bottom spectrum).

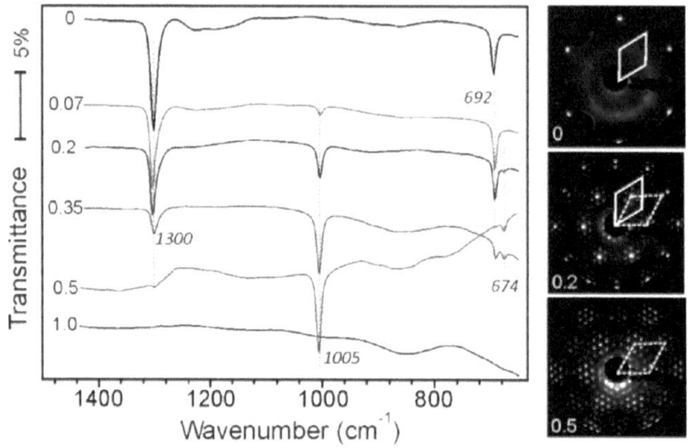

Figure 6.7 IRAS spectra and LEED patterns of the Fe-silicate films grown on Ru(0001) at different Fe:(Fe+Si) ratios shown adjacent to the spectra.

LEED inspection also revealed two unit cells implying two coexisting structures, see right panel in Fig. 6.7. The (2×2)-Ru(0001) diffraction pattern only visible at low Fe:Si ratios corresponds to the crystalline structure as in well-ordered, pure silicate films. Another unit cell with a lattice constant of 5.22 Å is rotated by 30° with respect to Ru(0001). In addition, six sharp satellite spots around each integer spots clearly point to a Moiré-like coincidence structure. (A careful reader may notice some resemblance of the above flower-like LEED pattern with that observed on the UHV-annealed monolayer films on Ru(0001), see Chapter 5).

The two phases, anticipated from the IRAS and LEED results, are clearly distinguished in STM images (Fig. 6.8a). Patches showing a long-range superstructure can straightforwardly be assigned to the Fe-containing phase since its area increases with increasing Fe:Si ratio. These regions appear in STM a bit higher (brighter) than the surrounding pure silicate phase, although the height difference may result from an electronic effect rather than reflecting the surface topography. Both phases show a honeycomb-like structure with the same periodicity and azimuthal orientation. Figure 6.8b zooms in the interface between two phases showing that both lattices are in perfect registry. Since the unit cell of Fe-silicate is rotated by 30° with respect to the Ru(0001) lattice, the rotation applies to pure silicate

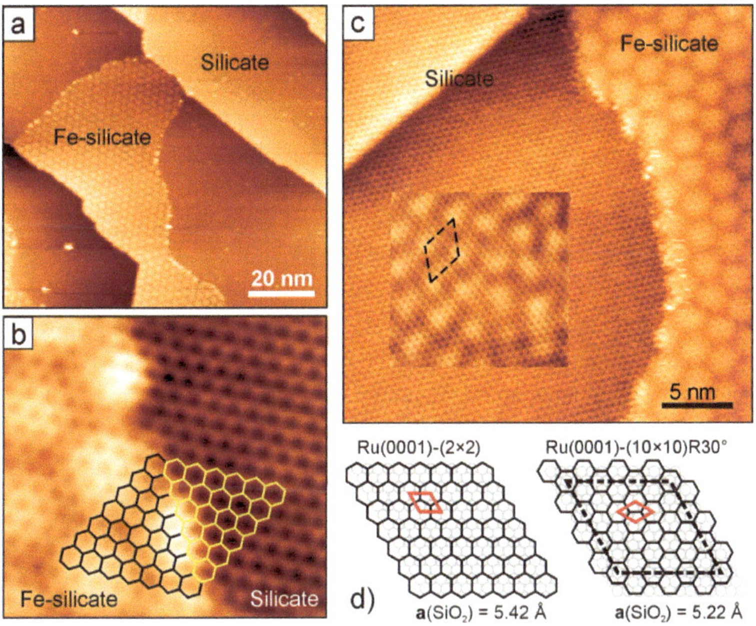

Figure 6.8 (a–c) STM images of a Fe-silicate film at relatively low Fe-Si atomic ratios. Fe-silicate regions show a Moiré structure. Panel (b) zooms in the boundary between two phases showing that both lattices follow the same registry. The 30° lattice rotation causes a very weak Moiré-like modulation on the pure silicate phase, highlighted with the enhanced contrast in panel *c*. (d) Schematic representations of the pure silicate film on Ru(0001) forming a (2×2) structure and of the Ru(0001)-(10×10)R30° coincidence structure formed by the 30°-rotated hexagonal layer with a 5.22 Å lattice constant.

phase as well, that causes a very weak surface modulation clearly seen in the image portion presented with the enhanced contrast in Fig. 6.8c.

In principle, any rotation of a hexagonal layer over the (0001) or (111) metal surface will result in a coincidence structure, with a long-range periodicity depending on the two lattices mismatch and the rotation angle. It is easy to show that the 30° rotation of a layer with 5.42 Å lattice constant over Ru(0001) yields a Ru(0001)-(7×7)R30° coincidence structure. Accordingly, reducing the lattice constant to 5.22 Å leads to a Ru(0001)-(10×10)R30° superstructure schematically shown in Fig. 6.8d. It is this structure that is observed on pure silicate phase (highlighted in Fig. 6.8c) showing a long-range surface modulation. Small deviations from a perfect Moiré pattern can be assigned to the O ad-atoms residing on the Ru surface in the "O-rich" films (see Chapter 4).

Actually a very small, in the range of a few picometers, modulation amplitude observed on pure silicate regions also supports the DFT-derived conclusion that bilayer silicate films weakly interact with the metal surface. Indeed, unsupported bilayer silicate sheets are expected to be perfectly flat. Moreover, our finding that the 1300 cm^{-1} band associated with the pure silicate phase in the 30° rotated films does not shift (see Fig. 6.7), provides solid evidence that vibrational characteristics of bilayer films are insensitive to the film orientation relative to the metal surface. Likewise, this explains previously mentioned insensitivity of IR frequencies to whether the bilayer silicate film is crystalline or amorphous, again, in full agreement with its weak interaction to the metal support. In contrast, a considerably larger modulation amplitude observed on the Fe-containing regions points to a much stronger interaction between a Fe-silicate film and a substrate.

It is fair to say that establishing the atomic structure of the Fe-silicate films would not be possible without DFT modeling, even though the experimentally found Moiré superstructure was not included in the calculations described below, because it is too large unit cell for rigorous calculations.[37] Starting from a crystalline "O-rich" bilayer silicate film on a Ru(0001)-(2×2) unit cell, several possible structures were created by Fe substitution of the Si atoms, resulting in $Fe_nSi_{8-n}O_{16}2O/Ru(0001)$ composition within the cell. The most stable structures obtained for $n = 1–4$ are displayed in Fig. 6.9(a–d). The first substitution, which takes place preferentially in the bottom layer, does not affect the principal structure of the film (Fig. 6.9a). The second Fe atom prefers to substitute the nearest neighbor site, still in the bottom layer, so that the bridging (Fe–O–Fe) oxygen atom moves closer to the metal surface (Fig. 6.9b). Even for the third Fe atom, the bottom layer remains energetically preferable (Fig. 6.9c). Note that, in all these models, the remaining Si–O–Si and newly formed Si–O–Fe linkages are normal to the surface as in the pristine bilayer film. It has turned out, however, that the structure

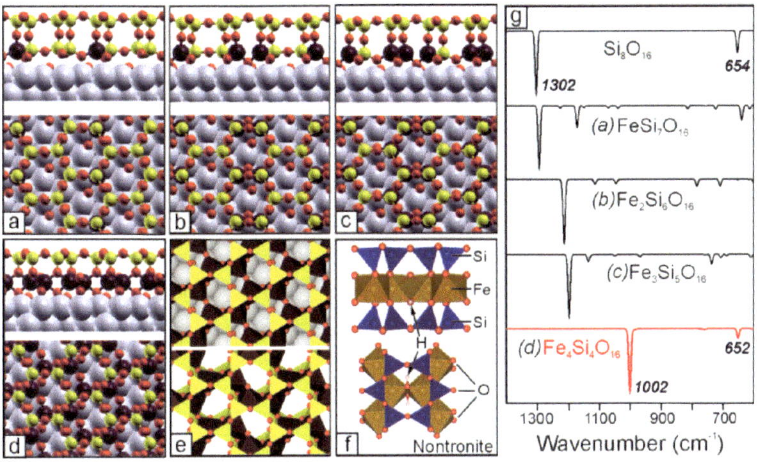

Figure 6.9 (a–d) Top and side views of the most stable DFT-optimized structures found by consecutive substitution of Si by Fe in the $Si_8O_{16}2O$/Ru(0001)-(2×2) unit cell (Si, yellow; Fe, brown; O, red; Ru, gray): (a) $FeSi_7O_{16}$; (b) $Fe_2Si_6O_{16}$; (c) $Fe_3Si_5O_{16}$; (d) $Fe_4Si_4O_{16}$. (e) Top-view comparison of the Fe-silicate film and of a sheet of dehydroxylated form of the clay mineral nontronite. (f) Top and cross views of ideal nontronite. (g) Scaled harmonic frequencies for IRAS active bands simulated for the structural models shown in (a–d). Adapted with permission from ref. 37.

with four Fe atoms in the bottom layer easily transforms into the considerably more stable one depicted in Fig. 6.9d. In this structure, the bottom layer is formed by a hexagonal layer of edge-sharing and corner-sharing FeO_5 square pyramids.

Vibrational spectra computed for DFT-optimized structures using harmonic approximation are shown in Fig. 6.9g. Although the spectra (a–c) exhibit many additional weak bands, the principal band falls in the range of 1200–1300 cm^{-1} since they originate from the interlayer Si–O–Si(Fe) linkages normal to the plane. Only the spectrum simulated for the most stable structure in Fig. 6.9d shows a strong band at 1002 cm^{-1} perfectly matching the experimentally observed one at 1005 cm^{-1} (Fig. 6.7). This band results from the stretching vibration of the Si–O bond oriented perpendicular to the surface and, in essence, does not involve vibrations of the Fe atoms. Therefore, the spectrum becomes similar to those on monolayer films on Mo(112) and Ru(0001), in the range of 1030–1060 cm^{-1}

(see Chapter 4). The low frequency band (654 cm^{-1} computed for the pure silicate film) is almost unperturbed by the Fe incorporation in the bottom layer (652 cm^{-1} computed) since it largely comes from bending of the Si–O–Si bonds in the topmost layer, which is only slightly affected by Fe in the bottom layer. Basically, the same behavior is observed in the experiment, although the red-shift is considerably larger (from 692 to 674 cm^{-1}).

Therefore, DFT analysis of thermodynamic stability and vibrational characteristics provides compelling evidence of that a Fe-silicate film consists of a single silicate layer on top of an iron oxide layer. The latter bears some similarity, but not the same as a FeO(111) monolayer film formed on metal surfaces like Pt(111) (see Fig. 2.16 in Chapter 2) and Ru(0001). As a result, the Fe-silicate film becomes strongly bound to the metal surface, which facilitates electron tunnelling in STM and explains a higher apparent height, although it is geometrically lower than the surrounding pure silicate surface, both imaged in Fig. 6.8.

Furthermore, the DFT analysis shed light on the experimentally observed phase separation into Fe-containing and pure silicate structures. To rationalize this finding, total slab energies were calculated to estimate the probability of the following "disproportionation" reactions:

$$2\mathrm{FeSi_7O_{16}\cdot 2O/Ru} \rightarrow \mathrm{Fe_2Si_6O_{16}\cdot 2O/Ru} + \mathrm{Si_8O_{16}\cdot 2O/Ru}, \quad \Delta E = -1\ \mathrm{kJ/mol} \tag{6.1}$$

$$3\mathrm{FeSi_7O_{16}\cdot 2O/Ru} \rightarrow \mathrm{Fe_3Si_5O_{16}\cdot 2O/Ru} + 2\mathrm{Si_8O_{16}\cdot 2O/Ru}, \quad \Delta E = +66\ \mathrm{kJ/mol} \tag{6.2}$$

$$4\mathrm{FeSi_7O_{16}\cdot 2O/Ru} \rightarrow \mathrm{Fe_4Si_4O_{16}\cdot 2O/Ru} + 3\mathrm{Si_8O_{16}\cdot 2O/Ru}, \quad \Delta E = -158\ \mathrm{kJ/mol} \tag{6.3}$$

Equation (6.1) describes the reaction when two identical slabs with one Fe atom in the unit cell separates into one slab containing two Fe atoms and one slab of a pure silicate. Equations (6.2) and (6.3) are for case of, respectively, three and four slabs each containing one Fe atoms. (NB: Although the models include "O-rich" structure with O ad-atoms adsorbed on the Ru surface, this does not alter the conclusions drawn.) A considerable energy gain is obtained for the reaction (3) where all four Si atoms in the bottom layer are replaced by Fe. Therefore, the phase separation is fully thermodynamically driven.

Even though the experimentally found 30°-rotation has not been included in the calculations, the proposed structure model nicely explains the experimental results. The Fe-silicate film bears a resemblance to the clay minerals, in particular of nontronite, representative of the smectites. Schematically shown in Fig. 6.9f, nontronite is formed by the two layers of silicate that sandwich a layer of Fe-hydroxide like in 2:1 sheet silicates (see Fig. 1.5 in Chapter 1). Since the films are prepared in water-free atmosphere and annealed at high temperatures, there seems to be no hydroxyl groups in the structure. Therefore, one should compare the Fe-silicate films with dehydroxylated nontronite depicted in Fig. 6.9e. The respective silicate sheets (so called T-(tetrahedral) layers in clays) are virtually identical in both structures. There are some differences in connections between FeO_5 square pyramids constituting the O-(octahedral) layer in clays. In particular, the six-membered rings are slightly shifted with respect to rings in the silicate layer in the film. Another point to mention is the oxidation states of iron in these two systems. According to the compositional stoichiometry of the Fe-silicate film ($FeSiO_4$) iron must be in the formal oxidation state 4+, whereas an oxidation state of iron in dehydroxylated nontronite $(Fe_2O_3)(SiO_2)_4$ is 3+. We should bear in mind, however, that our Fe-silicate film is strongly bound to a metal substrate, and the charge transfer from the metal leads to a lower oxidation state of iron. Despite some differences, the Fe-silicate film can be viewed as a single sheet of dehydoxylated nontronite, in which one silicate layer in the 2:1 clay is replaced by the metal layer.

Obviously, a Fe-silicate film is structurally different from an aluminosilicate film discussed in the previous section. While the Fe-silicate is strongly bound to the metal surface and exhibits phase separation resulting in a sharp interface between silicate and iron oxide layers, the Al-silicate film shows rather a random distribution of Al within the bilayer framework and remains weakly bonded to the metal support. Actually, the difference between the structural motifs in these two silicate films reflects the different behavior of naturally occurring Al- and Fe-silicate materials. In aluminosilicate films, Al^{3+} is present in fourfold coordination typical of natural zeolites (Fe-zeolites are not found in nature), whereas the Fe-silicate film adapts the layered structure characteristic for clay minerals.

6.4 (Fe, Al)-Silicate Films

In this section, we address ultrathin silicate films containing both Fe and Al. Among other reasons, the interest to such films was driven by attempts to deeper understand the mechanisms of catalytic reactions over Fe-modified zeolites like already mentioned ZSM-5. The nature of active species in these catalysts remains controversial due to a diversity of Fe species present in the system, so-called "in-frame" and "out of frame" and their combinations.[39]

The preparation including Fe deposition on top of a well-ordered Al-silicate film followed by high temperature annealing either in oxygen or vacuum usually results in disordered nanoparticulate film morphology. Further anneal at higher temperatures to improve ordering only causes film dewetting. Therefore, the preparation of ordered films involves co-deposition of all metals onto the O-precovered Ru(0001) surface to ensure homogeneous distribution of elements in the "as deposited" films. Then the films are heated in 10^{-6} mbar of O_2 to elevated temperatures until characteristic features of silicate films appear in IRAS and/or LEED. For comparison with previous Al- and Fe-silicate films, the three elements were deposited in total amounts equivalent to form a bilayer film.

Figure 6.10 summarizes structural characteristics of the Fe–Al-silicate films prepared at relatively low Fe concentrations. A sharp IRAS band at 1277 cm^{-1} is very similar in position and intensity to that of a pure aluminosilicate film (Fig. 6.6). Accordingly, the 1008 cm^{-1} signal bears close similarity to that of the Fe-silicate (Fig. 6.7). Also LEED inspection revealed a sharp (2×2) diffraction pattern like in pure Al-silicate films, while additional spots are coming from a shorter unit cell 30°-rotated and accompanied by a Moiré-like structure as in pure Fe-silicate films. Combined together the results suggest that the Fe–Al-silicate films at low Fe contents

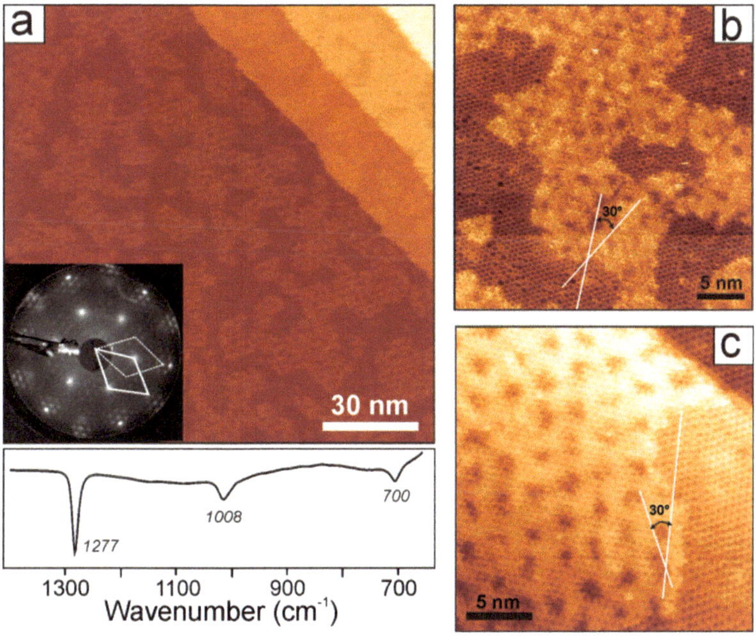

Figure 6.10 STM images of $Fe_{0.25}Al_{0.2}Si_{0.55}O_2$ film prepared by oxidation at 1150 K (a,b) and 1230 K (c). IRAS spectrum is shown below the image. Inset shows LEED pattern at 70 eV. The coexisting two unit cells are indicated. Both structures show a honeycomb-like lattices which are rotated by 30° with respect to each other in agreement with LEED. Adapted with permission from ref. 38.

form two separate structures, Al-silicate and Fe-silicate, rather than a mixed phase. It should be noted, however, that prolonged annealing or increasing oxidation temperature up to 1250 K results in all Fe-related features in IRAS, LEED and XPS disappeared as if Fe is no more present at the surface. In principle, iron oxides are known to sublime at these high temperatures. Alternatively, iron could migrate into the Ru crystal.

In full agreement with the prediction based on LEED and IRAS, STM images identified two different surface structures with irregularly shaped boundaries between them (Fig. 6.10a,b). One exhibits a long-range periodic structure, albeit not perfect, and the unit cell rotated by 30° with respect to the surrounding area that

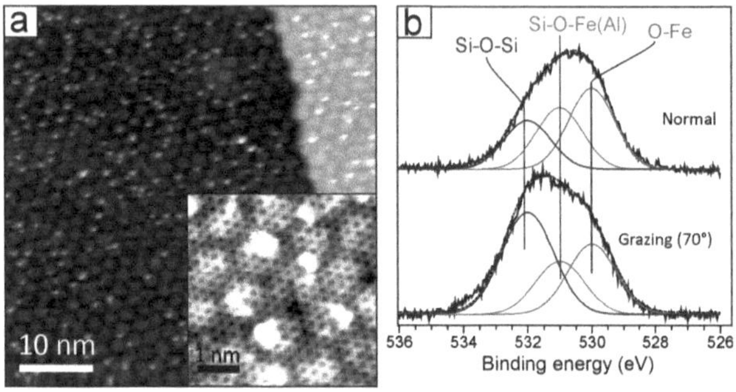

Figure 6.11 (a) STM images of the $Fe_{0.5}Al_{0.13}Si_{0.37}O_2$ film prepared by oxidation at 1250 K. (b) The normalized O 1s signals in XPS spectra measured at normal and grazing (70°) electron emissions. The spectra are deconvoluted using three O species having different environment as indicated. Adapted with permission from ref. 38.

constitutes the second phase. This latter looks virtually identical to the Al-silicate film (see Fig. 6.6 for comparison). In the same manner as we have explained STM contrast in the Fe-silicate films, the Fe-containing domains in our Fe-Al-silicate films appear higher than the Al-silicate domains because they are strongly bonded to the metal surface, through the FeO_x interfacial layer.

The situation changes, albeit not dramatically, at increasing Fe concentration. Neither pure Fe-silicate nor pure Al-silicate phases are formed in this case. In fact, spectral changes of the IRAS band around 1000 cm^{-1} (which is representative of the Fe-silicate phase) suggest some transformations occurring during oxidation at elevated temperatures. Also monitored by STM, at least two slightly different structures were found co-existing in such films before they become rather uniform at high temperatures as illustrated in Fig. 6.11a. The Fe–Al-silicate films prepared by annealing at high temperatures showed the IRAS fingerprints solely of the Fe-silicate film. It may well be that in the Fe-rich films Al is pushed out by Fe forming stronger Fe–O–Ru bonds to the metal surface, ultimately resulting in small alumina clusters observed on top of well-ordered Fe-silicate surface (Fig. 6.11a). However, at low oxidation temperatures, it appears that both Al and Fe constitute a mixed bottom layer.

To get more information on the element distribution in our Fe–Al-silicate films, XPS spectra were measured at normal and grazing electron emissions. As shown in Fig. 6.11b for the O 1s core level, a broad signal can be deconvoluted in three components: The one at around 532 eV is assigned to oxygen ions in Si–O–Si bonds, another one at 530 eV is typical for O^{2-} in iron oxides, and the one centered at 531 eV — to the mixed Si–O–Fe and/or Si–O–Al bonds. Compared to the spectrum at normal emission, the Si–O–Si associated component gains considerable intensity at grazing emission. It is, therefore, clear that the topmost layer is primarily formed by siloxane bonds, whereas Fe and Al seems to prefer the "sub-surface" layers closer to the metal surface.

To conclude, iron in Fe-silicates does not randomly substitute Si(Al) cations in the silicate framework, but segregates into a pure silicate (Al-silicate, respectively) phase and an Fe-silicate phase. At high Fe concentrations, the films may exhibit several phases depending on the annealing temperature and the Si/Al ratio. The surface preferentially exposes a silicate layer, whereas FeO_x dominates in the bottom layer. All in all, the results suggest that the formation of "in-frame" Fe species in silicalites and zeolites is thermodynamically unfavorable, and iron prefers to form iron oxide phase.

6.5 Ti-Silicate Films

Combining titania (TiO_2) and silica (SiO_2) into one composite exhibiting functional properties missing for individual substances has been considered from both material science and catalysis perspectives. Possible application of such materials in catalysis falls into three categories: (i) photocatalysis, associated with the support and the quantum-size effects; (ii) acid catalysis, related to the generation of new acid sites; and (iii) catalytic supports with enhanced thermal and mechanical stability of SiO_2 while preserving the catalytic performance of TiO_2.[40] Synergy effects imply intimate interaction of TiO_2 and SiO_2 which, in turn, critically depends on the preparation method and the conditions used. Based on available phase diagrams, truly mixed oxide phase can only be obtained at relatively low TiO_2 content (<15 wt %). At higher Ti/Si ratios, TiO_2 forms separate crystalline phase.

Perhaps the most prominent example of titania-silica systems in catalysis is a titanium derivative of silicalite ZSM-5, so called titanium silicalite-1 (TS-1), which received great attention thanks to its unique catalytic performance in the clean oxidative reactions of a variety of organic compounds at low temperature. The structure is believed to contain Ti atoms in lattice positions replacing Si atoms.[41] Structural properties of TS-1, such as crystallinity, particle size, and morphology, all play a role in the catalytic performance. TS-1 is commercialized in chemical industry and produced by hydrothermal crystallization from a batch containing proper Si and Ti sources and template. Currently, many studies focus on the preparation of TS-1 with superior catalytic properties by a cheaper and more environmentally friendly process.

Having a high dielectric constant and high refraction index, TiO_2 is attractive material for potential replacement of SiO_2 as the gate dielectric material in metal-oxide-semiconductor devices. Mixed titania-silicas are also considered for such purposes, even though their dielectric constant is expected to be lower than that of pure titania (~100), but still higher than that of SiO_2 (~4). In addition, titanium silicate films were investigated for optical applications such as passive or active waveguides, antireflective coatings, and notch filters.

Before addressing metal supported *ultrathin* Ti-silicate films, let us shortly survey here some preparations used for *thin* titanium silicate films. Such films can routinely be fabricated, for example, by pulsed laser deposition or by magnetron sputtering of Ti–Si or TiO_2/SiO_2 targets in a reactive gas atmosphere. Figure 6.12 shows a typical IR (not IRAS!) spectrum measured in the transparent mode on thin films about 100 nm in thickness grown on a Si(100) substrate kept at 300°C in 10^{-3} mbar of the reaction mixture (O_2 and Ar).[42] In addition to the characteristic absorption bands originated from the Si–O–Si vibrations (i.e., at around 400, 800, and 1050–1200 cm^{-1}) and associated, respectively, with the Si–O–Si asymmetric rocking, bending, and stretching, a new band at 930 cm^{-1} was observed which was missing in pure SiO_2 and TiO_2 and hence assigned to stretching vibrations in Si–O–Ti species.

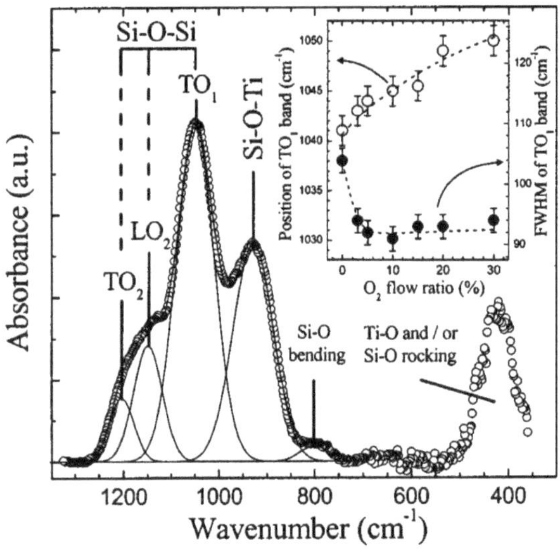

Figure 6.12 Typical IR spectrum (in transmission mode) of the $TiSi_xO_y$ films prepared on Si(001) by reactive magnetron sputtering (~100 nm in thickness). The inset shows the variation of the position and width of the principal TO_1 mode of the Si–O–Si stretching vibrations with the oxygen flow rate ratio. Reproduced with permission from ref. 42.

Interestingly, dielectric characteristics of such films were found to depend on the O_2/Ar ratio. Good dielectric properties (means, high dielectric constant and low leakage current) were only obtained at high oxygen flow. The effect was rationalized in terms of oxygen vacancies formation in the films prepared under O_2 lean conditions.

Another fingerprint of the titanium silicate phase formation comes from XPS measurements indicating the presence of Ti–O–Si bonds. In a simple way to prepare titania-silica interfaces, Ti was vapor-deposited in the presence of oxygen onto a fused quartz or a thermally oxidized Si wafer. In addition to significant shifts of the Ti 2p and O 1s core levels, the observed dependence of modified Auger parameter (see the basics in Chapter 2) on the Ti/Si ratio was attributed to the TiO_x/SiO_2 interface formed by cross-linking bonds. The interface altered the electronic properties of the titania overlayer as well. Ultraviolet absorption spectra revealed a significant blue shift indicating a larger band-gap as compared to pure TiO_2.[43]

Less popular is the preparation based on CVD where Ti silicate thin films grow via pyrolysis of individual Ti- and Si-containing precursors like, for example, di-acetoxi–di-butoxisilane, tetrabutoxysilane, and titanium (4+)*n*-butoxide.[44] Although the films grown on Si(100) remained amorphous even after annealing at temperatures as high as 950°C, IR and Raman spectra indicated the formation of small TiO_2 crystallites in anatase-like structure embedded into a SiO_2 matrix.

In case of phase separation into TiO_2 and SiO_2 the synergetic effects observed in such systems is governed by the titania/silica interface. To get more insight into interfacial structures, we address the DFT calculations[45] performed for the atomically sharp interface formed by an anatase $TiO_2(101)$ surface and a single layer of silica forming a two-dimensional network of one eight-, one six-, and one four-membered ring per surface cell as shown in the top and side views in Fig. 6.13. In case of hydroxylated silica overlayer, three Si atoms in the unit cell yield a (2Ti–O–Si) group, and the other

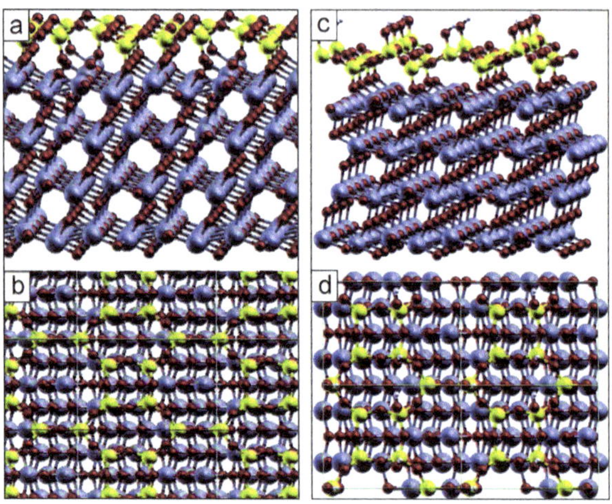

Figure 6.13 DFT-derived models of SiO_2/TiO_2 interfaces in perspective (a,c) and top (b,d) views. Hydroxylated (a,b) and H-free (c,d) silica monolayers adsorbed on the anatase $TiO_2(101)$ surface. (Ti, gray; Si, yellow; O, red; H, white.) Adapted with permission from ref. 45.

three Si atoms bind to an OH group pointing into the vacuum. In the case of H-free termination, three Si atoms participate in a (Ti–O–Si) group and the other three form a (2Ti–O–Si) group each, with no dangling bond pointing outwards. Analysis of electronic density of states (DOS) showed that an interface between TiO_2 and SiO_2 gives a large contribution to the states near the valence and conduction band edges which, in turn, determine the photo-physical properties of such materials.

In the above-mentioned experimental studies of titania-silica films the atomic structures were not established, most likely because of its amorphous nature. Goodman's group seems to be the first which was aimed at preparation of *well-ordered* TiO_x–SiO_2 thin films.[46,47] In their approach, Ti was vapor-deposited at room temperature onto a monolayer silicate film on Mo(112), then oxidized at 950 K, and finally annealed at 1150 K in UHV. At low Ti coverages, STM images revealed additional atomic-size protrusions which were interpreted as Ti atoms substituting Si in the silicate frame (Fig. 6.14a,c). Upon increasing the Ti content, the elongated rectangular islands appeared (Fig. 6.14b,d) which were tentatively assigned to a TiO_x layer.

The annealing temperature was crucial for the prepared surfaces. TiO_x disperses and covers the silica surface after oxidation and annealing below 800 K, whereas TiO_x–SiO_2 interfaces undergo significant restructuring when annealed above 1000 K, forming Si–O–Ti linkages fingerprinted in high-resolution electron energy loss spectra (HREELS). The inertness of the prepared films towards water vapor points to that Ti is most likely not at surface, but located below the silicate layer and bonded via Si–O–Ti–O linkages to Mo(112). (As a side note, vacuum annealing at 1400 K of the high Ti coverage samples completely removed Si from the surface, leaving behind a well-ordered TiO_3 film in a (8×2)-Mo(112) symmetry.)

Ultrathin Ti-silicate films were succesfully prepared in our own laboratories by Ti and Si co-deposition onto Ru(0001).[48] It should be mentioned first, that in contrast to Fe for which the 4+ oxidation state is not common, Ti readily accommodates the 4+ state. Therefore, one could anticipate random substitution of Si^{4+} by Ti^{4+} ions in the silicate framework. It has turned out, however, that structural characteristics of the Ti-silicate films closely resemble the

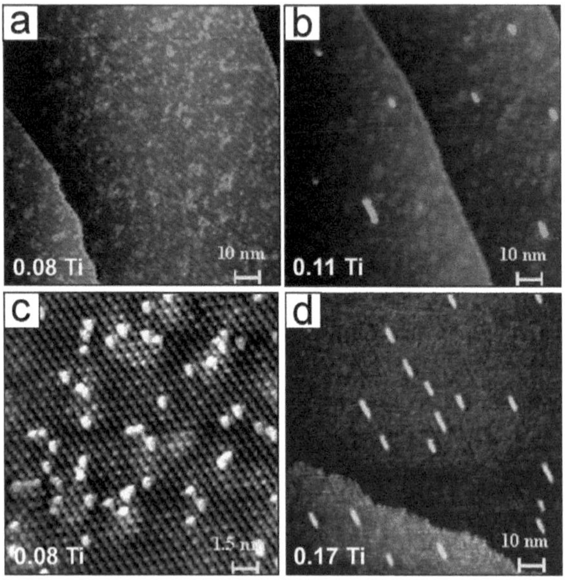

Figure 6.14 STM images of a TiO_x-SiO_2 thin films at Ti content of 8% (a,c), 11% (b), and 17% (d). Adapted with permission from ref. 46.

Fe-silicate films discussed above. Firstly, LEED showed a crystalline film with the lattice constant of ~5.25 Å which is rotated by 30° with respect to the Ru(0001) surface, thus resulting in a Moiré-like structure also seen in STM images via a long-range surface modulation (Fig. 6.15). Secondly, a new IRAS band at 1022 cm^{-1} emerges from the onset in the Ti-silicate films, which is only slightly blue-shifted as compared to Fe-silicate films (1005 cm^{-1}, see Fig. 6.7). This similarity points to a similar structural motif.

The atomic structure of a Ti-silicate film has been studied in detail by DFT.[48] About 40 different structures at different Ti/Si ratios between 1/7 and 1 were examined. Among all possible structures formed by Ti substitution of one of the eight Si atoms present in the silicate bilayer placed onto the Ru(0001)-(2×2) slab, the Ti atom prefers Si in the bottom layer (Fig. 6.16a). Although the structures might introduce certain strain, because of longer bond distances for Ti–O than for Si–O, the bilayer structure is basically preserved upon sequential substitution of up to three Si atoms

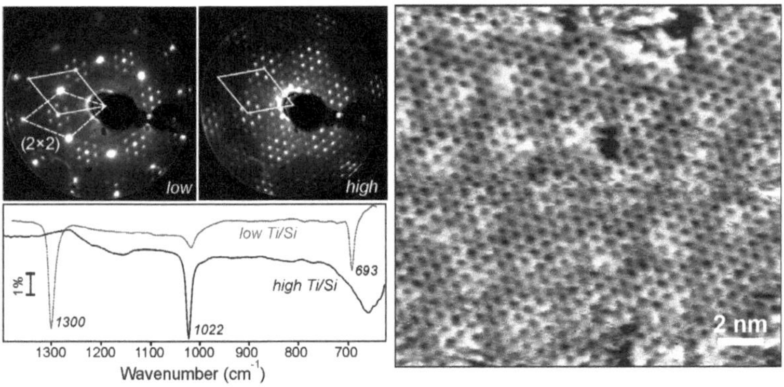

Figure 6.15 LEED patterns (60 eV) and IRAS spectra of the Ti-silicate films at low and high Ti/Si ratios. High-resolution STM image of the Ti-silicate surface. A long-range surface modulation as suggested by LEED is also visible. Adapted with permission from ref. 48.

in the cell. At higher Ti/Si ratios, i.e., in $Ti_3Si_5O_{16}$ and $Ti_4Si_4O_{16}$ films, the lowest energy structures are qualitatively different. The energies per unit cell of the structures shown in Fig. 6.16c,d, where all the substituted Ti atoms are in the bottom layer, are found to be 68 and 217 kJ/mol lower than their counterparts with the bilayer structure. Therefore, the Ti-silicate film must be described as a silicate monolayer on top of a monolayer formed by edge-sharing TiO_6 octahedra. The layers are connected by shared oxygen atoms at the corners of the SiO_4 tetrahedra and TiO_6 octahedra. The TiO_6 octahedra are also connected to the Ru substrate via oxygen atoms shared by two Ti atoms and one Ru atom. Fairly good agreement between the calculated within harmonic approximation frequency (1003 cm^{-1}, scaled) and experimentally measured (1022 cm^{-1}) frequency validates the proposed model. Like in Fe-silicates, the principal IR band manifests the Si–O bond vibrations in the Si–O–Ti linkages and is, therefore, close to the typical frequencies in monolayer silicates.

Furthermore, DFT calculations predicted Ti segregation and phase separation into pure silicate and Ti-silicate that even occurs at low Ti content. The energy gain computed for disproportionation reactions below, resulting in two, "Ti-rich" and pure silica, phases is particularly large for the formation of a Ti-silicate phase with a

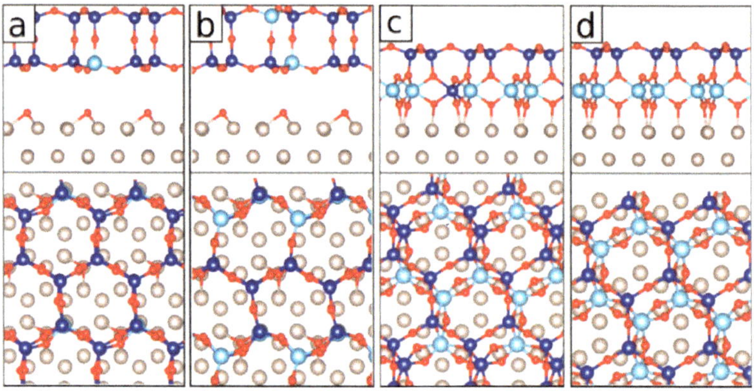

Figure 6.16 Top and side views of the most stable structures found by DFT at various Ti/Si ratios. (a) $TiSi_7O_{16} \cdot 2O/Ru$, (b) $Ti_2Si_6O_{16} \cdot 2O/Ru$, (c) $Ti_3Si_5O_{16} \cdot 2O/Ru$, and (d) $Ti_4Si_4O_{16} \cdot 2O/Ru$. Si, blue; Ti, light blue; O, red; Ru, gray. Reproduced with permission from ref. 48.

Si:Ti = 1:1 ratio (eq. 6.6). A large energy gain is even observed for the segregation of the Ti_3Si_5 film into Ti_4Si_4 and pure bilayer silicate phases described by Eq. (6.7).

$$2TiSi_7O_{16}\cdot 2O/Ru \rightarrow Ti_2Si_6O_{16}\cdot 2O/Ru + Si_8O_{16}\cdot 2O/Ru$$
$$\Delta E = -11\ \text{kJ/mol} \quad (6.4)$$

$$3TiSi_7O_{16}\cdot 2O/Ru \rightarrow Ti_3Si_5O_{16}\cdot 2O/Ru + 2Si_8O_{16}\cdot 2O/Ru$$
$$\Delta E = -73\ \text{kJ/mol} \quad (6.5)$$

$$4TiSi_7O_{16}\cdot 2O/Ru \rightarrow Ti_4Si_4O_{16}\cdot 2O/Ru + 3Si_8O_{16}\cdot 2O/Ru$$
$$\Delta E = -227\ \text{kJ/mol} \quad (6.6)$$

$$4Ti_3Si_5O_{16}\cdot 2O/Ru \rightarrow 3Ti_4Si_4O_{16}\cdot 2O/Ru + Si_8O_{16}\cdot 2O/Ru$$
$$\Delta E = -390\ \text{kJ/mol} \quad (6.7)$$

In principle, layered titanosilicates such as bafertisite or nafertisite appear in nature as a class of minerals called heterophyllosilicates, albeit rare.[49] They have "T–O–T" stacking like in all 2:1 sheet silicates, but the hetero atoms are primarily substituting Si in the *T-layer*. In contrast, the TiO_6 octahedra constitute the *O-layer* in our films.

6.6 Controlling Silica Film Crystallinity via Metal Doping

As shown above, the ultrathin silicate films incorporating (or, using semiconductor terminology, "doped" with) other metals (Al, Fe, Ti) all exhibited high degree of crystallinity even being prepared at considerably lower temperatures, whereas pure bilayer silicate films are often amorphous, and it is quite difficult to prepare perfectly ordered films. In the first example, Al substitutes Si in the silicate framework so that the bilayer structure is maintained, although the replacement seems to begin within the bottom layer. The Al-silicate films exhibit a (2×2) symmetry, the same as in a crystalline form of a pure silicate, resulting in the lattice constant 5.42 Å. In contrast, Fe- and Ti-containing films showed Fe (Ti) segregation within the bottom layer forming a FeO_x (TiO_x) layer underneath a silicate layer. The unit cell in the Fe- and Ti-silicates has a shorter lattice constant (~5.22 Å), and it is rotated by 30° with respect to the unit cell of Ru(0001) resulting in Moiré-like coincidence structures. (Henceforth, we will refer to these two ordered structures as "(2×2)" and "30°-rotated," for brevity.) Furthermore, all doped films undergo phase separation into the metal-silicate phase and "all Si"-silicate phase from the onset. For the "mixed" Fe–Al-silicate films, the process leads to "30°-rotated" Fe-silicate and "(2×2)" Al-silicate phases, respectively. Based on the DFT calculations, the structures are thermodynamically stable, and the segregation process is thermodynamically, not kinetically driven.

As discussed in the last sections of Chapter 4 while comparing films prepared on different metal substrates, a strong interaction with the metal surface and the lattice mismatch, both controls film crystallinity and its azimuthal orientation to the metal surface. Therefore, we can admit that Fe- and Ti-silicate films are crystalline thanks to the formation of the interfacial FeO_x (TiO_x) layer strongly bonded to the metal surface that drives the film to be 30°-rotated. Surprisingly, however, the films showed good crystallinity even at the lowest metal concentration studied, when these two phases coexist. Moreover, the pure silicate phase surrounding the "30°-rotated" Fe(Ti)-silicate domains is crystalline and 30°-rotated, too.

STM images (see Fig. 6.8b) clearly show that the silicate layer follows the same registry upon crossing the border from the Fe-silicate to the pure silicate region which is perfectly crystalline. The same is observed for the Al-silicate films, where both phases form "(2×2)" structure (see Fig. 6.6). Non-doped films under similar preparation conditions show predominantly the amorphous structure.

To rationalize such behavior, we recall the basics of the crystal growth. Typically the process follows an initial stage of nucleation, i.e., the formation of a small nucleus containing the newly forming crystallites. After the nucleation, the crystal homoepitaxially grows outwards from the nucleating site. However, the nucleation process is relatively slow in homogeneous systems. To accelerate the growth, a small crystallite is added as a seed. In principle, there is nothing against proposing that similar processes occur in two-dimensional, thin film systems as well. Accordingly, our metal modifier acts as the seed.

To verify this hypothesis, we look into the initial stages of the Fe-silicate film formation using the sample containing the smallest amount of Fe, but still allowing to monitor the Fe-silicate phase via the corresponding IRAS feature at 1005 cm^{-1}. Note that the sample had to be only shortly annealed, in this case to 1050 K; otherwise the film fully crystallizes. LEED inspection showed complex pattern indicating coexistence of several surface structures (Fig. 6.17a). The "30°-rotated" unit cell with flower-like satellite spots is a clear fingerprint of the Fe-silicate phase. Relatively strong (2×2) spots come from the crystalline portions of pure silicate phase, and a faint diffraction ring is indicative of the amorphous silicate. In full agreement with the LEED results, STM images observed slightly protruding Fe-silicate domains (clearly distinguished in the large-scale images thanks to a long-range Moiré structure) which are surrounded by a pure silicate. However, a closer look revealed that the "30°-rotated" structure is only found in the regions which are in direct contact to the Fe-silicate domains (Fig. 6.17b). Farther away, the film exhibits, in essence, the structure of a pure silicate film formed in the absence of any dopant, where small (2×2) domains coexist with amorphous phase as shown in Fig. 6.17c. Obviously, once the Fe-silicate domain has been formed, it triggers

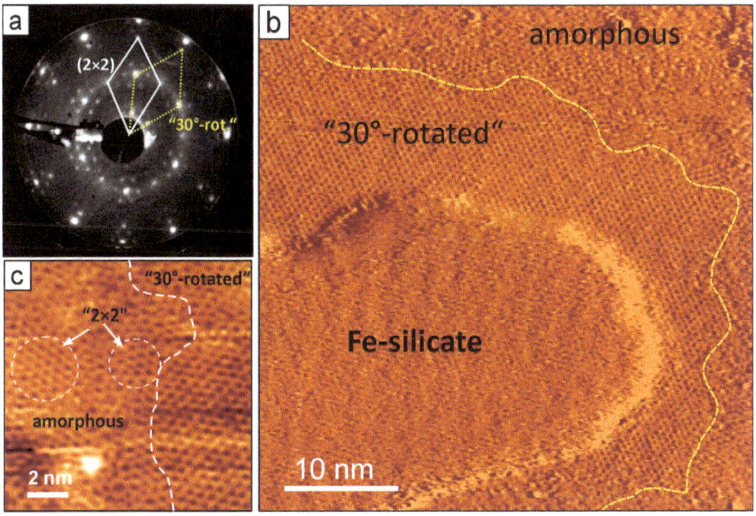

Figure 6.17 (a) LEED pattern (at 70 eV) of the Fe-silicate film prepared by short annealing at 1050 K. The unit cells of the "30°-rotated" and "2×2" structures are indicated. In addition, a faint diffraction ring is seen, which is characteristic for amorphous bilayer film. (b) STM image presented in the differentiated contrast. A Fe-silicate domain is surrounded by a "30°-rotated" crystalline silicate layer, further zoomed in in panel (c). Dashed lines guide the eye to separate crystalline and amorphous phases. Small (2×2) domains within the amorphous regions are outlined by circles. Adapted with permission from ref. 50.

the formation of the crystalline structure that propagates as a "crystallization wave" outwards the domain.

The proposed scenario is schematically shown in Fig. 6.18a. On heating of an "as deposited" Fe–Si-oxide overlayer, small Fe-silicate domains readily form at relatively low temperatures, as evidenced by appearance of the Si–O–Fe related band, while the Si–O–Si bands coming from a bilayer structure develop at higher temperatures.[50] The Fe-silicate domains crystallize into the "30°-rotated" structure and concomitantly make a surrounding more weakly bonded silicate bilayer grow in the same azimuthal orientation until the entire film becomes crystalline. Therefore, the introduced metal atoms may reasonably be referred to as "crystal formers" in analogy to the term "glass formers" commonly used for the glassy systems. In fact, using

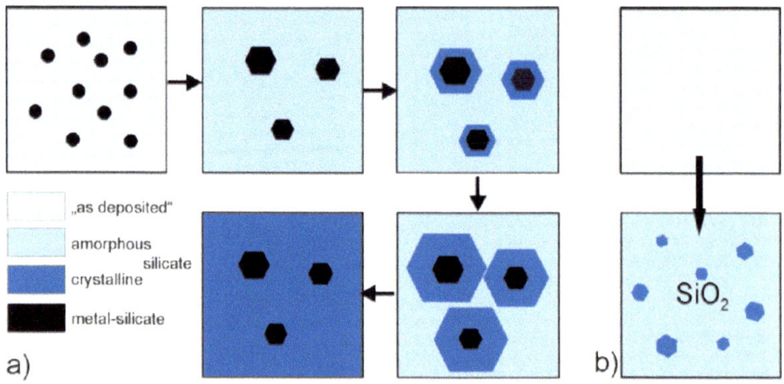

Figure 6.18 (a) Proposed scenario for the growth of crystalline metal (Al, Fe, Ti)-silicate films for comparison with pure silicate film (b) during heating to elevated temperatures.

transition metal as "a seed" one can substantially improve silica film crystallinity and, in addition, lower the preparation temperature, which may be important for technological applications based on the ultrathin silica films, yet remained to be explored.

In case of the mixed Fe–Al-silicate films, segregating into Fe-silicate and Al-silicate phases, the Fe-silicate phase is 30°-rotated like in pure Fe-silicate films, whereas the Al-silicate follows the same orientation as in pure Al-silicate films. Perhaps, the above-presented scenario does not hold true for the Al-containing films. On the other hand, the origin of the long-range order in the Al-silicate films remains unclear. The experimental results (in particular, IRAS) and DFT calculations available thus far indicate that bilayer motif is preserved in the Al-silicate films, hence implying rather weak interaction with the metal support. Therefore, one expects that strongly bound and rotated Fe-silicate domains will push the surrounding Al-silicate phase to grow in the same (30°-rotated) orientation. However, it keeps the (2×2) symmetry as in pure Al-silicates. We may only speculate that Si substitution by Al imposes restrictions on flexibility of Si–O–Al bonds in the silicate framework, resulting in the behavior as observed.

As mentioned in the end of Section 4.5.1 in Chapter 4, Altman and co-workers suggested an ellegant way to control film crystallinity via tuning the lattice parameter in metal alloy as a substrate. This has been exemplified by the growth of a well-ordered silica film on the NiPd(111) surface.[51] In the light of the experimental results presented in this chapter, one had to further examine whether the effect observed for the NiPd systems is really caused by a better lattice match, or it originates solely from having Ni at the alloy surface which may form NiO-like layer that induces film crystallinity in the same manner as shown above for Fe- and Ti-silicates. Indeed, the same group using IRAS and DFT provided later strong evidence for the formation of interfacial NiO_x layer which stabilizes a crystalline silicate layer on top.[52] It appears that many other thin silicate film systems involving 3d-metals will show a similar structural motif.

References

1. Chiang, A. S. T., and K.-J. Chao, Membranes and films of zeolite and zeolite-like materials. *Journal of Physics and Chemistry of Solids*, 2001. **62**(9–10): pp. 1899–1910.
2. Tsapatsis, M., Toward high-throughput zeolite membranes. *Science*, 2011. **334**(6057): pp. 767–768.
3. McLeary, E. E., J. C. Jansen, and F. Kapteijn, Zeolite based films, membranes and membrane reactors: Progress and prospects. *Microporous and Mesoporous Materials*, 2006. **90**(1–3): pp. 198–220.
4. Ozin, G. A., A. Kuperman, and A. Stein, Advanced zeolite, materials science. *Angewandte Chemie International Edition*, 1989. **28**(3): pp. 359–376.
5. Lew, C. M., R. Cai, and Y. S. Yan, Zeolite thin films: From computer chips to space stations. *Accounts of Chemical Research*, 2010. **43**(2): pp. 210–219.
6. Cundy, C. S., and P. A. Cox, The hydrothermal synthesis of zeolites: Precursors, intermediates and reaction mechanism. *Microporous and Mesoporous Materials*, 2005. **82**(1–2): pp. 1–78.
7. Yan, Y., S. R. Chaudhuri, and A. Sarkar, Synthesis of oriented zeolite molecular sieve films with controlled morphologies. *Chemistry of Materials*, 1996. **8**(2): pp. 473–479.

8. Pham, T. C. T., H. S. Kim, and K. B. Yoon, Growth of uniformly oriented silica MFI and BEA zeolite films on substrates. *Science*, 2011. **334**(6062): pp. 1533–1538.
9. Hedlund, J., B. Schoeman, and J. Sterte, Ultrathin oriented zeolite LTA films. *Chemical Communications*, 1997(13): pp. 1193–1194.
10. Bein, T., Synthesis and applications of molecular sieve layers and membranes. *Chemistry of Materials*, 1996. **8**(8): pp. 1636–1653.
11. Davis, S. P., E. V. R. Borgstedt, and S. L. Suib, Growth of zeolite crystallites and coatings on metal surfaces. *Chemistry of Materials*, 1990. **2**(6): pp. 712–719.
12. Valtchev, V., and S. Mintova, The effect of the metal substrate composition on the crystallization of zeolite coatings. *Zeolites*, 1995. **15**(2): pp. 171–175.
13. Jansen, J. C., et al., Zeolitic coatings and their potential use in catalysis. *Microporous and Mesoporous Materials*, 1998. **21**(4–6): pp. 213–226.
14. Mintova, S., et al., Continuous films of zeolite ZSM-5 on modified gold surfaces. *Chemical Communications*, 1997. **1**: pp. 15–16.
15. Yang, H., et al., Synthesis of oriented films of mesoporous silica on mica. *Nature*, 1996. **379**(6567): pp. 703–705.
16. Tolbert, S. H., et al., A new phase of oriented mesoporous silicate thin films. *Chemistry of Materials*, 1997. **9**(9): pp. 1962–1967.
17. Mintova, S., and T. Bein, Microporous films prepared by spin-coating stable colloidal suspensions of zeolites. *Advanced Materials*, 2001. **13**(24): pp. 1880–1883.
18. Kirschhock, C. E. A., et al., Zeosil nanoslabs: Building blocks in nPr_4N^+-mediated synthesis of MFI zeolite. *Angewandte Chemie International Edition*, 2001. **40**(14): pp. 2637–2640.
19. Doyle, A. M., et al., Ultra-thin zeolite films prepared by spin-coating silicalite-1 precursor solutions. *Chemical Physics Letters*, 2003. **382**(3–4): pp. 404–409.
20. Varoon, K., et al., Dispersible exfoliated zeolite nanosheets and their application as a selective membrane. *Science*, 2011. **334**(6052): pp. 72–75.
21. Rangnekar, N., et al., 2D zeolite coatings: Langmuir–Schaefer deposition of 3 nm thick MFI zeolite nanosheets. *Angewandte Chemie International Edition*, 2015. **54**(22): pp. 6571–6575.
22. Kumar, P., et al., Quantification of thickness and wrinkling of exfoliated two-dimensional zeolite nanosheets. *Nature Communications*, 2015. **6**: p. 7128.

23. Kestell, J. D., et al., Studying two-dimensional zeolites with the tools of surface science: MFI nanosheets on Au(111). *Catalysis Today*, 2017. **280**, Part 2: pp. 283–288.

24. Cho, M.-H., et al., Annealing effects of aluminum silicate films grown on Si(100). *Journal of Vacuum Science & Technology A*, 2002. **20**(3): pp. 865–872.

25. Němec, P., et al., Pulsed laser deposited alumino-silicate thin films and amorphous chalcogenide/alumino-silicate structures. *Thin Solid Films*, 2010. **519**(4): pp. 1341–1345.

26. Katada, N., and M. Niwa, Silica monolayer solid-acid catalyst prepared by CVD. *Chemical Vapor Deposition*, 1996. **2**(4): pp. 125–134.

27. Apblett, A. W., L. K. Cheatham, and A. R. Barron, Chemical vapour deposition of aluminium silicate thin films. *Journal of Materials Chemistry*, 1991. **1**(1): pp. 143–144.

28. Bhachu, D. S., et al., Zeolite films: A new synthetic approach. *Journal of Materials Chemistry A*, 2013. **1**(4): pp. 1388–1393.

29. Böszörményi, I., et al., Model silica-alumina acid catalysts for surface science and catalysis studies prepared by argon ion beam sputter deposition using HY-zeolite targets. *Catalysis Letters*, 1991. **10**(5): pp. 343–355.

30. Gründling, C., J. A. Lercher, and D. W. Goodman, Preparation of mixed Al_2O_3/SiO_2 thin films supported on Mo(100). *Surface Science*, 1994. **318**(1–2): pp. 97–103.

31. Stacchiola, D., et al., Synthesis and structure of ultrathin aluminosilicate films. *Angewandte Chemie International Edition*, 2006. **45**(45): pp. 7636–7639.

32. Boscoboinik, J. A., et al., Modeling zeolites with metal-supported two-dimensional aluminosilicate films. *Angewandte Chemie International Edition*, 2012. **51**(24): pp. 6005–6008.

33. Dempsey, E., Acid strength and aluminum site reactivity of Y zeolites. *Journal of Catalysis*, 1974. **33**(3): pp. 497–499.

34. Schroder, K. P., and J. C. Sauer, Preferred stability of Al–O–Si–O–Al linkages in high silica zeolite catalysts-theoretical predictions contrary to Dempsey's rule. *Journal of Physical Chemistry*, 1993. **97**(25): pp. 6579–6581.

35. Stucki, J. W., Properties and behaviour of iron in clay minerals, in *Developments in Clay Science*. 2013. pp. 559–611.

36. Van De Leest, R. E., and F. Roozeboom, Iron-silicate glassy films by sol-gel conversion induced by rapid thermal processing. *MRS Online Proceedings Library Archive*, 1998. **525**: p. 351 (8 pages).

37. Włodarczyk, R., et al., Atomic structure of an ultrathin Fe-silicate film grown on a metal: A monolayer of clay? *Journal of the American Chemical Society*, 2013. **135**(51): pp. 19222–19228.
38. Tissot, H., et al., Preparation and structure of Fe-containing aluminosilicate thin films. *Physical Chemistry Chemical Physics*, 2016. **18**(36): pp. 25027–25035.
39. Zecchina, A., et al., Structure and nuclearity of active sites in Fe-zeolites: Comparison with iron sites in enzymes and homogeneous catalysts. *Physical Chemistry Chemical Physics*, 2007. **9**(27): pp. 3483–3499.
40. Gao, X., and I. E. Wachs, Titania–silica as catalysts: Molecular structural characteristics and physico-chemical properties. *Catalysis Today*, 1999. **51**(2): pp. 233–254.
41. Lamberti, C., et al., Ti location in the MFI framework of Ti-silicalite-1: A neutron powder diffraction study. *Journal of the American Chemical Society*, 2001. **123**(10): pp. 2204–2212.
42. Brassard, D., et al., High-k titanium silicate thin films grown by reactive magnetron sputtering for complementary metal–oxide–semiconductor applications. *Journal of Vacuum Science & Technology A*, 2004. **22**(3): pp. 851–855.
43. Lassaletta, G., et al., Spectroscopic characterization of quantum-sized TiO_2 supported on silica: Influence of size and TiO_2–SiO_2 interface composition. *The Journal of Physical Chemistry*, 1995. **99**(5): pp. 1484–1490.
44. Hodroj, A., J.-L. Deschanvres, and U. Gottlieb, Growth of amorphous Ti–Si–O thin films by aerosol CVD process at atmospheric pressure. *Journal of The Electrochemical Society*, 2008. **155**(2): pp. D110–D114.
45. Seriani, N., et al., Titania–silica interfaces. *The Journal of Physical Chemistry C*, 2012. **116**(20): pp. 11062–11067.
46. Min, B. K., W. T. Wallace, and D. W. Goodman, Synthesis of a sinter-resistant, mixed-oxide support for Au nanoclusters. *The Journal of Physical Chemistry B*, 2004. **108**(38): pp. 14609–14615.
47. Chen, M. S., and D. W. Goodman, An investigation of the TiO_x–SiO_2/Mo(112) interface. *Surface Science*, 2005. **574**(2–3): pp. 259–268.
48. Fischer, F. D., et al., Ultrathin Ti-silicate film on a Ru(0001) surface. *The Journal of Physical Chemistry C*, 2015. **119**(27): pp. 15443–15448.
49. Ferraris, G., A. Bloise, and M. Cadoni, Layered titanosilicates—A review and some results on the hydrothermal synthesis of bafertisite. *Microporous and Mesoporous Materials*, 2008. **107**(1): pp. 108–112.

50. Li, L., et al., Transition metal induced crystallization of ultrathin silica films. *Chemistry of Materials*, 2017. **29**(3): pp. 931–934.

51. Hutchings, G. S., et al., Epitaxial $Ni_xPd_{1-x}(111)$ alloy substrates with continuously tunable lattice constants for 2D materials growth. *ACS Applied Materials & Interfaces*, 2017. **9**(12): pp. 11266–11271.

52. Zhou, C., et al., Structure of a two-dimensional silicate layer formed by reaction with an alloy substrate. *Chemistry of Materials*, 2019. **31** pp. 851–861.

Chapter 7

Adsorption on Ultrathin Silicate Films: Surface and "Sub-Surface" Reactions

In this chapter, we address the reaction of ultrathin silicate films toward typical gases present in the atmosphere, except water that will separately be discussed in Chapter 8. Chemical reactions on silica surfaces are important topics in many scientific disciplines and applications ranging from weathering and erosion to catalysis and sensors. Due to a limited thickness of our films, the processes accompanied by the loss of a material such as dissolution and sublimation can hardly be studied on these systems. However, the initial stages of gas/silica interaction and related fundamental questions can be investigated at the molecular level using structurally well-defined films. Besides the reactions taking place directly at the silica surface, on weakly bound bilayer silicate films small molecules can penetrate the films and intercalate the silica/metal interface. Therefore, chemical reactions can additionally occur on a metal surface in confined space underneath the film. Yet poorly understood the chemical reactions "under cover" have recently received considerable attention in particular for graphene-related systems.[1] Further, we discuss adsorption of metal atoms on ultrathin

Introduction to Ultrathin Silica Films: Silicatene and Others
Shamil Shaikhutdinov

ISBN 978-981-4877-28-2 (Hardcover), 978-1-003-08862-2 (eBook)
www.jennystanford.com

films. Metal deposition is an important process in fabrication of electric junctions on silica in nanoelectronic devices. In addition, the formation of catalytically active metal species on silica-based catalysts is the subject of interest in heterogeneous catalysis. As in the case of gas adsorption, metal intercalation and its effect on functional properties should also be taken into account.

7.1 Interaction with Ambient Gases

With respect to the typical gases in the ambient atmosphere oxide surfaces are usually less reactive than metals, albeit it is a case-sensitive. It is this difference that underlies the metal passivation via the formation of oxide layer suppressing further oxidation and corrosion. Carrying out gas adsorption experiments over metal-supported ultrathin films, one has to be sure that the film under study fully covers the metal support, otherwise the obtained information, in particular by spatially averaging techniques, can be dominated by reactivity of the uncovered metal surface rather than of the oxide film itself. To examine the integrity of the silicate films, one can use titration of a metal surface using probe molecules like CO or H_2, which is well documented in the literature for many metal surfaces. Typically, only a few percent of the sample surface area exposes a bare substrate in the well-prepared films. Solely such films should be used in adsorption studies.

7.1.1 Oxygen

In principle, ultrathin silica films on metals are all prepared by oxidation at high temperatures. Therefore, additional exposure to O_2 only heals adventitious oxygen vacancy type defects in silica, if any. Mostly, oxygen in the ambient atmosphere alters the amount of O ad-atoms directly adsorbing on a metal substrate forming so-called "O-rich" structures. This takes place both for monolayer and bilayer films. However, the process reversibility depends on oxygen affinity to the metal. On Mo-supported monolayer films, "extra" oxygen can be easily added to the O-poor structure, but the reverse process is practically impossible due to very strong bonding of O atoms to Mo that, in turn, requires very high temperatures often leading to the film decomposition. In contrast, more weakly bound O ad-atoms present in monolayer films on Ru can, indeed, be removed by heating in UHV. In this case, however, the transformation is accompanied by substantial film restructuring (see Chapter 5). Surprisingly, the process is fully reversible. As far as bilayer silicate films on Ru and Pt are concerned, heating of the "O-rich" films in vacuum only reduces the amount of O ad-atoms at the interface, while the principal structure of the film retains (Chapter 4). The "O-poor" film can be transformed back to "O-rich" by oxidation. Reversible transformation between "O-poor" and "O-rich" structures implies oxygen diffusion to/from the interface and its adsorption/desorption on/from a metal support, thus suggesting that O_2 molecules readily penetrate the silicate film.

Before discussing a more complex interaction of our silicate films with oxygen, it is instructive to recall oxygen adsorption on the Ru(0001) surface which has been a subject of numerous studies in the "surface science" community.[2] It is well established that chemisorbed oxygen on Ru(0001) forms several ordered structures such as (2×2)-O, (2×1)-O, (2×2)-3O, and (1×1)-O. Exposure to high pressures of O_2 at high temperatures results in a well-ordered $RuO_2(110)$ film of the thickness depending on exposure conditions. Its formation can be monitored with low-energy electron diffraction (LEED). In addition, the $RuO_2(110)$ surface shows a characteristic O_2 desorption peak at around 400 K in the temperature programmed desorption (TPD) spectrum (dashed line in Fig. 7.1a), which is associated with O atoms sitting on top of otherwise under-coordinated Ru surface atoms. The most prominent signal centered at 1000 K showing zero-order desorption kinetics is assigned to oxide decomposition. It is fair to say that O ad-atoms, "sub-surface" oxygen species, surface oxide layers, and even "buried" oxides might all be formed upon oxygen interaction with Ru(0001). The structural diversity explains, to some extent, a complex chemistry taking place on the Ru surfaces.

Let us now look at oxygen interaction with a bilayer silicate film grown on Ru(0001) at the conditions which, in the film absence, would result in $RuO_2(110)$ (in this case, exposure to 10^{-5} mbar of O_2 at 700 K for 45 min). Again, TPD revealed two O_2 desorption

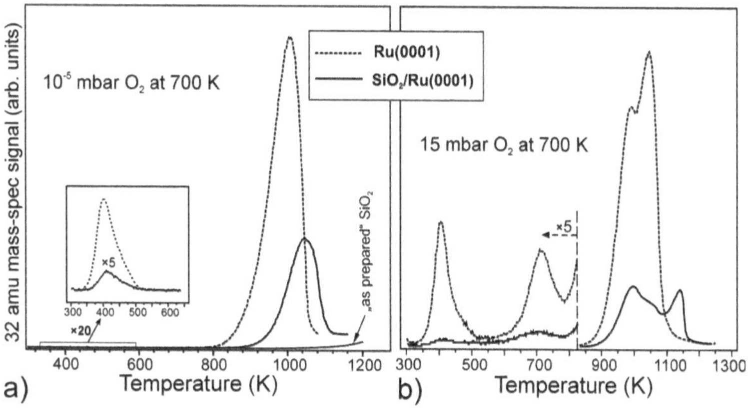

Figure 7.1 O_2 (32 amu) desorption signals in TPD spectra of Ru(0001) and SiO_2/Ru(0001) after exposure to 10^{-5} mbar O_2 at 700 K for 45 min (a) and to 15 mbar O_2 at 700 K for 45 min (b). Panel (a) also shows the spectrum for the "as prepared" SiO_2 film before O_2 exposure. Note that *y*-axis in panel (b) is reduced by a factor of 4. Adapted with permission from ref. 3.

signals, at 400 and 1045 K (Fig. 7.1a), which resemble those of the silica-free Ru(0001) sample. Since the "as prepared" film does not show oxygen desorption in this temperature range (as shown in the same panel for comparison), it is clear that both desorption peaks come from a Ru oxide layer formed underneath the silicate film. Based on the integral intensity of the high-temperature peak, which obviously scales with the Ru oxide thickness, the silica film definitely inhibits metal oxidation. Interesting is, however, that this peak is shifted towards higher temperatures by about 40 K, whereas the low-temperature peak stays non-shifted. In addition, the signal at 400 K is much smaller, by a factor of 25, as compared to the case of bare Ru, and it falls in the range of the sample surface area that is uncovered by silica and measured by CO titration. It is, therefore, most likely that these bare Ru regions react with O_2 and transform into the RuO_2(110) phase as on pristine Ru(0001). Meanwhile, the Ru-oxide phase formed underneath the silicate film is poorly defined and, based on LEED, amorphous. However, the silicate film keeps showing vibrational characteristics of a bilayer, i.e., two sharp bands at 1300 and 692 cm^{-1} in infrared reflection absorption spectroscopy (IRAS) spectra, indicating that the bilayer silicate film itself remains intact after such treatment. It seems plausible that the Ru-oxide layer preferentially forms in the proximity to the largest pores in the silicate film which allow oxygen to penetrate and react with the metal surface.

More recent in situ x-ray photoelectron spectroscopy (XPS) measurements on similarly prepared films showed, however, no formation of the Ru-oxide on the films exposed to 10^{-4} mbar O_2 at 820 K.[4] The conclusion has been drawn based on that the O 1s signal associated with the O ad-atoms on Ru underneath the silicate only gained in intensity (by a factor of 2.5), but did not shift as expected if the Ru oxide would form. Apparently, the discrepancy with the TPD-based results originates from the differences in the film preparation as well as exposure conditions.

In principle, the passivation effect of silica depends on the oxygen pressure. Figure 7.1b compares TPD spectra of the two surfaces, Ru(0001) and SiO_2/Ru(0001), after exposure to 15 mbar, instead of 10^{-5} mbar used in the previous case. Not surprisingly, the total amount of O_2 desorbing from the silica-free sample is increased substantially (by factor of 4), indicating progressive oxidation of a Ru substrate at elevated pressures. However, the absolute intensity of a weakly bound state (at 400 K), which is a fingerprint of the RuO_2(110) surface, is substantially smaller than on the sample oxidized at low pressure. In addition, a new desorption feature appeared at 710 K. Apparently, several Ru-oxide phases and/or surfaces form upon oxidation at near atmospheric pressures. Basically, a similar desorption profile is observed on the silicate-covered sample. A much lower intensity suggests that the silicate film effectively passivates the Ru surface at high oxygen pressures as well, even though the silicate film is obviously "transparent" to oxygen. Moreover, the silicate bilayer film remains practically unchanged after high pressure treatments, as revealed by IRAS. Perhaps, this finding is not surprising since the silicate film is terminated by fully saturated oxygen layers on either side and should be stable in oxygen. Importantly, the silicate film remains decoupled from the support which is no more a metal, but oxide. Further experiments on the films prepared using $^{18}O_2$ isotope and then exposed to high pressures of $^{16}O_2$ revealed desorption exclusively of $^{16}O_2$ and no traces of $^{18}O_2$ or

$^{16}O^{18}O$.[3] Therefore, oxygen in the silicate film does not exchange with oxygen in the gas phase, and the Ru surface interacts solely with the ambient O_2 molecules penetrating the film. Direct comparison of the silica-covered and silica-free surfaces shows, however, that surface reactions on metals, in this case oxidation, may proceed very differently in confined space formed by a weakly bound silicate.

Even stronger passivation effect is observed for the aluminosilicate bilayer films. One possible explanation could be a smaller distance between the film and the Ru surface as compared to pure silicate that would, in turn, suppress oxygen intercalation. Another factor that might influence the reaction is that charge imbalance caused by Al^{3+} substitution of Si^{4+} in the framework is likely compensated via the charge transfer from the metal support. The additional electrostatic forces between the Al-silicate and the metal surface could also affect the reaction.

To conclude, bilayer silicate films is very stable in the oxidizing atmosphere and effectively passivate the metal surface. The results also indicate a considerable confinement effect on the metal oxidation.

7.1.2 Hydrogen

Since the silicate films are terminated by fully saturated siloxane bonds, hydrogen adsorption shows essentially no effect on surface structures. In particular, no hydroxyls are formed even by exposure to 10 mbar of hydrogen, which is in the "surface science" community considered as a "high pressure." In fact, all changes observed upon H_2 exposure only relate to adsorption on the metal substrate.

In order to study interaction of hydrogen with bilayer films in details, in particular to see whether hydrogen intercalates the interface like oxygen discussed above, one first needs to get rid of O ad-atoms, which are always present in some amounts in the "as prepared" films. In principle, they can desorb, at least partially, upon vacuum annealing that usually requires high temperatures depending on the metal support. In fact, a more efficient way to remove O from metal surfaces at mild conditions is to expose it to hydrogen or carbon monoxide (CO) which react with O to form water (respectively, CO_2), desorbing at much lower temperatures

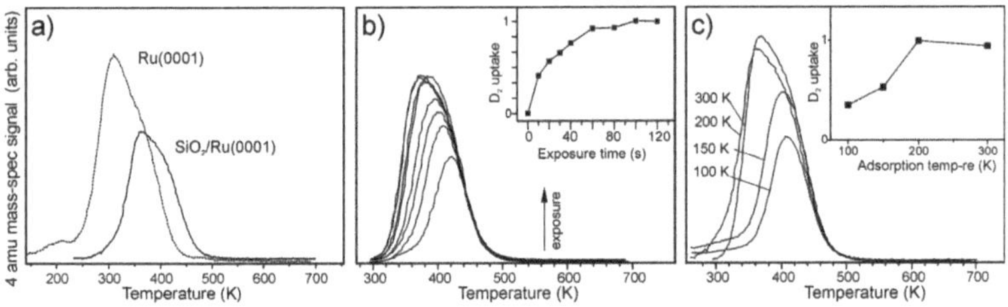

Figure 7.2 (a) TPD spectra of D_2 at saturation coverage measured on the bare Ru(0001) surface and on SiO_2/Ru(0001) film. (b,c) TPD spectra measured on SiO_2/Ru(0001) after exposure to 10^{-6} mbar of D_2 as a function of exposure time all at 300 K (b), and of adsorption temperature all at 2 min exposure time (c). Integral intensities (D_2 uptakes) are shown in the insets. Adapted with permission from ref. 5.

than O_2. Interestingly, exposure of bilayer films to hydrogen at the UHV-compatible pressures (typically, 10^{-6} mbar) showed no effect. Either H_2 molecules cannot penetrate the silicate and hence reach the metal surface at these pressures or, most likely, the O ad-atoms block H_2 dissociation necessary for its reaction with O. Only at elevated pressures (0.1–1 mbar) is it possible to get rid of O species from the interface and actually replace them by H ad-atoms, which, in turn, desorb as H_2 upon thermal flash leaving the metal surface clean.

For studying interaction of hydrogen with solids, it is common to use deuterium (D_2) which is better suited for TPD and IRAS measurements, but the conclusions hold true for hydrogen as well. Figure 7.2a compares two TPD spectra measured after prolonged exposure of bare Ru(0001) and SiO_2/Ru(0001) film surfaces to D_2 until saturation. Although the absolute intensities considerably differ, the desorption profiles are quite similar. In addition, desorption from the silica films looks like shifted to higher temperatures by ~30 K. In this respect, it is worth mentioning that O_2 desorption signal at around 1000 K coming from the decomposition of Ru-oxide underneath the film also showed a shift to higher temperatures (see Fig. 7.1a).

As expected, the amount of hydrogen intercalating the interface increases with exposure time as shown in Fig. 7.2b. The uptake at saturation is practically independent of the hydrogen pressure used, i.e., 1 mbar vs. 10^{-6} mbar. The adsorption kinetics follows the one observed on bare Ru(0001). Therefore, all findings point out that hydrogen can easily intercalate the interface as if the silicate layer is transparent for the H_2 molecules. However, adsorption experiments at low temperatures showed that the hydrogen uptake considerably, by a factor of 2, *decreases* (Fig. 7.2c), whereas on the clean Ru(0001) surface the uptake *increases*. Apparently, the permeation of $H_2(D_2)$ molecules through the silicate film depends on temperature. On the other hand, one cannot exclude the possibility that the $H_2(D_2)$ dissociation and diffusion of H(D) atoms on the Ru surface is affected by silicate layer above, thus manifesting the confinement effects.

The results shown above are obtained for the Ru-supported films. Similar experiments were also carried out for Pd-supported films (see Section 7.1.4 below).

7.1.3 Carbon Monoxide

It becomes clear now that one should take into account gas permeation through the film in order to rationalize the environmental effects on metal-supported silicate bilayer films. Since the siloxane terminated surface is inert, a bilayer silicate film behaves, in essence, as a membrane or a molecular sieve having a certain pore size distribution. The film reaction to the ambient gases is primarily determined by adsorption and reactions on the metal support which, in turn, may be affected by the silicate layer. In case of hydrogen, it is quite difficult to detect and differentiate H ad-species by conventional electron and vibrational spectroscopies. In this respect, CO is a molecule that is better suited for studying permeability of our films by making use of the metal surface as a CO "detector." Of course, the metal surface must first be cleaned from any adsorbates formed during the film preparation, either by UHV annealing or by reaction with hydrogen as discussed above.

To monitor CO adsorption kinetics, IRAS spectra were continuously recorded immediately after introducing CO into the vacuum chamber. Since the intensity of an IRAS band not always correlates with the surface species population for several reasons, it is recommended to complement the IRAS results with TPD measurements which allow to determine total amount of CO on the sample surface. The IRAS and TPD experiments were performed at different adsorption temperatures and pressures, as illustrated in Figs. 7.3 and 7.4 for Ru(0001)-supported films. The results on SiO_2/Pd(111) will be presented in the next section for comparison.

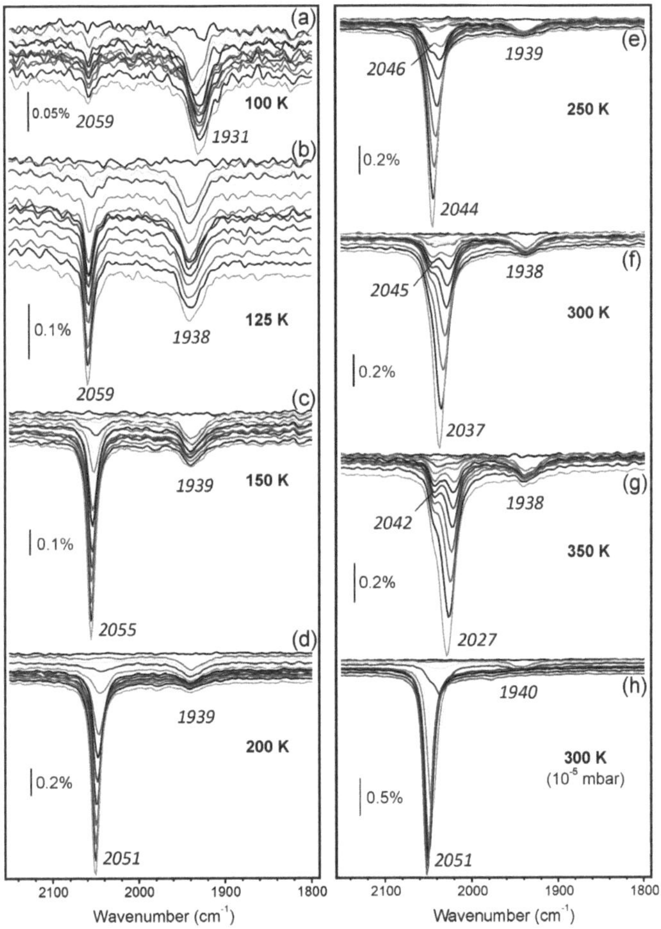

Figure 7.3 Consecutively recorded IRAS spectra (from top to bottom in every panel) on SiO_2/Ru(0001) exposed to 2×10^{-6} mbar CO (a–g) and 10^{-5} mbar CO (h) at the indicated temperatures. Each spectrum takes ca 12 sec. Spectra are offset for clarity. Reproduced with permission from ref. 5.

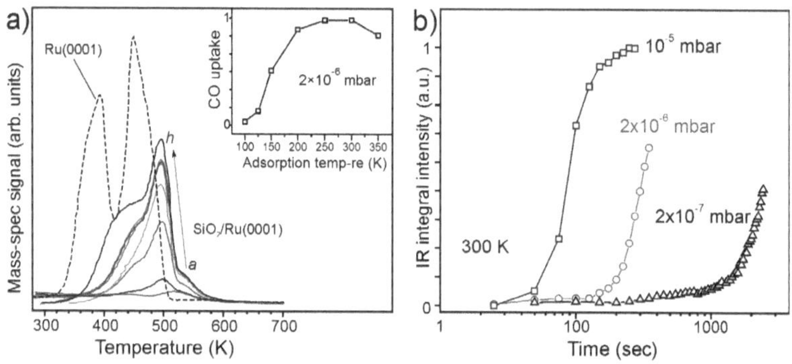

Figure 7.4 (a) TPD spectra of CO measured after taking the last IRAS spectra shown in Fig. 7.3(a–h). (The spectra are cut off at 300 K). The inset shows CO uptake as a function of adsorption temperature at constant pressure. CO TPD spectrum on the clean Ru(0001) surface at saturation is shown for comparison. (b) Kinetics of CO adsorption measured by IRAS at three different CO pressures as indicated. Note the logarithmic scale for the exposure time. Adapted with permission from ref. 5.

A weak band at around 1940 cm^{-1} appearing immediately at all temperatures most likely originates from CO adsorption on the silica uncovered spots and will be omitted in the following discussion. The spectra are dominated by the band(s) in the 2030–2060 cm^{-1} region depending on the exposure conditions. The band bears close similarity to CO adsorption on the clean Ru(0001) surface. In particular, it was established that the band position almost linearly correlates with CO coverage, i.e., the higher the coverage the higher the frequency.[6] At low temperatures, the signal at 2059 cm^{-1} is weak, although the frequency itself suggests the highest CO coverage. The band gains in intensity with increasing adsorption temperature, but shifts to the lower frequencies as if the CO coverage diminishes, which is at a variance to the TPD-measured CO uptake increasing (see inset in Fig. 7.4a). The discrepancy can be rationalized by that the band position reflects a *local* CO density rather than average CO coverage determined by TPD. In fact, the band only approaches the highest frequency at the highest CO coverage upon exposure to 10^{-5} mbar CO at 300 K as shown in Fig. 7.3h.

Basically, TPD spectra recorded after each IRAS experiment look similar to those obtained for bare Ru(0001) (Fig. 7.4a). However, desorption traces are shifted towards higher temperatures, by about 25 K if measured at the descending desorption edge. Without any knowledge of the system under study, one would assign such a shift to a stronger CO bonding. Recall, however, that such shifts were observed for O_2 and D_2 dissociative adsorption as well (see Figs. 7.1 and 7.2), and as such it becomes common for desorption processes from the intercalated silicate/metal interface. It appears more plausible that the observed shifts in TPD spectra originate from the time delay that is needed for a molecule to find the pore in the silicate film to escape into the vacuum.

Figure 7.4 shows that the CO uptake decreases at low temperatures and drops almost to zero at 100 K, in nice agreement with IRAS results. However, this is purely a kinetic effect. As depicted in Fig. 7.4b, CO adsorption exhibits the induction period. The time needed for the adsorption to set in inversely correlates with the pressure: The higher the pressure, the shorter the induction period. This finding can hardly be explained if CO penetrates the film uniformly through all N-membered silica rings. Apparently, only large size rings are suited for CO penetration.

CO adsorption on a metal-supported silicate bilayer film was further studied using density functional theory (DFT) calculations.[7] As the dispersion forces play a critical role in this system, various dispersion corrections at the DFT level have been examined for a crystalline film only. Figure 7.5(a,b) shows optimized geometries for CO adsorbing in different positions on Ru(0001) at the lowest coverage, i.e., 1/16 ML. As in the case of bare metal surface, CO prefers the on-top sites with the adsorption energy -2.3 eV, which is almost equal to that calculated for bare Ru(0001) (-2.26 eV) and much higher than for CO in the hcp site (-1.43 eV). Trapping CO in the silicate cage is thermodynamically unfavorable: The energy gain is small, only about 0.1 eV. At high coverages, the adsorption energy is considerably reduced (-1.84 eV), and the CO stretching frequency becomes comparable which that measured in the experiment. Note that, at this high coverage, the interface gap increases dramatically, from ~260 to ~590 pm, measured between the Ru and O layers for the CO-free and CO intercalated films, respectively. In addition, this

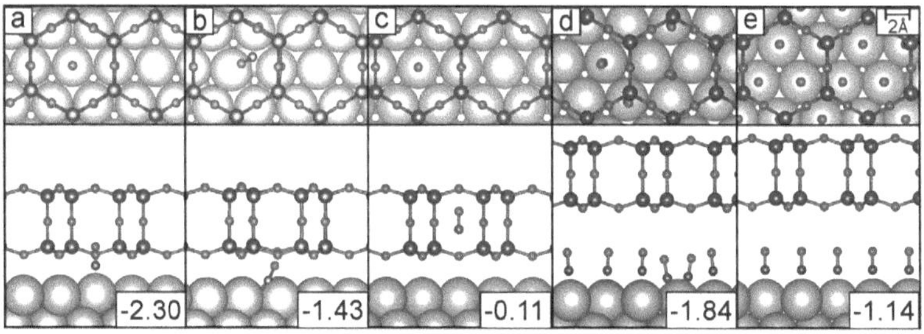

Figure 7.5 (a,b) Optimized geometry of CO at low coverage (1/16 ML) in top (a) and hcp (b) positions on Ru(0001). (c) CO inside the silica cage. (d,e) CO ad-layer at high coverages: 5/8 ML (d) and 1 ML (e). 1 ML (monolayer) corresponds to the surface density of Ru atoms. The calculated adsorption energies (in eV) are indicated. Reproduced with permission from ref. 7.

may explain the experimentally observed small ($\sim$5 cm^{-1}) blue-shift of a phonon band at 1300 cm^{-1} due to further decoupling of a film from a Ru substrate covered by CO. Finally, the authors estimated barrier for diffusion of CO through the unsupported crystalline silica bilayer ($\sim$0.5 eV) computed using a potential energy curve obtained from a constrained optimization, not a real barrier for diffusion.

7.1.4 SiO_2/Pd(111) vs. SiO_2/Ru(0001)

Adsorption experiments over bilayer silicate films prepared on Pd(111) revealed some similarities as well as differences if compared to the films grown on Ru(0001). Note that both CO and H_2 readily adsorb on Pd(111), and the adsorption mechanism is well documented in the literature that allows to identify adsorbed species by IRAS and/or TPD.

Figure 7.6a presents a series of IRAS spectra continuously recorded at room temperature in CO at pressures increasing from 1×10^{-7} to 5×10^{-6} mbar. At low exposures (*ca.* 10 Langmuir (*L*), 1 $L = 10^{-6}$ Torr$\times$sec), no CO adsorption is found, suggesting that the film fully covers the metal surface. At increasing CO dosage, a broad feature develops at 1930 cm^{-1}, which gains in intensity and shifts to 1935 cm^{-1}. In order to accelerate CO uptake, the pressure was increased to 10^{-6} mbar. At 200 L of CO, an additional band

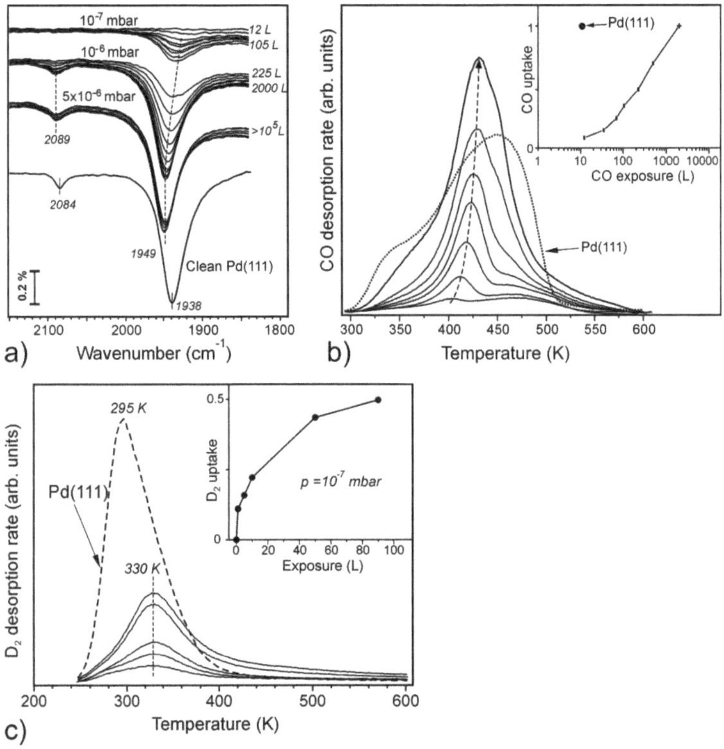

Figure 7.6 (a) Consecutive IRAS spectra measured at 300 K on a SiO_2/Pd(111) film at CO pressures as indicated. The spectra are offset for clarity. The total CO exposures (in L) are shown adjacent to the spectra. The spectrum for the clean Pd(111) surface exposed to 10 L is also shown for comparison. (b) TPD spectra at increasing CO dosage at 300 K. The spectrum for bare Pd(111) at 10 L is shown for comparison. Inset shows CO uptake as a function of CO exposure (in the logarithmic scale). (c) TPD spectra of D_2 on SiO_2/Pd(111) as a function of exposure at 250 K. The inset shows D_2 uptake normalized to the one on bare Pd(111) obtained after dosing 1 L onto clean Pd(111), also shown for comparison. Adapted with permission from ref. 8.

appears at 2089 cm^{-1}. The bands grow and saturate at very high exposures (10^4 L) reached by increasing CO pressure. At saturation, a strong band at 1949 cm^{-1} and a weak signal at 2089 cm^{-1} are observed. The peak positions and the intensities are almost identical to those (1938 and 2084 cm^{-1}) obtained on saturation of the clean Pd(111) surface even at the lowest CO exposure studied.

The bands are associated with CO adsorbed in bridge and top sites, respectively. Therefore, we can conclude that CO readily intercalates the interface between the silicate and the Pd surface. The conclusion is fully supported by TPD results shown in Fig. 7.6b. The CO uptake at saturation is almost equal to that measured on bare Pd(111). However, the process is rather slow at room temperature as it needs relatively high exposures, again suggesting that CO penetrates most likely through the largest silica pores. Similarly to Ru-supported films, the uptake vs. exposure curve exhibits the induction period on Pd(111) supported films as well.

TPD spectra of CO on Pd(111) are well understood.[9] On heating of the saturated CO adsorbed layer, CO molecules first desorb from the most weakly bound, on-top sites followed by desorption from the bridge sites (at $T < 400$ K). The remaining CO molecules diffuse and occupy the threefold hollow sites, from which they desorb as a peak at 450 K (see the spectrum in Fig. 7.6b). At initially very low coverage, CO only occupies the hollow sites, and the desorption peak shifts to lower temperatures at increasing coverage (up to 1/3 ML) due to repulsive CO–CO interaction.

Although CO desorbs from SiO_2/Pd(111) in the same temperature range, the desorption profiles showed certain difference to bare Pd(111) (Fig. 7.6b). In addition, the most prominent desorption peak at around 400 K shifts to *higher* temperatures at increasing CO dosage, whereas the desorption maximum on pure Pd(111) shifts to *lower* temperatures.[9] The observed peak shift may, again, be associated with the time needed for the CO molecules underneath the film to find an appropriate pore to escape.

As far as adsorption of hydrogen is concerned, Fig. 7.6c collects TPD spectra of D_2 at increasing exposure. In principle, hydrogen desorbs in the same temperature range as for the clean Pd(111) surface, but there are some differences between two systems. In contrast to Pd(111), D_2 desorbs from SiO_2/Pd(111) at high temperature (330 K) independently of the coverage, whereas the maximum peak on Pd(111) gradually shifts to the lower temperatures at increasing dosage.[10] The D_2 uptake increases non-linearly and approach saturation which is remarkably, by a factor of 2, smaller than on bare Pd(111).

On the Ru(0001)-supported films, the D_2 uptake correlates well with the CO uptake. However, on the Pd(111) supported films the D_2 uptake is substantially lower than the CO uptake. (In both cases, the comparative CO and D_2 measurements were performed using the same sample in order to minimize any instrumental effects.) It is tempting to link the difference between Pd and Ru to the well-known specific H_2/Pd interaction resulting in almost spontaneous migration of H(D) atoms into the Pd bulk and formation of Pd hydrides, which does not occur on Ru. Apparently, some D ad-atoms diffusing underneath the silicate film migrate into the Pd bulk on heating and hence become unavailable for desorption into the vacuum, thus reducing the total amount of D_2 measured by TPD. If so, then this finding provides one more example of confinement effects.

DFT calculations of CO adsorption on silica covered Pd(111) revealed rather small effects on adsorption energies.[8] The hollow sites on Pd(111) are most favorable for both systems, although the adsorption energy per CO molecule decreases with increasing coverage (Table 7.1), i.e., from −2.17 eV (0.25 ML), to −1.87 eV (0.5 ML), and to −1.59 eV (0.75 ML). The adsorption energies for CO underneath the silica follow the same trend and are −1.91, −1.73, and −1.51 eV, respectively. Accordingly, the differences in adsorption energies between clean and silica covered Pd(111) become smaller at high coverages. As in the case of Ru(0001) supported films, intercalation of CO causes the interface distance to remarkably increase, from 3 to 6 Å. Although performed solely on crystalline silica overlayer the calculations showed fairly good agreement with experimental results.

Table 7.1 Adsorption energy E_{ads} (in eV) per CO molecule adsorbed in threefold hollow sites on Pd(111) and SiO_2/Pd(111)

Coverage (ML)	1/4	1/2	3/4
Pd(111)	−2.17	−1.87	−1.59
SiO_2/Pd(111)	−1.91	−1.73	−1.51
ΔE_{ads}	0.26	0.14	0.08

Source: Taken from ref. 8.

7.1.5 General Mechanism

We may now speculate about what happens when a bilayer silicate film is exposed to "simple" gases like oxygen, hydrogen, and carbon monoxide. Obviously, the silicate layer itself is inert towards such gases. However, gas molecules penetrate through the film and react with the metal surface underneath. Let us first consider the case of CO for which the experimental and theoretical database is the most complete. We neglect macroscopic "holes" exposing bare metal surface in the discontinuous films since CO immediately adsorbs there and the related spectral features remain unchanged during the experiment.

We first assume that CO penetrates the silica film without any barrier. CO adsorbs on the metal surface sites right behind the pore and then can either desorb back through the same pore or diffuse across the metal surface underneath the silicate. CO diffusion even on the bare metal surfaces is not trivial issue. First, the diffusion coefficient is difficult to measure, and, second, it often depends not only on temperature, but also on coverage due to the lateral CO–CO interaction. Nevertheless, the estimated diffusion length for the time-scale of our IRAS experiments (>10 sec) is relatively large, about 10^3–10^4 nm at room temperature, but becomes comparable with the interatomic distances at cryogenic temperatures. Certainly, diffusion underneath the film is, at least, not as fast as on bare surface. Therefore, upon exposure at low temperatures, first adsorbed CO molecules hardly diffuse across the surface, and the next incoming CO molecules cannot penetrate through the same pore as the adsorption site is already occupied. Therefore, very small CO uptake measured at low temperatures probes solely the largest silica rings present either at line defects or in amorphous portions of the film (see Chapter 5). The low diffusivity of CO, in turn, results in high local concentration of adsorbed CO molecules near the pore they entered. Indeed, the 2059 cm^{-1} band observed at low temperatures on Ru-supported films (see Fig. 7.3) suggests CO coverage of about 0.5 ML, whereas the total coverage measured by TPD is an order of magnitude lower. In addition, this finding suggests that CO does not penetrate every silica ring, but only the largest ones. Based on the kinetic diameter of CO (~3.8 Å), there are must be relatively large silica rings for molecules to go through. Note that in the adsorption experiments, gas phase molecules possess the kinetic energy corresponding to the room temperature. Therefore, the low temperature effect on CO uptake, shown in Fig. 7.4a, cannot be explained in terms of ballistic flow through the pores. Most likely, the temperature effect reflects the activated process.

The induction period and the following-up self-acceleration of the CO adsorption demonstrated by Figs. 7.4b and 7.6b are intriguing. In principle, such a behavior is indicative of non-linear processes. For instance, CO surface diffusivity on metals considerably enhances at increasing coverage due to the CO–CO repulsive interaction. In our case, the increased local CO coverage right behind the pore may lead to the situation when CO diffusion rapidly increases above the critical coverage. Another explanation is based on the theoretical prediction, yet unproven experimentally, that intercalated CO at high coverages pushes the silicate layer away from the metal surface (Fig. 7.5). Accordingly, CO diffusion underneath the silica layer will be certainly enhanced once this locally happened. Such "gate lifting" model has to be further examined, for example, by scanning tunneling microscopy (STM), particularly at various temperatures.

Certainly, the adsorption kinetics is more involved in case of dissociating molecules like hydrogen and oxygen. Despite the limited database of experimental results, the H(D) atoms on the bare metal surfaces are theoretically predicted to diffuse very fast at our temperatures studied. Therefore, they can migrate far away from the place they penetrated through the film and dissociated, thus resulting in their fairly uniform distribution on the metal surface. On heating during the TPD run, the H ad-atoms desorb as H_2 via recombinative mechanism. As the desorption temperature and time are directly linked in TPD through the heating rate (β), $T = T_0 + \beta \times$time, any time delay immediately causes the desorption profile shifting to the higher temperatures. For the heating rate 3 K/s applied in the above-shown TPD experiments, the obtained shift of ~30 K (see Fig. 7.2a) corresponds to about 10 sec of delay for CO desorption from the silicate covered surface. In other words, ad-species spend about 10 sec, on average, to find the right pore

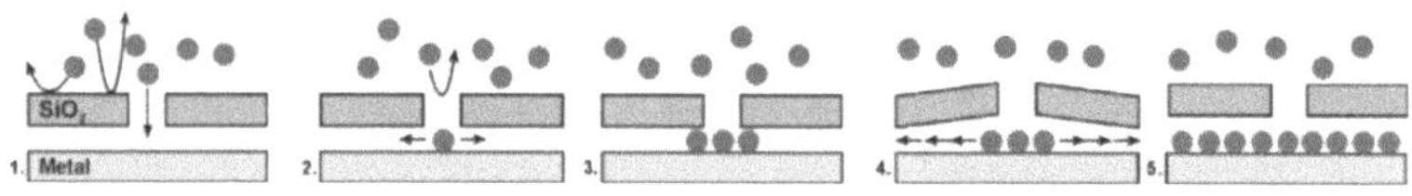

Figure 7.7 General scenario of gas adsorption on bilayer silicate films.

to escape into the vacuum, which is enormously long for the flat surfaces! Therefore, the H ad-atoms on the silicate-covered Pd(111) surface have enough time to migrate into the Pd bulk rather than to desorb into vacuum through the film, resulting in a lower hydrogen uptake measured by TPD. The hydride formation is well-known phenomenon for Pd, and not for Ru, that may explain the difference between the two systems.

A general scenario of what may happen to the metal-supported bilayer silicate films exposed to ambient gases is schematically shown in Fig. 7.7. For simplicity, it only considers non-dissociative adsorption on a metal substrate. (1) A molecule penetrates through the large pores present in the amorphous portions of the film and line defects most likely via an activated process. The crystalline layer is either impermeable or exhibits large activation barrier. (2) The molecule adsorbs on the metal surface right behind the pore. Next molecules only penetrate through the same pore if the ad-molecule makes the site free by hoping into adjacent site if available. At low temperatures, the process ends with (3). The metal surface around the pore exhibits a high local density of adsorbed species, whereas the total coverage is low. If, however, the diffusion across the surface becomes possible (e.g., at increasing temperature), then such area expands. Concomitantly, the silicate layer lifts up (4), thus making the surface diffusion facile and accelerating the adsorption until it reaches saturation (5). Apparently, a critical coverage and/or area size exists, which must be formed during the induction period, before the accelerated migration takes place through the "open gate."

7.2 Bilayer Film as the Thinnest Silica Membrane

In previous sections, we discussed primarily experimental results showing that ambient gas molecules migrate through a silicate film

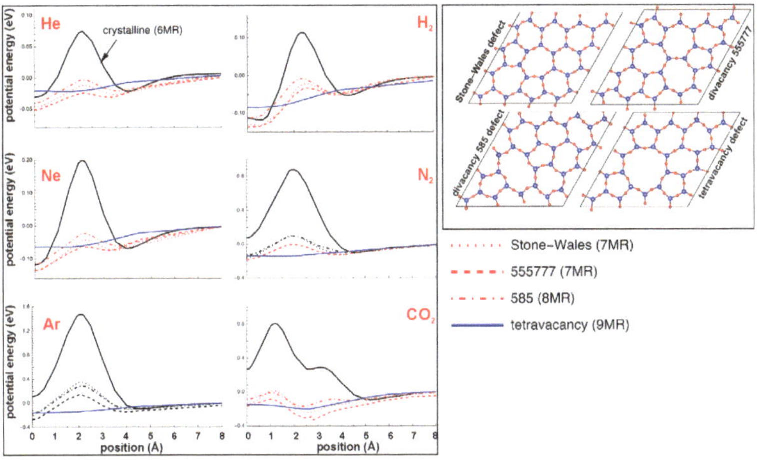

Figure 7.8 On the right: Top views of silica bilayer geometries used to mimic different N-membered rings. On the left: Potential energy profiles for different gases as a function of position, measured from the mirror plane in the bilayer to the center of the molecule. Only the unique portion of the symmetric potential energy is plotted. The curves are labeled for all plots. Adapted with permission from ref. 11.

like a membrane. In fact, bilayer silicate constitutes the thinnest possible stable structure for a silica membrane. If one includes the van der Waals radii of the surface oxygen atoms (1.52 Å), it is only 7.2 Å in thickness. Although its practical realization in the membrane technologies is hardly possible in the very near future, we address here some fundamental aspects related to the gas permeation through such nanoporous membranes. For this, we refer to the DFT calculations[11] for free-standing silica bilayers of different geometries comprising crystalline (i.e., of six-membered rings) as well as the seven-, eight-, and nine-membered rings. The models are depicted in Fig. 7.8 also showing the potential energy profiles computed at the DFT + D (dispersive corrections) level for gas passage through these various pore types for several molecules.

Basically, three different types of potential energy profiles are found, depending on the relative sizes of the gas and silica ring. Consider the example of He. It shows a minimum on the silica surface corresponding to a physisorbed molecule about 4 Å from the center, a maximum as it enters the silica ring (at ~2 Å), and another minimum (at 0 Å) exactly between the silicate layers, thus corresponding to a trapped molecule inside bilayer. This Type I is observed when the size of the gas molecule is larger than or comparable to the size of the pore. For He and Ne, this occurs only for passage through the crystalline layer. For the larger Ar atom, this type of repulsive barrier also occurs for all of the seven- and eight-membered rings.

Like the Type I, the Type II consists of two local minima (one on the exterior surface and one inside the bilayer) and a local maximum. The difference is that the local maximum has an attractive (negative) potential energy as compared to the non-interacting gas–silica system (see, for example, dashed lines for He and Ne). Thus, the Type II has no barrier for ballistic gas transport (because the entering gas molecule will have a positive kinetic energy), but does have an effective barrier for gases attempting to enter the pore after being physisorbed on the surface. This occurs when the pore is slightly larger than the gas molecule, e.g., for He and Ne with all of the seven- and eight-membered ring defects, but never for Ar or larger molecules. Both Type I and Type II profiles have stable states for the gas molecule within the bilayer. For He and Ne, this interior trapped states are thermodynamically more stable than the physisorbed surface state. For Ar transport through the crystalline six-membered ring, this interior state may be kinetically trapped by the large potential energy barrier for exit, despite being significantly higher in energy than the free gas. This interior state is increasingly stabilized for the various seven- and eight-membered rings, due to the surrounding attractive dispersion interactions.

Finally, Type III profiles have no barrier, and consist of approximately only a single local energy minimum within the bilayer. This type occurs when the gas molecule is significantly smaller than the pore. Nine-membered ring structures arising from the tetravacancy defect have no energy barrier for any of the gases considered here, and may therefore serve as the most possible candidate for the pores responsible for gas penetration in the metal-supported films discussed in the previous sections.

The energy profiles of H_2 are qualitatively similar to that of the comparably sized He atom. The N_2 profiles are qualitatively similar to that of Ar. The larger triatomic CO_2 molecule has additional features. First, CO_2 has a lower potential energy barrier for entry than N_2 for all of the defects. If one considers both molecules as cylinders, CO_2 has a smaller effective diameter than N_2, due to the electron density distribution. The lower potential energy barrier for entry of CO_2 is especially pronounced for the seven- and eight-membered rings. Second, the potential energy surface has an additional local minimum and local maximum, as the C and O atoms in the CO_2 molecule interact with distinct layers of the silica structure. Finally, for the nine-membered ring, the minima occur at 2.5 Å, and are separated by 0.05 eV at the geometric center of the bilayer. This is qualitatively similar to the Type III for N_2, but with a larger central barrier.

The DFT results provide a basis for calculating relative permeances, which in turn allow to estimate the selectivity of the different pore structures for the different molecules in gas mixtures. In general, the transmission process consists of several intermediates and transition states, which can be used to compute the barrier crossing rates by applying classical transition state theory. However, in a crude approximation, these multistep processes can be described in terms of the energetic span, δE, which is the largest energy difference between an intermediate (local minimum) and any subsequent transition state (local maximum) during the barrier crossing process. The energetic span serves as apparent activation energy of the process, and thus a smaller energetic span leads to a faster reaction. Table 7.2 shows the computed energetic span values for the gases transiting through the various silica bilayer.

The selectivity for two gases, X and Y, is the ratio of the transport rates, r_X and r_Y, respectively, each of which in turn can be expressed in terms of the classical Arrhenius rate equation $S_{X/Y} = \frac{r_X}{r_Y} = \frac{A_X e^{-\delta E_X/k_B T}}{A_Y e^{-\delta E_Y/k_B T}}$. For simplicity, one can assume pre-exponential factor to be equal for both molecules. In (more realistic) case of amorphous silicate film having a certain ring size distribution, the selectivity can be obtained by weighting the selectivities by the transmission of each ring type by the probabilities (p_i) at

Table 7.2 Energetic span, δE (in eV), for several gases transiting through bilayer silica

Structure	He	Ne	H_2	CO_2	Ar	N_2	CO	Kr
Crystalline (6)	0.104	0.316	0.232	0.900	1.576	0.977	0.962	2.554
Stone–Thrower-Wales	0.046	0.136	0.132	0.254	0.520	0.259	0.259	0.882
555777	0.051	0.136	0.137	0.330	0.406	0.184	0.211	0.710
585	0.039	0.114	0.113	0.170	0.444	0.224	0.205	0.572
Tetravacancy	0.022	0.065	0.085	0.202	0.160	0.143	0.142	0.245

Note: The corresponding structures are shown in Fig. 7.8 (from ref. 11).

which each ring type is observed: $S_{X/Y} = \frac{\sum_i p_i e^{-\delta E_{X,i}/k_B T}}{\sum_j p_j e^{-\delta E_{Y,j}/k_B T}}$. The calculations using the energy span (δE) values, presented in Table 7.2, and experimentally observed ring size distribution in Chapter 5 showed that the crystalline film has sufficiently high permeances and selectivities, which allows to consider it even for practical applications in industrially important H_2/CO and He/CH_4 separations. Not surprisingly, the amorphous silicate increases the permeabilities of all gases due to the smaller potential energy barriers for the larger rings. As a result, transport through the larger rings dominates, but the selectivity falls off.

It is interesting that DFT results suggest a non-zero probability for small molecules to be trapped inside the silica bilayer, either thermodynamically, i.e., showing the lowest energy in potential energy profiles, or kinetically, due to the large barrier to exit. This would allow to use certain inert gases as selective probes for particular ring size to measure the ring size distribution in entire films, which was thus far investigated only locally using microscopy techniques like TEM and STM. Moreover, such systems would be interesting for atomistic studies of gas absorption on nanoporous materials like zeolites. To date, our own attempts to experimentally detect trapped molecules by TPD were unsuccessful. However, a recent study[12] demonstrated that Ar might, indeed, be trapped in bilayer films during XPS measurements in Ar atmosphere. Tentatively, the effect has been assigned to the ionization of Ar atoms by X-rays.

7.3 Chemical Reactions under Bilayer Silica Films

The results presented in previous sections clearly demonstrate that environmental effects on structure and properties of metal-supported bilayer silicate films primarily involve adsorption and reaction on a metal substrate underneath. The system can be viewed as a nanoporous silica membrane coating the metal surface, on which chemical reactions take place in space limited in one direction to a few Angstroms. In addition to that the silicate film controls both the access of reactants and the removal of products and hence the reaction equilibrium according to Chatelier's principle, the spatial confinement can also alter the reaction kinetics through the modification of the transition state and/or by changing the reaction pathway. Therefore, surface chemistry in these kind of materials becomes more complicated. On the other hand, it may result in new functional properties anticipated for such hybrid systems.

To illustrate the complexity of processes occurring on silicate covered metal surfaces, we address a reaction of oxygen removal from the SiO_2/Ru(0001) interface by hydrogen. In the experiments described below, the "O-rich" film containing O ad-atoms at the interface was exposed to 10^{-6} mbar of H_2 at elevated temperatures (540–650 K). Based on XPS results, all O atoms are ultimately removed from the interface upon this treatment. Obviously, hydrogen diffuses to the interface, dissociates there to produce atomic H which react with O to form water that desorbs. Since water was not detected (neither by XPS nor by IRAS) during the reaction, it probably desorbs through the silicate layer immediately upon formation.

The process was visualized in situ by low-energy electron microscopy (LEEM).[13] LEEM movies during sample heating monitor the reaction via a change in image intensity (local darkening). Figure 7.9 displays a series of snapshots showing the reaction propagating from the bottom right to the upper left. Combined LEEM and XPS results of bright and dark regions identify the support underneath as "O-rich" and clean Ru(0001) surfaces, respectively. Although the reaction front showed no preferential directions, step bunches (highlighted by arrows in the images) seem to hamper its propagation.

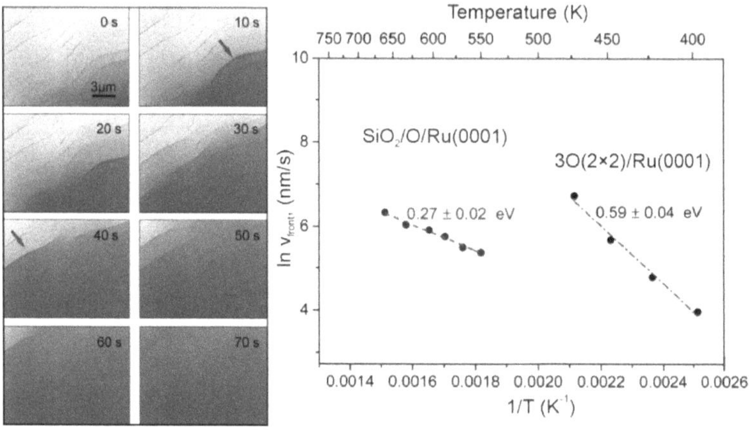

Figure 7.9 On the left: Snapshots from the LEEM movie of the reaction propagating across the "O-rich" SiO_2/Ru(0001) surface in 10^{-6} mbar H_2 at 550 K. The arrows highlights step bunches decelerating the reaction front. On the right: Arrhenius plots of the front velocities on "O-rich" SiO_2/Ru(0001) and 3O(2×2)/Ru(0001). Adapted with permission from ref. 13.

The same set of experiments performed on the silica-free 3O(2×2)-Ru(0001) surface also revealed the reaction front propagation which sets in at a much lower temperatures (400–500 K) and therefore proceeds too fast to be precisely measured at the temperatures used for the silica samples. This is not surprising, since hydrogen can freely access the Ru surface and also water desorbs easier. The temperature dependence of the front velocity showed the Arrhenius behavior, from which the apparent activation energy (E_{act}) was determined, i.e., 0.59 ± 0.04 eV, in good agreement with previously reported values determined from TPD measurements. The H_2 + O/Ru(0001) reaction is commonly described in several elementary steps:

$$H_{2(gas)} \rightleftharpoons 2H_{(ads)} \tag{7.1}$$

$$H_{(ads)} + O_{(ads)} \rightleftharpoons OH_{(ads)} \tag{7.2}$$

$$OH_{(ads)} + H_{(ads)} \rightleftharpoons H_2O_{(ads)} \tag{7.3}$$

$$2OH_{(ads)} \rightleftharpoons H_2O_{(ads)} + O_{(ads)} \tag{7.4}$$

$$H_2O_{(ads)} \rightleftharpoons H_2O_{(gas)} \qquad (7.5)$$

As the first step includes H_2 dissociation, the reaction can only proceed within the "dilute" O ad-layer having empty sites for dissociate. Apparently, the reaction starts at some structural defects or where local fluctuations allow H_2 dissociation to occur. The associative reaction (7.2) is believed to be the rate determining step.

The temperature dependence on reaction over silicate covered surface showed the Arrhenius behavior with $E_{act} = 0.27 \pm 0.02$ eV, which is almost two times smaller than on the bare Ru surface. In addition, the pre-exponential factor is substantially smaller, by a factor of 10^4. The origin of the difference between the activation energies remains not fully understood. As discussed in previous sections, H_2 readily penetrates through defects in bilayer film, thus leaving the step (7.1) unaltered or only slightly affected by the silica layer. In contrast, water desorption (7.5) that occurs spontaneously on uncovered Ru is most likely hampered by silica overlayer. Theoretical considerations of water diffusion in silica, performed on small silica cluster models, suggested about 1.5 eV for the activation barrier for the six-membered ring, but decreases down to 0.3 eV for the seven-membered cage.[14] Since water formed by reactions (7.3) and (7.4) has to be removed for the reaction front to propagate further, the reaction (7.5) probably becomes the rate limiting step on the silica-covered surface. Thus, the prerequisite of highest E_{act} for the rate determining step can be overruled by a lower pre-exponential factor for the process, here interpreted as the probability for water to find the right pore in the silica film to desorb. On the other hand, the accumulation of water due to its slow removal results in enhanced probability of reaction (7.4) providing an additional reaction pathway for OH formation. It is, therefore, tempting to assign the difference between two systems to confinement of water molecules leading to enhancement of self-dissociation reaction.

Certainly, further elaborative studies have to be done in order to understand chemical reactions at the bilayer silicate/metal interface, which may open new perspectives for rational design of catalytic hybrid materials combining molecular sieve and confinement effects.

7.4 Metal Deposition

In the next parts of this chapter, we consider adsorption of metals on ultrathin silica films. It is fair to say that the preparation of thin silica films on metals in early days was to a larger extent motivated by making them as suitable oxide supports for metal nanoparticles to model highly dispersed metal catalysts. It is also worth mentioning that first experiments on metal deposition onto ultrathin silica films were carried out even before their atomic structure has been established. Although metal deposits on ultrathin oxide films often showed adsorption characteristics identical to metal nanoparticles supported on oxide single crystals, it was realized that individual metal atoms may migrate into the ultrathin films. There is a growing body of experimental results showing that this phenomenon is related to the thickness of an oxide film rather than its nature. It is needless to say that metal deposition onto ultrathin silica may become a critical issue while considering possible applications of such films in nanoelectronics.

7.4.1 Metal Adsorption on a Monolayer Silicate

The initial stages of metal deposition on single layer silicates have been studied for Pd, Ag and Au on $SiO_{2.5}$/Mo(112) films. In order to minimize diffusion processes, the deposition and STM imaging were carried out at low temperatures. Figure 7.10 displays typical STM images of Pd and Au deposited and recorded at 10 K. The Au atoms form small clusters decorating antiphase domain boundaries in the film. In contrast, Pd-associated species are found randomly distributed, and they are imaged as weakly protruding hexagons on the atomically flat honeycomb-like surface. The absence of large single "blobs" suggests that the Pd atoms most likely reside below the surface, implying their migration into the silica layer even at cryogenic temperatures.

The ability of metal atoms to penetrate the silicate layer was examined by DFT by Pacchioni and co-workers.[16,17] It has turned out that a Pd atom easily, in essence barrierless, goes through the six-membered silica ring and strongly binds to the Mo atoms as depicted in Fig. 7.10. A driving force is obviously a much higher adsorption

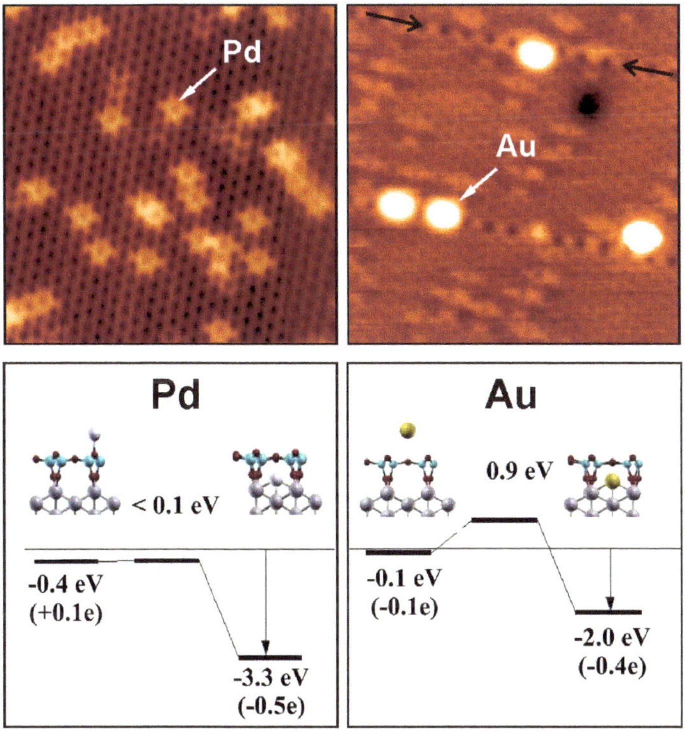

Figure 7.10 Low-temperature STM images of Pd (on the left) and Au (on the right) atoms deposited onto $SiO_{2.5}$/Mo(112) film at 10 K. Black arrows mark antiphase domain boundary. Adapted with permission from ref. 15. Below the images are the DFT-derived energies for Pd and Au atoms on and inside the silicate film with respect to a gas-phase atom. The associated energy barrier is also given. The values in parenthesis show the charge transfer from corresponding atom. Mo, dark gray; Si, cyan; O, red spheres; Pd, light gray; Au, yellow. Adapted with permission from ref. 16.

energy on metal than on silica, i.e., 3.3 and 0.4 eV, respectively. The potential energy profiles are found similar for both the six- and eight-membered rings, thus explaining uniform distribution of the Pd atoms on the films as observed by STM. Basically, every Pd atom that sticks to the surface becomes trapped in the silica ring. For comparison, an Au atom exhibits a considerably high activation barrier (0.9 eV) for a perfect silicate layer, despite a large energy gain (2 eV), albeit smaller than for Pd (3.3 eV). The barrier is, however, substantially reduced for the eight-membered ring. Therefore, at low temperatures, the Au atoms only penetrate large rings available on antiphase domain boundaries consisting of alternating four- and eight-membered rings. Interestingly, further calculations revealed the eight-membered rings large enough to accommodate more Au atoms, thus forming dimer and trimer. To distinguish those by STM is not a trivial task, and required imaging at different tunneling conditions in combination with DFT-based image simulations.[17]

Intuitively, one may think of the observed "molecular sieve" effect is related to the metal atom size, that is, the smaller the atom, the more easily it penetrates the silica ring. Indeed, van der Waals radius of Pd (163 pm) is smaller than of Ag (172 pm). However, the calculated barrier for Au penetration (0.9 eV) is substantially higher than for Ag (0.3 eV), although its radius (165 pm) is smaller (because of the relativistic contraction of Au orbitals) and comparable with that of Pd. Therefore, the penetration probability does not seem to correlate directly with the effective size of the atom, but depends on its electronic structure. The calculations showed that the half-filled Ag 5s and Au 6s orbitals during penetration experience a substantial repulsion with the surface O 2p orbitals, thus leading to a higher barrier. In contrast, the unfilled Pd 5s orbital produces only a small repulsion. The much lower barrier for Ag in comparison with Au is attributed to a positive charging of the Ag atom in the transition state while passing the ring plane. The Ag atom transfers some charge into the Mo states, thereby reducing the electron–electron interaction with the oxide states. Being strongly electro-negative, the Au atom remains neutral when passing the silicate ring, and the repulsive interaction with the oxide surface electrons cannot be weakened. The Pd atom remains neutral during penetration, but becomes partly negatively charged in its final binding configuration.

The above scenario holds true solely for the "O-poor" films. The energetics considerably alters in the presence of O ad-atoms on the Mo support in the "O-rich" films as they occupy the sites the metal atom would otherwise adsorb on. In this case, the Pd adsorption energy decreases substantially, from 3.3 eV on the "O-poor" film to 0.7 eV on the "O-rich" film, although the activation barrier is almost unaffected. Therefore, to minimize or even suppress metal migration one should expose the films to oxygen prior to the metal deposition.

Once a metal atom has been migrated to the interface, it may act as an anchor for the next incoming atoms to form metal clusters. The deposition temperature and flux often influence the nucleation-and-growth process. At relatively high amounts of metal deposited, hemi-spherical nanoparticles form, as shown in Fig. 7.11a for the case of Pd deposited at room temperature. Adsorption of CO as a probe molecule for Pd revealed TPD and IRAS fingerprints of the clean Pd surface. Therefore, relatively large Pd aggregates formed on monolayer silica films are similar to those obtained on "thick" silica films and crystals, unless the sample is not heated to elevated temperatures inducing Pd migration into the film.

Interestingly, the particle size distribution considerably alters if Pd deposition is performed in the CO atmosphere. STM image in Fig. 7.11b shows that Pd particles deposited at 90 K and exposed

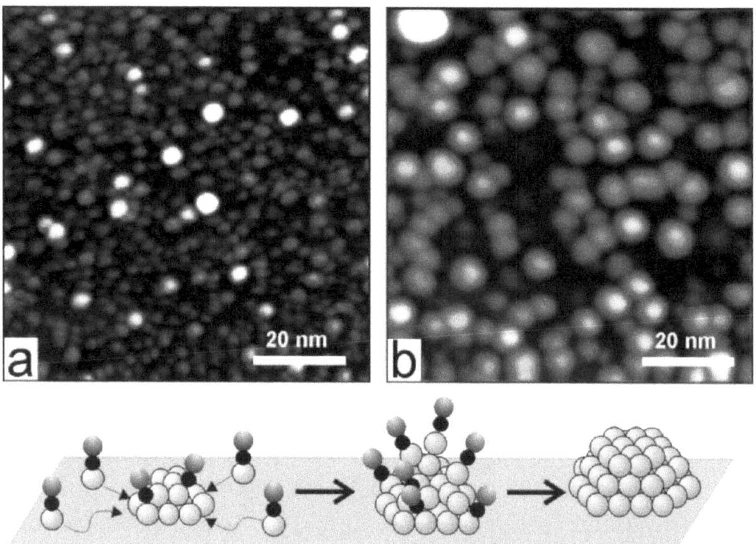

Figure 7.11 STM images of Pd particles on $SiO_{2.5}$/Mo(112). For image (a), Pd was deposited at 300 K; (b) The same amount of Pd was deposited at 90 K following adsorption of 20 L CO and heating to 300 K. CO-induced growth of Pd particles: CO adsorbs on Pd ad-atoms forming Pd carbonyls which diffuse and stick together forming a large agglomerate which decomposes upon heating and transforms into a metallic Pd particle. Adapted with permission from ref. 18.

immediately to CO and then heated to 300 K in UHV are substantially larger than the particles directly formed in UHV at 300 K (Fig. 7.11a). The effect can reasonably be rationalized by that CO molecules strongly adsorb on the Pd ad-atoms and small clusters formed at low temperatures as well as on the Pd atoms migrated into the film. Based on DFT predictions, the carbonyl PdCO complexes can even escape the silica cage, diffuse on the surface, and form aggregates. Monitored by TPD, CO desorbs from Pd carbonyl aggregates at about 200 K, resulting in larger Pd nanoparticles with CO still residing in multi-coordinated sites on the particle facets. Those species desorb on heating to 500 K.

Finally, the interaction of alkali metals with monolayer silicate films on Mo(112) has been studied using Li as an example.[19] As in case of Pd, the Li atoms readily penetrate the silicate layer and bind to the metal surface. Moreover, an ordered superstructure was found at increasing coverage seen as alternated stripes with high and low density of Li atoms along the particular direction on Mo(112). On the basis of the DFT calculations, the effect was explained in terms of the anisotropic screening response of the Mo(112) surface to Li adsorption. The Li atoms become positively charged and substantially lower the work function, which was measured by conductance spectroscopy.

7.4.2 Metal Adsorption on Bilayer Silica Films

In the same manner, the initial stages of metal adsorption were investigated for bilayer silicate films. It was reasonably expected that deposition at cryogenic temperatures (5 K) will only result in "monomeric" ad-species. Indeed, STM inspection revealed single atoms, however, not as individual protrusions, but as spots of a butterfly (crescent) shape (see Fig. 7.12). Again, there are some differences how Pd and Au deposits look in STM images. The Pd-related protrusions are more or less randomly distributed across the surface, whereas Au-related species are only found within the amorphous portions of the film and/or at domain boundaries, which obviously possess larger silica rings than in the crystalline film.

As in case of Pd on monolayer films, solely STM appearance of Pd species favors its assignment to an atom located underneath

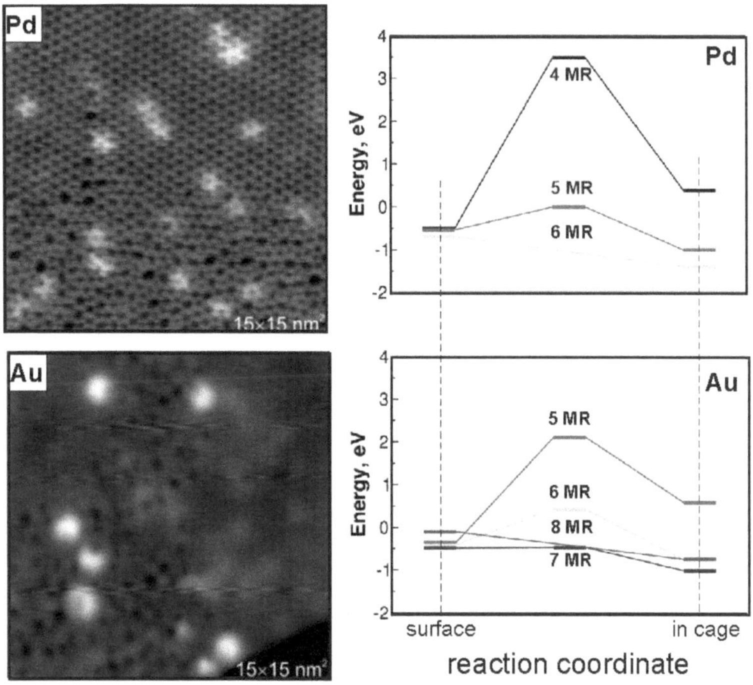

Figure 7.12 On the left: STM images of Pd and Au atoms deposited onto a bilayer silica film at 5 K. On the right: Energy profiles for Pd and Au penetrating the bilayer silica rings of different size (n = 4–8) computed for unsupported cluster models. Energies are calculated with respect to the metal atom in the gas phase. Adapted with permission from ref. 20.

or inside the film rather than on top. Indeed, DFT calculations predict the Pd atom to penetrate the hexagons without any barrier, thus making no significant difference for Pd adsorbed either on crystalline or amorphous portions. In contrast, the Au atoms only diffuse through the rings larger than heptagons existing in amorphous films and domain boundaries between the crystalline domains.

To gain more insight into metal interaction with silicate films, in particular whether metal atoms do migrate to the metal support, XPS measurements were carried out.[20] The idea behind is quite simple. As previously shown, adsorption of gas molecules on the metal surface underneath the film is accompanied by simultaneous

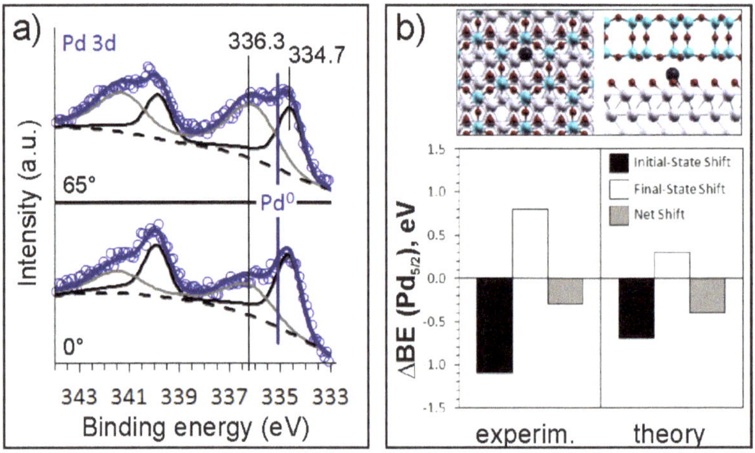

Figure 7.13 (a) Pd 3d region in XP-spectra measured for 0.05 ML Pd at two emission angles (65° and 0°) relative to the surface normal. Pd^0 marks the position of the $3d_{5/2}$ line for the Pd(111) crystal surface. (b) Experimentally detected change to the 3d BE, relative to bulk, (in gray) is partitioned into two contributions attributed, respectively, to changes in the initial- (black) and final-state (white) configurations of the photoemission process. Inset shows top and side views of a Pd atom bound at the interface between the bilayer film and the O-covered Ru(0001) substrate. Adapted with permission from refs. 20, 21.

shifts of the Si 2p and O 1s core levels due to the band re-alignment caused by the work function changes (see Chapter 4). Therefore, one anticipates such a shift also in case of metal atoms migrating to the support. Indeed, the measurements observed both levels shifted to higher binding energies (BE), by 0.3 and 1.2 eV after deposition of 0.05 and 0.24 ML of Pd, respectively. These shifts are actually equal to the work function changes (−0.3 and −1.2 eV, respectively) directly measured in the same experiments. However, the Pd 3d level showed a broad signal which was deconvoluted into two Pd states having BE higher and lower than that of measured on a Pd(111) single crystal, marked as Pd^0 in Fig. 7.13a. Relative intensity of the high BE component (at 336.3 eV) considerably increases at grazing emission, pointing to its surface origin. It is also qualitatively consistent with BE shifts typically observed for metal nanoparticles supported on thick oxide films and single crystals and attributed to the “final state” effect caused by lack of screening of an electron hole in the photoemission process (see Chapter 2). Taking into account the relatively high deposition temperature (150 K) in these experiments that favors metal penetration through the film, the low BE signal in spectra (at 334.7 eV) can accordingly be assigned to the Pd atoms at the interface.

Of course, evaluation of XPS shifts associated with supported metal clusters or atoms on ultrathin oxide films is not straightforward because of a number of initial and final state effects such as charge transfer, size-dependent screening, electrostatic charging, lattice contraction which may even cause shifts in opposite directions and hence cancel each other. As introduced in Chapter 2, some of these contributions can be disentangled using the Auger parameter, which consists of the difference between the kinetic energies (KE) of XPS and Auger peaks from the same atom in a given material. In fact, more informative are the changes in Auger parameter when comparing the spectra for a given atom in different environment, in particular for the so-called modified Auger parameter ($\Delta\beta$), where the Auger measurement is limited to core-core-core lines with final states consisting of two holes in the same orbital probed by the XPS measurement. In our case of Pd, one should measure Pd 3d photoelectron lines and the $L_3M_{45}M_{45}$ Auger transition lines. Accordingly, analysis of the silica states includes $K_1L_{23}L_{23}$ Auger transition to complement the Si 2p photoelectron line. As the energy needed to excite these Auger transitions (1600–3200 eV, depending on the element) exceeds that of conventional x-ray sources, the inherent Bremsstrahlung radiation was used to probe the specified Auger peaks.

Applying Auger parameter by comparing the change in the KE of the Si $K_1L_{23}L_{23}$ Auger peaks to determine energy contributions for the final-state relaxation showed that the latter is negligible. In fact, the O 1s and Si 2p BE shifts solely follow the work function. Any changes to the system that result in a decreased (increased) work-function will always decrease (increase) the energy of the band structure within the ultrathin silica film, i.e. in the same manner as was suggested for the “O-rich” versus “O-poor” film systems. The analysis applied to the Pd adsorption, summarized in Fig. 7.13b, showed that the negative shift of BE by 0.3 eV, relative to the Pd 3d peak position in the bulk, should be attributed to large variations in the initial-state orbital energies of the single atoms (1.1 eV towards the Fermi level), which is large enough to overcome decreased final-state relaxation contributions (0.8 eV), ultimately yielding the net-negative 0.3 eV shift measured by XPS. The conclusion was nicely supported by theoretical calculations, also presented in Fig. 7.13b.

Theoretical considerations were further expanded to the cases of adsorptions of Li, Na, K, and Mg, not yet experimentally studied. Basically, the idea was to check whether alkali and alkaline-earth metal atoms and ions can be stabilized in the small cages provided by the bilayer Al-silicate representing a two-dimensional model of zeolites. To reduce computational costs, the unsupported cluster models were only employed, all shown in Fig. 7.14. According to the calculations, no charge transfer takes place between silica and adsorbed atom which remains metallic. The interaction is dominated by dispersive forces. As the ring size increases, the stability of the atoms inside the cages enhances, but is still unfavorable for all the ring sizes studied. Interestingly, the adsorption energy does not follow any regular trend like in the Periodic Table, e.g., going from Li, to Na and to K. In fact, for the hexagonal ring, it is largest for K, and smallest for Na, with Li in an intermediate position (−0.53, −0.37, and −0.45 eV, respectively). This is the result of two different factors affecting the bonding. On the one hand, there is the atomic polarizability that is largest for heavier atoms (K > Na > Li). On the other hand, the atomic dimension favors a stronger interaction for the lighter atoms (Li > Na > K).

The interaction becomes stronger when one Si atom in the cluster model (indicated by the arrow in top panel of Fig. 7.14) is replaced by Al to mimic zeolitic frames. This has an enormous effect on the adsorption energies of Na atom, which increases by an order of magnitude, i.e., from 0.3 to 4 eV. As a result, Na is positively charged (Na^+), and electrostatic interaction starts to dominate the adsorption process, albeit it still depends on the ring size. For large rings, the Na ion ultimately occupies sites in vicinity to the Al ions in the framework.

Not surprisingly, the situation changes substantially for the metal supported films. The alkali metals all become thermodynamically

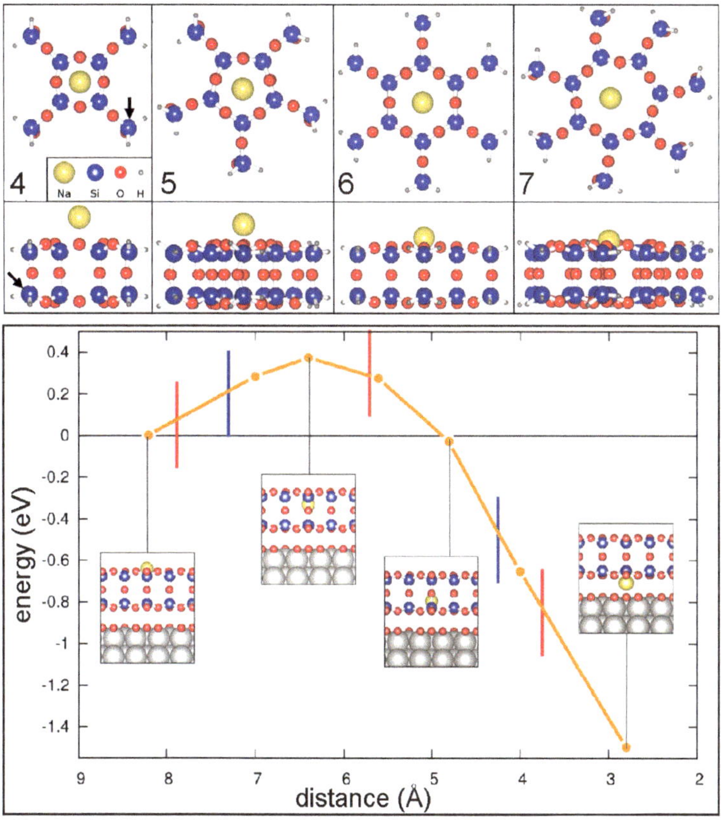

Figure 7.14 Top panel: Top and side views of the optimized geometry of a Na atom interacting with a cluster model mimicking N-membered rings ($N = 4$–7). The arrow marks the Si atom that was replaced by Al to mimic aluminosilicate frame in separate calculations. Bottom panel: Potential energy curve for the diffusion of a Na atom into the six-membered ring of the $SiO_2/3O(2\times2)/Ru(0001)$ films. The zero of energy corresponds to Na adsorbed on the surface of the film. The vertical bars indicate the position of the O (red) and of the Si (blue) layers. Adapted with permission from ref. 22.

the most stable if intercalated the interface. The energy profile only computed for six-membered ring is shown in the bottom panel of Fig. 7.14. Although the barrier (0.4 eV) does not correspond to a real transition state, it appears that the energy required for penetration can readily be overcome at moderate temperatures.

For larger rings present in the amorphous portions of the film, the barrier is presumably smaller. Upon reaching the metal surface the Na atom experiences charge transfer and become Na^+. Noteworthy, the distance between the silica film and the metal support upon alkali metal adsorption does not change, as it was the case for CO and oxygen gas adsorption studied in the previous sections (see Fig. 7.5).

As a consequence of the metal diffusion to the interface, the calculated work function decreases substantially, i.e., by 1.1 eV (for Na) and 1.3 eV (for Mg). Note, however, that the work function changes may even be by 1–1.5 eV higher, if Na(Mg) ions could be stabilized at the surface. Such a strong reduction of the work function by alkali metal adsorption may have a strong impact on the subsequent adsorption of gaseous molecules. This is interesting issue that has to be studied in the future.

References

1. Sutter, P., J. T. Sadowski, and E. A. Sutter, Chemistry under cover: Tuning metal–graphene interaction by reactive intercalation. *Journal of the American Chemical Society*, 2010. **132**(23): pp. 8175–8179.
2. Over, H., and M. Muhler, Catalytic CO oxidation over ruthenium—bridging the pressure gap. *Progress in Surface Science*, 2003. **72**(1–4): pp. 3–17.
3. Emmez, E., et al., Oxidation of the Ru(0001) surface covered by weakly bound, ultrathin silicate films. *Surface Science*, 2016. **646**: pp. 19–25.
4. Zhong, J.-Q., et al., Oxidation and reduction under cover: Chemistry at the confined space between ultrathin nanoporous silicates and Ru(0001). *The Journal of Physical Chemistry C*, 2016. **120**(15): pp. 8240–8245.
5. Emmez, E., et al., Permeation of a single-layer SiO_2 membrane and chemistry in confined space. *The Journal of Physical Chemistry C*, 2014. **118**(50): pp. 29034–29042.
6. Pfnür, H., et al., High resolution vibrational spectroscopy of CO on Ru(001): The importance of lateral interactions. *Surface Science*, 1980. **93**(2–3): pp. 431–452.
7. Schlexer, P., et al., CO adsorption on a silica bilayer supported on Ru(0001). *Surface Science*, 2016. **648**: pp. 2–9.

8. Tissot, H., et al., Ultrathin silica films on Pd(111): Structure and adsorption properties. *Surface Science*, 2018. **678**: pp. 118–123.
9. Guo, X., and J. T. Yates, Dependence of effective desorption kinetic parameters on surface coverage and adsorption temperature: CO on Pd(111). *The Journal of Chemical Physics*, 1989. **90**(11): pp. 6761–6766.
10. Gdowski, G. E., T. E. Felter, and R. H. Stulen, Effect of surface temperature on the sorption of hydrogen by Pd(111). *Surface Science*, 1987. **181**(3): pp. L147–L155.
11. Yao, B., et al., Gas separation through bilayer silica, the thinnest possible silica membrane. *ACS Applied Materials & Interfaces*, 2017. **9**(49), 43061–43071.
12. Zhong, J.-Q., et al., Immobilization of single argon atoms in nano-cages of two-dimensional zeolite model systems. *Nature Communications*, 2017. **8**: p. 16118.
13. Prieto, M. J., et al., Water formation under silica thin films: Real-time observation of a chemical reaction in a physically confined space. *Angewandte Chemie International Edition*, 2018. **57**(28): pp. 8749–8753.
14. Kostinski, S., et al., Diffusion of water molecules in amorphous silica. *IEEE Electron Device Letters*, 2012. **33**(6): pp. 863–865.
15. Baron, M., et al., Adsorption of Au and Pd atoms on thin SiO_2 films: The role of atomic structure. *The Journal of Physical Chemistry C*, 2008. **112**(9): pp. 3405–3409.
16. Ulrich, S., et al., Realization of an atomic sieve: Silica on Mo(112). *Surface Science*, 2009. **603**(8): pp. 1145–1149.
17. Ulrich, S., et al., Evidence for a size-selective adsorption mechanism on oxide surfaces: Pd and Au atoms on SiO_2/Mo(112). *ChemPhysChem*, 2008. **9**(10): pp. 1367–1370.
18. Lu, J. L., et al., Structure, thermal stability, and CO adsorption properties of Pd nanoparticles supported on an ultra-thin SiO_2 film. *Surface Review and Letters*, 2007. **14**(5): pp. 927–934.
19. Jerratsch, J. F., et al., Lithium incorporation into a silica thin film: Scanning tunneling microscopy and density functional theory. *Physical Review B: Condensed Matter and Materials Physics*, 2009. **80**(24): p. 245423.
20. Büchner, C., et al., Adsorption of Au and Pd on ruthenium-supported bilayer silica. *The Journal of Physical Chemistry C*, 2014. **118**(36): pp. 20959–20969.

21. Kaden, W. E., et al., Understanding surface core-level shifts using the Auger parameter: A study of Pd atoms adsorbed on ultrathin SiO_2 films. *Physical Review B*, 2014. **89**(11): p. 115436.
22. Schlexer, P., L. Giordano, and G. Pacchioni, Adsorption of Li, Na, K, and Mg atoms on amorphous and crystalline silica bilayers on Ru(0001): A DFT study. *The Journal of Physical Chemistry C*, 2014. **118**(29): pp. 15884–15891.

Chapter 8

Water Adsorption on Silica Films

The interaction of water and silica, the two most abundant compounds on our planet, has always been a subject of research interest in the course of the formation and development of natural sciences, and it covers a wide variety of scientific disciplines. Water/silica reactions play a decisive role in natural processes such as weathering and dissolution. The performance of silica in catalysis largely depends on the density and spatial distribution of hydroxo species at surface. Everyone opening boxes with new household items has seen small bags filled by silica gel granules as drying agent.

Since this book highlights the "surface science" approach to understanding the structure and properties of silica and related materials, it is fair to mention here Thiel and Madey's review paper from 1987[1] summarizing experimental studies of water interaction with metal and oxide surfaces primarily carried out on single crystals. Although the review has been updated and extended by Henderson in 2002,[2] these reviews provide, in fact, not so much information about silica. It is because of (i) well-known difficulties to prepare clean surfaces of silica single crystals, and (ii) insulating properties of silicas, that makes the experimental studies quite difficult and renders the results often uncertain.

Spectroscopic techniques such as nuclear magnetic resonance spectroscopy, infrared and Raman spectroscopy are primarily used

Introduction to Ultrathin Silica Films: Silicatene and Others
Shamil Shaikhutdinov

ISBN 978-981-4877-28-2 (Hardcover), 978-1-003-08862-2 (eBook)
www.jennystanford.com

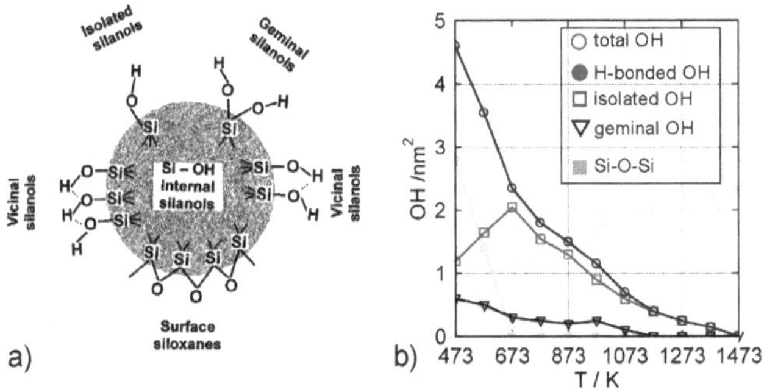

Figure 8.1 (a) Types of silanol groups on amorphous silica. Adapted with permission from ref. 3. (b) Distribution of the surface OH groups as a function of the treatment temperature. Adapted with permission from ref. 4.

for investigating hydrated silicas. Numerous results from these studies suggested the presence of isolated silanols (i.e., single silanols, Si–OH, and geminal silanols, Si–$(OH)_2$) and hydrogen-bonded silanols at surface (Fig. 8.1). Statistical analysis of hydrated silica materials by Zhuravlev[3] led him to conclude that the maximum density of silanol groups on a hydroxylated surface amounts to about 5 OH per nm^2 irrespectively of the preparation method and silica morphology.

It is generally accepted that water chemisorbs by opening strained siloxane (Si–O–Si) bonds at surface resulting in two silanol groups as schematically shown below. Subsequent water molecules bind to existing silanols via hydrogen bonds rather than adsorb molecularly on the pristine siloxane surface.

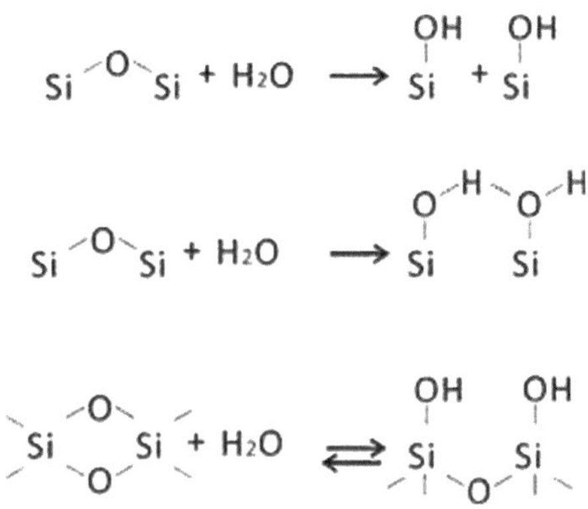

Early theoretical works addressing water/silica interfaces have been performed using cluster calculations, in particular to understand the atomic structures of various hydroxyl species. Significant progress has been achieved by employing density functional theory (DFT). Note, however, that weak interactions, such as hydrogen bonds, are notoriously difficult for DFT methods to implement accurately. The accuracy of a given functional may depend on the choice of basis set. For obvious reasons, calculations have first been performed on the crystalline surfaces, mainly of quartz.

Dissociative adsorption at the (0001), (10-10), (10-11), and (10-1-1) surfaces of α-quartz is energetically favorable, as it results in the large adsorption energies and lowering the surface energies.[5] However, water molecules do not adsorb on the reconstructed, siloxane (Si–O–Si)-terminated α-quartz(0001) surface (for its surface structure, see Fig. 1.12 in Chapter 1). DFT analysis of layer-by-layer water adsorption on the clean reconstructed as well as hydroxylated α-quartz (0001) surfaces showed that, except the case of monolayer adsorption, the interaction between water molecules and bare quartz surfaces with Si–O–Si termination is relatively weak. The interaction with the hydroxyl-terminated surface is much stronger. A stable water bilayer may be formed on both surfaces as illustrated in Fig. 8.2b.

In experiments, however, the initially OH-free silica surfaces are usually obtained by dehydroxylation of silicas via vacuum annealing at elevated temperatures. DFT analysis of this process for the (001), (101), and (111) surfaces of β-cristobalite[7] suggested complex surface reconstruction involving silica units migration. The latter was required to reach the experimentally measured values for hydroxyl coverage (3.7–5.2 OH/nm^2 in standard conditions, see Fig. 8.1b) despite obvious topological differences between ordered model structures used in calculations and amorphous silica samples used in experiments.

The results for the Si–O–Si reconstructed quartz surface can, in principle, be adopted for water adsorption on natural sheet silicates. Molecular dynamics simulations on mica (muscovite) showed that, at monolayer coverage, water forms a fully connected two-dimensional H-bond network like a "two-dimensional ice" depicted in Fig. 8.3.[8] The structure is predicted to be stable at

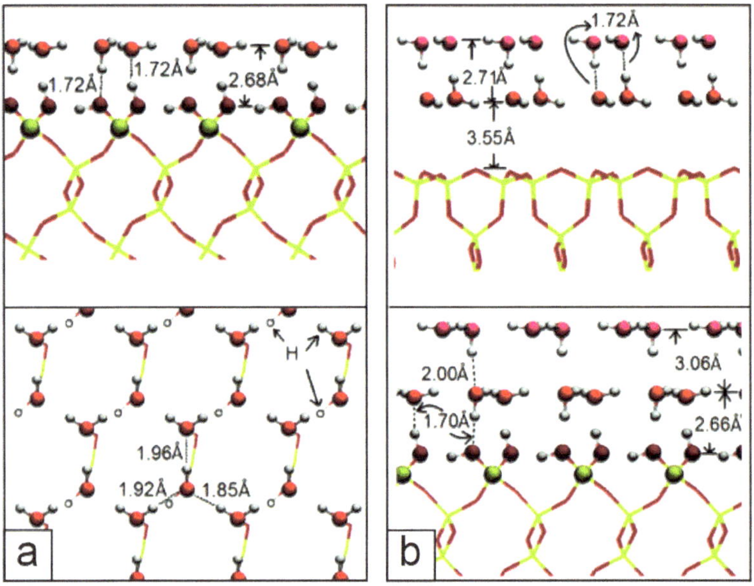

Figure 8.2 (a) Side and top views of water monolayer on fully hydroxylated quartz (0001) surface. (b) Cross views of water bilayer formed on the siloxane-terminated (top) and hydroxylated (bottom) quartz (0001) surfaces. Adapted with permission from ref. 6.

300 K, and as such it may explain the ordered structure observed by atomic force microscopy (AFM) on the hydrated mica surface at room temperature.[9] Also, AFM images of mordenite (one of the most abundant high-silica zeolite minerals) indicated an alignment of hydroxyl groups on both the (010) and (100) surfaces.[10]

In the following sections, we will discuss water adsorption on thin silica and related films primarily by infrared spectroscopy, which remains the principal technique for studying water interaction with silica.

8.1 Water Adsorption on a Monolayer Film

First studies of water adsorption on single-layer silicate films grown on Mo(112) were performed using high-resolution electron

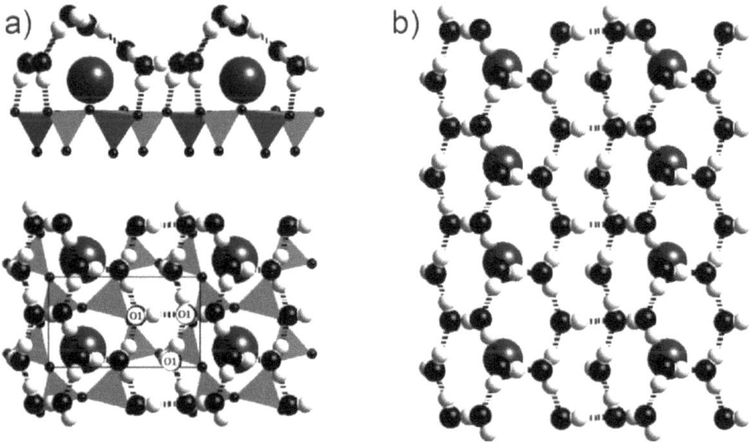

Figure 8.3 (a) The optimized structure of water monolayer on mica in side and top views. Dashed lines are hydrogen bonds. The unit cell of the 2D-ice structure, which coincides with the unit cell of mica, is indicated. There are six water molecules per unit cell; half of them (denoted by O1) are bound to the surface oxygen atoms. The other water molecules solvate the potassium ions, and form bonds to the underlying water molecules and to each other. (b) Top view of the 2D-ice structure, where only the K ions of the mica surface are shown. The H-bonds network consists of distorted hexagons. Reproduced with permission from ref. 8.

energy loss spectroscopy (HREELS).[11] The results suggested the formation of three-dimensional water clusters at low temperatures which desorb intact upon heating, leaving no surface hydroxyls behind. Our own infrared reflection absorption spectroscopy (IRAS) studies[12] confirmed growth solely of so-called "amorphous solid water" (ASW) film. (NB: One should discriminate ASW from ice, which is a crystalline form of water, although the "ice" is frequently used for brevity.) Accordingly, water desorption follows zero-order desorption kinetics showing the common leading edge in temperature programmed desorption (TPD) spectra, with a maximum gradually shifting to the higher temperature at increasing coverage. However, dosing H_2O at 140 K, i.e., very close to the onset of water desorption (~145 K), results in a pseudo-first order kinetics (i.e., the signal intensity scales with the coverage, but the maximum peak does not shift). Interestingly, water dosed at 140

K desorbed at temperatures by 7 K higher than the same amount of water dosed at 100 K. Even a larger difference (18 K) was observed for deuterated water (D_2O). It should be mentioned that for systems dominated by weak H-bonding such differences are quite substantial.

The IRAS spectra recorded during slow heating of the water film from 100 to 140 K considerably alter, even though water is not desorbing at these temperatures. This phenomenon is well-documented for other substrates and commonly interpreted in terms of the ASW-to-ice transformation.[13] Looking in the region of silica phonons, a slight blue-shift (9 cm^{-1}) of the principal band at 1060 cm^{-1} was observed upon water adsorption at 100 K as the ASW film, i.e., with the water molecules randomly oriented with respect to the silica surface. The formation of the ice-like layer (at 140 K) results in a much larger (by 20 cm^{-1}) shift. The experiments with the silica film prepared with the ^{18}O isotope confirmed that this effect originates from the water/silica interface. Since water is a polar molecule having a large dipole moment, the observed shifts are most likely caused by polarization of the water/silica interface. As the shift is found larger for the ice-like layer, it suggests water reorientation with respect to the silica surface and hence some sort of ordering.

Indeed, temperature-programmed LEED measurements (i.e., recording LEED patterns during sample heating) revealed additional diffraction spots (Fig. 8.4a) shortly appearing at around 145 K, i.e., at temperature close to the desorption maximum in TPD spectra, suggesting that the corresponding ordered structure forms at relatively low water coverage. The observed pattern corresponds to the water ad-layer in $c(2\times2)$ symmetry with respect to the silicate layer, which is in turn form $c(2\times2)$ structure to Mo(112). That the same pattern is observed by water dosing directly at 145 K points to that water ordering is thermodynamically driven, not kinetically. (Certainly, in all LEED experiments one has to take precautions to ensure that the effect is not induced by the incident electron beam).

The formation of water ordered structures has been reported for noble metal surfaces using STM at low temperatures.[14,15] On the basis of the DFT calculations, it was assigned to a water layer like in the basal plane of the hexagonal *Ih* phase of ice. Basically, the

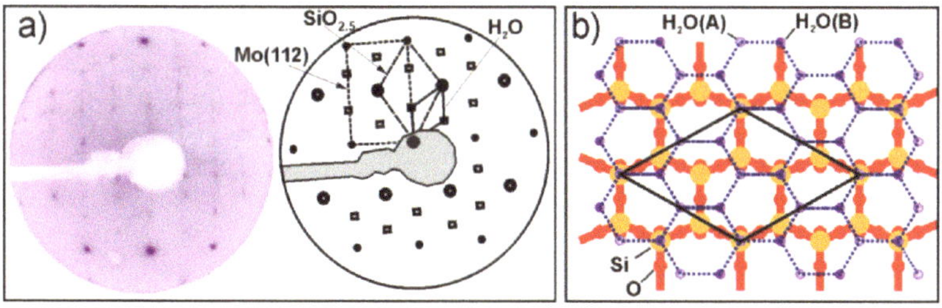

Figure 8.4 (a) LEED pattern of ordered water ad-layer on a $SiO_{2.5}$/Mo(112) film (in negative contrast) and its schematic presentation. The unit cells are only indicated for one rotational domain. (b) Top view of a structural model. H-bonding network of two differently oriented water molecules (A, B) within the bilayer structure of ice forms a c(2×2) superstructure (marked as rhomb) with respect to the silicate layer.

weakly adsorbed water molecules form a honeycomb-like network stabilized by hydrogen bonds (see also Fig. 8.2a). In the same manner, one can place the ice layer on top of a silicate film such that it fits the c(2×2) symmetry. The resulted model is depicted in Fig. 8.4b.[12] In this structure, there is a relatively small difference between the O—O distances in the ice *Ih* (2.61 Å, if projected onto the (0001) plane) and in the topmost O-layer of silicate (2.73 Å).

Such a possibility has been examined by DFT,[16] although the calculations were carried out for free-standing silicate bilayer in a slab containing a (1×1) unit cell, whereas water layer forms the c(2×2) structure. At low coverage (Fig. 8.5a), water molecules are isolated. The oxygen atom in the water molecule sits on top of the Si atoms, whereas the water protons are directed toward the siloxane oxygen atoms, but without forming a hydrogen bond. At higher coverages, two stable structures can be formed (Fig. 8.5b,c) in a H-bonding network resembling an ice-like arrangement. For three H_2O molecules per unit cell, the optimized structure contains water molecules coordinated in a distorted tetrahedral fashion, making interconnected six-membered and three-membered H-bonded water rings. The latter is absent in known forms of ice. In case of 2 H_2O/unit cell, the layer derived from the *ice XI* phase fits better the hexagonal rings of the silicate film. This latter model is a good candidate to explain the experimental observations. (Note again, however, that the calculations were performed for

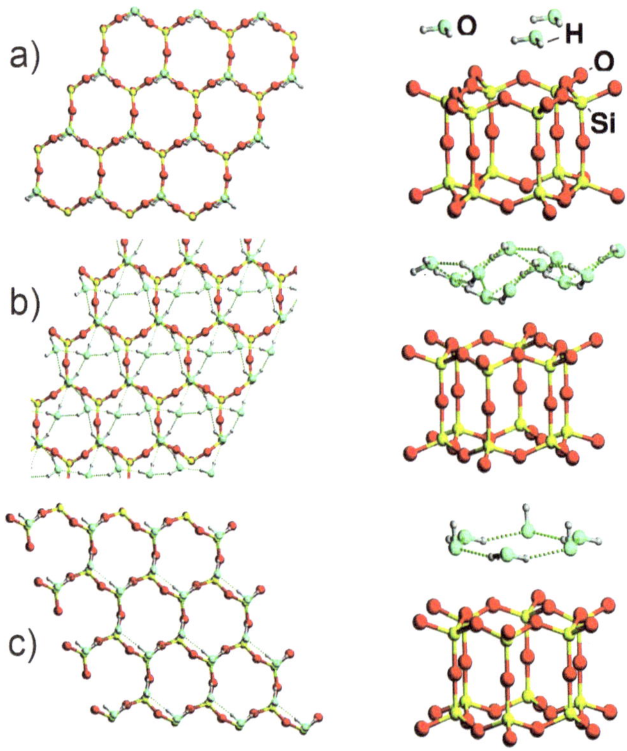

Figure 8.5 DFT-optimized structures of H_2O adsorbed on the free-standing silicate bilayer at low (a) and high (b,c) coverages. Adapted with permission from ref. 16.

the (1×1) unit cell). Nevertheless, the DFT results are consistent with the experimental finding of that water does not dissociate on the siloxane-terminated, silicate monolayer. Apparently, the water/water interaction is much stronger than the water/silica interaction.

We continue discussion of water interaction with a single silicate layer prepared on Ru(0001). Such films are considerably more defective than the crystalline films grown on Mo(112) (see Chapter 4). Therefore, one anticipates different reactivity of these two films towards hydroxylation. Indeed, IRAS inspection of the Ru-supported films exposed to water vapor at 100 K and then heated to 300 K to desorb the ASW film revealed a sharp band at

2760 cm^{-1} (see inset in Fig. 8.6b) reflecting O–D stretch vibration in silanols (Si–OD). (Henceforth, we primarily use D_2O instead of H_2O). Concomitantly, isolated protrusions about 1 Å in height appear in STM images, which can straightforwardly be assigned to surface hydroxyls. Only a few such species are observed on crystalline domains, while most of those are located at domain boundaries and at the rim of adventitious holes (Fig. 8.6a). Obviously, the surface hydroxylation under these conditions only takes place on defects. The latter are present in abundance on the same films annealed in UHV, forming a periodic defect structure (see Fig. 5.7 in Chapter 5). As a result, surface hydroxyls and their aggregates decorate the periodic defects (Fig. 8.6b). Trimers are reminiscent of triangular (T-) defects, while quatromers are definitely related to rectangular (R-) defects, both described in Chapter 5. Therefore, patterning defects on the silica surface would allow to control spatial distribution of surface hydroxyls on silica. The latter is important for anchoring catalytically active species in so-called "single atom catalysts".

Analysis of high-resolution STM images allowed to localize the position of silanols at surface as illustrated in Fig. 8.6c. The water-related protrusions are preferentially located above the Si atoms at the nodes formed by non-equivalent polygons, for instance, a pentagon, hexagon, and heptagon. Water adsorption experiments on the film prepared with $^{18}O_2$ showed no shift for the ν(OD) band at 2760 cm^{-1}, indicating that oxygen in silanols is only coming from the water, and no oxygen scrambling takes place. Therefore, the STM observation of hydroxyl on top of the Si atoms in defects suggests breaking of the SiO–Ru linkage which is followed by Si atom popping up to strongly bind OD from the water molecule that dissociates. In this scenario, the second D in the water should form another hydroxyl with the O atom in silica. However, the related band (that would definitely experience a shift in ^{18}O-labeled films) was never detected. The fate of the second H remains unknown. Either the respective hydroxyl is oriented parallel to the metal surface and hence becomes invisible due to the metal surface selection rules (Chapter 2) or this H atom migrates onto the metal surface.

One may argue that water adsorption on a single silicate layer is affected by its strong bonding to a metal substrate. In this respect, a bilayer silicate film weakly bonded to the metal surface is expected

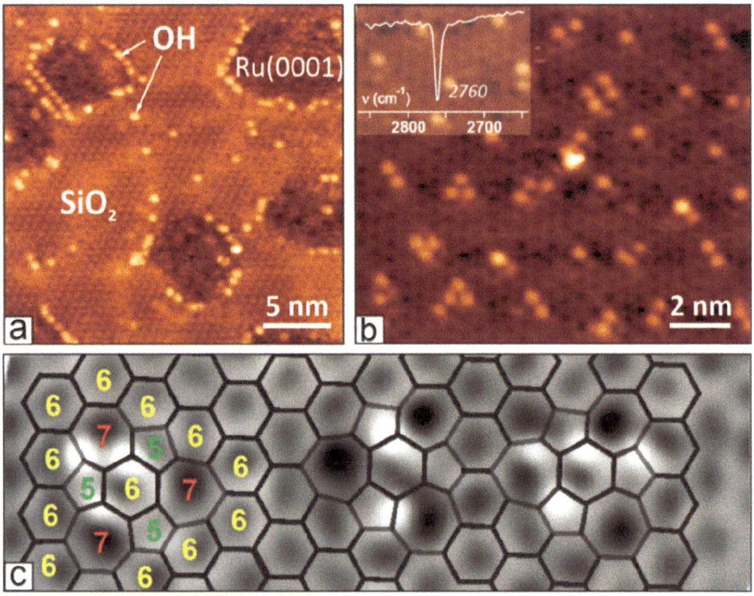

Figure 8.6 STM images of hydroxylated monolayer silicate films on Ru(0001) formed by water adsorption onto: (a) "as prepared" film; (b) UHV-annealed film showing periodic defects (see Figure 5.7 in Chapter 5). The inset in panel (b) shows typical IRAS spectrum in the OD stretch vibration region. (c) STM image superimposed with polygonal representation of silica rings. Adapted with permission from ref. 17.

to show some difference. In general, one could consider bilayer film as the more adequate system to study reactions of silica with water.

Water exposure under UHV-compatible conditions did not result in surface hydroxylation. Only tiny amounts of isolated hydroxyls are detected by IRAS via the 2763 cm^{-1} band.[18] Note that the band is only slightly blue-shifted when compared to a single silicate, suggesting that the OD vibration in surface silanols is sensitive to local Si coordination and not to surface ordering. The band disappears at heating at temperatures as high as 1000 K indicating a good thermal stability of silanols. Based on the signal intensity, the silanol density is in the order of 0.1 nm^{-2}, or about one hydroxyl per 40 silica rings. Such a small concentration points to the defect-mediated hydroxylation.

8.2 Electron Beam Assisted Hydroxylation

The results presented in the previous section are well consistent with anticipated inertness of the siloxane-terminated silica surfaces. However, to study the chemistry of silanols using planar model systems it is desirable to have them in substantial amounts. Moreover, one has to control the degree of hydroxylation in order to establish the relationship with their chemical activity. To promote the hydroxylation of our silicate films one needs to create defects. Under vacuum conditions this could be done, for example, by electron or ion bombardment. However, usage of ion bombardment is very tricky in the case of ultrathin films because of the obvious sputtering effects. On the other hand, electron irradiation of the pristine films showed negligible effects on subsequent water adsorption due to the presence of a metal support which efficiently quenches possible excitations induced by electron beam. Instead, we proposed another approach which is based on using electron irradiation of the silicate films covered by "ice" (Fig. 8.7).[19] The underlying idea is fairly simple: If high-energy electrons create short-living surface defects in silica, these can react immediately with surrounding water molecules more efficiently as compared to water post-dosing.

Electron bombardment of relatively thick ASW films causes, in essence, no effect on the amount of surface hydroxyls formed after subsequent heating of the bombarded sample to the room temperature to desorb unreacted water molecules. Apparently, ice

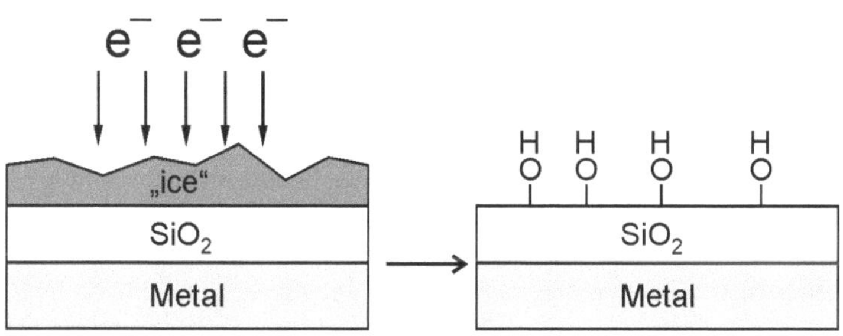

Figure 8.7 Schematic illustration of the silica hydroxylation assisted by electron irradiation of a solid water film.

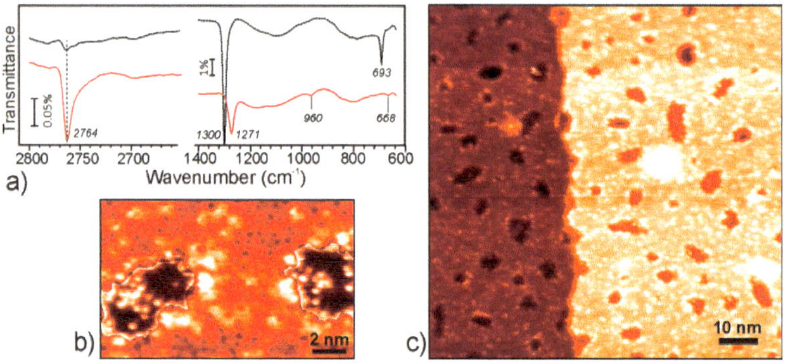

Figure 8.8 (a) IRAS spectra of the SiO_2/Ru(0001) surfaces hydroxylated without (dash line) and with (solid line) low-energy electron irradiation of the ASW film at 100 K. Both samples were heated in UHV to 300 K before recording the spectra at 100 K. Adapted with permission from ref. 19. (b,c) STM images of hydroxylated surfaces prepared with electron irradiation. The dark spots in these images are monolayer silicate domains initially presented in the sample as shown in image (b) presented in differentiated contrast.

overlayer must be of thickness comparable with the electron mean free path and/or exciton diffusion length in ice. In fact, low-energy electrons (100–500 eV) efficiently sputter the ice film, since the related IRAS bands attenuate at increasing beam exposure time. In practice one exposes the sample to water at low temperatures to form a relatively thick ASW film solely to ensure its uniform coating, and then irradiates with electrons until the water film is almost sputtered. After desorption of remaining water molecules by thermal flash in vacuum, IRAS spectra show a much higher concentration of hydroxyls (see Fig. 8.8a) as compared to the silica film similarly treated, but without electron irradiation. At such high concentration even a much weaker band at 960 cm^{-1} associated with Si–O bond stretch vibrations in silanols becomes clearly observed.

Using the results on a hydroxylated single layer silicate film for calibration, with the silanol density being directly measured by STM (see Fig. 8.6), the observed IRAS signals correponds to the amount of silanols occupying about 10% of all the Si atoms in the top silicate layer. (Note that such an estimate does not account for hydroxyls, if any, oriented parallel to the surface and hence "invisible" for IRAS due to the metal selection rules). The silanols are thermally very stable and disappear from the surface via water desorption on heating to 900 K and above, i.e., at considerably higher temperatures as compared to the monolayer silicate (~700 K). It appears that strong interaction of monolayer films with a metal support provides additional pathway for hydroxyls desorption as water.

The surface hydroxylation leads to strong attenuation of the silica phonon bands which in addition shift to the lower frequencies by a value roughly proportional to the degree of hydroxylation, but about 30 cm^{-1}, at most. Therefore, it is clear that the bilayer structure of the film remains. This is also confirmed by STM inspection of the hydroxylated films. Numerous white protrusions seen in Fig. 8.8c can be assigned to surface hydroxyls, which randomly cover the surface. There are in addition adventitious spots exposing the single-layer silicate (Fig. 8.8b) which look very similar to those observed upon hydroxylation of a monolayer silicate film (Fig. 8.6). The STM results could not yet precisely identify silanols position.

Following siloxane bond opening mechanism for hydroxylation (see the scheme in introduction), two hydroxyl species are formed. One includes oxygen from water (O_W) and another one incorporates oxygen from silica (O_S), both can be spectroscopically distinguished using isotopic labeling of oxygen in the water and/or the silica. The experiments on the films prepared with $^{18}O_2$, indeed, showed two bands, at 2764 and 2746 cm^{-1} (as shown in Fig. 8.9) which can be assigned to $^{16}O_WD$ and $^{18}O_SD$ species solely on the basis of a reduced mass analysis (the oscillator with the heavier mass exhibits the lower frequency). Also, the formation of silanols by atomic D, if the latter is produced by radiation, can be excluded as this would only result in a single O_SD band at 2746 cm^{-1}. Although the observation of the two OD bands agrees with the siloxane bond opening mechanism, resulting in two virtually identical hydroxyls, the bands behave differently on heating. Up to 700 K, the integral intensities of the bands stay fairly constant, but the bands become sharper. At further heating, the 2764 cm^{-1} signal attenuates, whereas the 2746 cm^{-1} band gains in intensity so that it becomes even larger than the 2764 cm^{-1} band. Both bands ultimately disappear upon heating to 1240 K. In addition, during heating the bands experience some

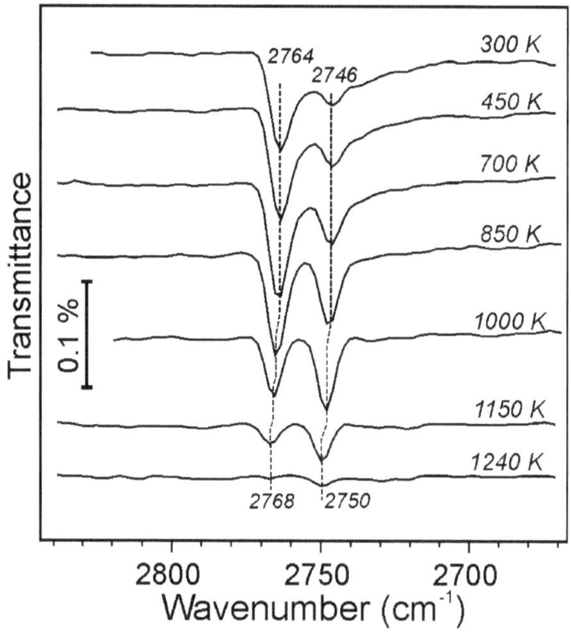

Figure 8.9 The ν(OD) spectra of the $Si^{18}O_2$ bilayer film, hydroxylated by using $D_2^{16}O$ water, as a function of annealing temperature. The hydroxylation was obtained with 150 V electron irradiation of the ASW film deposited at 100 K. Adapted with permission from ref. 19.

shift, albeit small. As already mentioned above, one has to bear in mind that the band intensity in IRAS depends also on the dipole orientation with respect to the metal surface and gets smaller when it becomes parallel to the surface. Therefore, the observed spectral changes may also reflect thermally induced structural relaxations.

Interestingly, the same two bands and the similar spectral evolution were observed for "thermally" hydroxylated films, i.e., without irradiation. The difference is only a substantially smaller signal intensity that implies a common mechanism for the surface hydroxylation. It appears that electron irradiation aids in overcoming the activation barrier which cannot be achieved thermally simply because water molecules desorb intact before the reaction could occur.

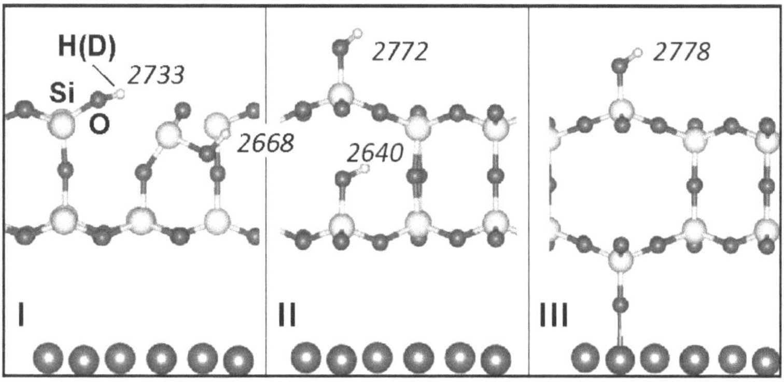

Figure 8.10 Side view of three possible structures formed upon hydroxylation of a silicate bilayer film. The DFT-computed (harmonic and scaled) frequencies for the OD stretch vibrations are shown next to hydroxyl species.

The atomic structure of the hydroxylated silicate surface was analyzed by Sauer and co-workers employing DFT calculations on a crystalline film.[19] One can anticipate several possible structures, all schematically shown in Fig. 8.10, which can be formed through breaking siloxane bonds either *in* the surface layer (structure I) or *between* the silicate layers (structure II). One more structure (III) is formed from the structure II by flipping of internal silanol towards the metal with simultaneous formation of the Si–O–Ru linkage and hydrogen migration onto the metal surface.

Actually, the relative stability of the structures depends on the amount and nature of ad-atoms adsorbed directly onto the Ru surface. If the Ru surface is covered by D atoms (that is likely based on the experimental TPD results showing large D_2 desorption signal), then structure I is most stable. Preferential breaking of the siloxane bonds in the surface plane is also consistent with the IRAS results (Fig. 8.8a) showing that the 693 cm^{-1} band, which involves bending vibrations of in plane O–Si–O bonds, is strongly suppressed, at least to a much larger extent than the high-frequency band at 1300 cm^{-1} associated with vertical Si–O–Si bonds.

The calculated (and properly scaled) frequencies are shown in Fig. 8.10 next to corresponding hydroxyl species. The terminal hydroxyls formed in structures II and III exhibit the frequencies

(2772 and 2778 cm^{-1}), which fit much better the experimentally observed 2764 cm^{-1} than the one in structure I (at 2733 cm^{-1}), albeit thermodynamically the most favorable. Note, however, that terminal hydroxyl in structure I is oriented nearly parallel to the surface and hence hardly detectable by IRAS. It seems plausible that tilted towards the surface silanol in structure I (2733 cm^{-1}) manifests itself as a weaker and broad feature (shoulder) at the low-frequency side of the principal peak at 2764 cm^{-1} (Fig. 8.8a). Second silanols formed in structures I and II are strongly red-shifted (2668 and 2640 cm^{-1}) because of hydrogen bonding to the nearest oxygen which in addition may cause band broadening. Again, these silanols are oriented parallel to the surface and hence IRAS silent. However, its presence in structure II is indirectly affirmed through its transformation into structure III at high temperatures, accompanied by release of the D atoms desorbing as D_2 in TPD spectra. In addition, the progressive formation of Si–O–Ru linkages in structure III explains why dehydroxylation of the samples by UHV annealing never recovers the original phonon spectrum of the film. Apparently, recombination of different types of hydroxyls is a stochastic process accompanied by structural transformations resulting in a complex spectral evolution of the related hydroxyl IRAS bands on heating (Fig. 8.9). Dehydroxylation of silicate films is not the reverse process of hydroxyl formation. Repeated hydroxylation over the once dehydroxylated film ultimately leads to disintegration (de-wetting) of the silica film after few cycles.

Finally, we speculate about origin of IR band splitting in above-presented isotopic experiments (Fig. 8.9). The low frequency 2746 cm^{-1} band has tentatively been assigned to $^{18}O_SD$ hydroxyl, whereas according to the DFT calculations such hydroxyls should show frequencies well below 2668 and 2640 cm^{-1} computed for "lighter" ^{16}OD species (Fig. 8.10). If, however, this band also corresponds to terminal, but ^{18}OD–Si silanol, then one should admit that ^{16}O oxygen in the water exchanges with ^{18}O oxygen in the silicate film during water $D_2^{16}O$ dissociation. One possible mechanism includes the exchange in the transition state involving both oxygens followed by rotation of the one of the two ODs resulting in the final state such that the formation of two (O_WD and O_SD) terminal hydroxyls are equally possible. Still, the precise mechanism of hydroxylation, in particular under electron irradiation remains to be further investigated.

Among the electron beam parameters varied, ice film thickness provides the largest correlation to changes in subsequent water/hydroxyl coverage. Since almost no increase of the water/hydroxyl coverage was observed if the film thickness was in the range of one nanometer, corresponding to the inelastic mean free path the electrons, it seems that that the siloxane bonds are activated for water dissociation by water radiolysis products, e.g., hydroxyl ions, formed in the ice layer during electron bombardment.

It is worth mentioning that electron-stimulated reactions have been studied on ASW films formed on metals like Pt(111).[20] In particular, electron bombardment resulted in the evolution of molecular oxygen and hydrogen. Most likely, mobile electronic excitations (excitons) are responsible for the reactions, although the mechanism remains poorly understood. Electron irradiation of physisorbed water layer can also cause surface oxidation as demonstrated for the Si(111) surfaces irradiated with low-energy (<20 eV) electrons.[21] As a side note, radiation induced chemical reactions on cosmic ice dust and dust grains mainly composed of silicates may also be involved in chemical evolution in interstellar space.

8.3 Hydroxylation of Metal-Doped Silicate Films

Hydroxyls in the aluminosilicates, in particular zeolites, are probably the most important entities underlying their application in chemical industry which is primarily determined by so-called "bridging" hydroxyls, with oxygen linking the Si^{4+} and the Al^{3+} ions in the silicate framework (Si–OH_{br}–Al). Therefore, once the preparation and atomic structure of the Al-silicate films has been established (see Chapter 6), the follow-up studies were focused on the formation, identification, and adsorption properties of OH_{br} species in the films.

IRAS studies of the Ru-supported aluminosilicate films showed a band that appears at 3594 (2650) cm^{-1} upon H_2O (D_2O) adsorption. The observed frequency falls in the range typical for the bridging hydroxyl groups measured on zeolitic systems. These bands only appear at relatively high Al:Si molar ratios in the films. The effect is explained by that Al first substitutes Si in the bottom silicate layer and only then in the topmost layer. The hydroxyls are fairly stable: The corresponding band attenuates on heating above 500 K. However, they are substantially less stable than the silanols in pure silicate films which only disappear above 900 K. OH_{br} can be replaced by OD_{br} upon exposure to D_2O and vice versa, thus indicating H–D exchange, which is also a well-known phenomenon in zeolite chemistry.

Being used as a solid acid, the important characteristic of a zeolite is their acidity, i.e., how strong is a proton bonded to the oxygen ion. Actually, measuring the acidity of zeolitic materials is not trivial.[22] A commonly used approach is based on its reaction with a weak base molecule, such as CO and ethylene. The probe molecule forms adduct with OH without breaking the O–H bond that causes a red-shift of the OH stretching frequency. This shift is believed to correlate with the ability of the proton to be transferred.

Adsorption of weak and strong bases on aluminosilicate films has been studied by IRAS. In case of CO, the spectra (see Fig. 8.11a) showed a strong red-shift for the OH stretch and a blue-shift for the CO stretch when compared to CO in the gas phase ($=2143\ cm^{-1}$), both shifts indicating CO coordination to OH. Such large shifts, about 380 and 245 cm^{-1} for $\nu(OH)$ and $\nu(OD)$ bands, respectively, and about 40 cm^{-1} for the $\nu(CO)$ band point out a high acidity of the prepared Al-silicate film surfaces, which is even comparable with the most acidic zeolites known so far.

Exposure to strong bases, such as ammonia NH_3 and pyridine C_5H_5N, even leads to a full transfer of acidic proton to the base, ultimately forming NH_4^+ and $C_5H_5NH^+$ surface ions, respectively. This is illustrated in Fig. 8.11b for the case of ammonia. The ammonium ion is located above the center of the hexagon, with three H atoms pointing toward the surface forming H-bonds to the O atoms in the topmost layer. There are two vibrational modes of NH_4^+ with strong components perpendicular to the surface plane in this configuration: A N–H stretch for the H pointing outward from the surface, and an "umbrella" mode of other three H atoms vibrating as one against N. The good match between experi-

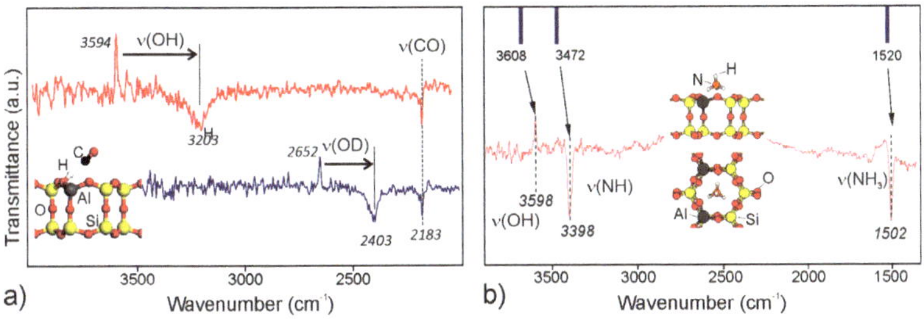

Figure 8.11 (a) IRAS spectra of the hydroxylated (OH- and OD-) aluminosilicate films grown on Ru(0001) recorded in 2×10^{-5} mbar of CO. (b) Spectrum of NH_3 adsorbed on a OH-exposing aluminosilicate film. Side and top views of DFT-optimized, free-standing bilayer model are shown in inset. The DFT-computed (scaled) frequencies for corresponding bands are presented as vertical bars. In both panels (a) and (b), each spectrum is divided by a spectrum taken before gas exposure. The positive transmittance shows the corresponding band disappearing upon the adduct formation. Adapted with permission from refs. 23, 24.

mental and theoretical values supports the proposed adsorption structure.

The adsorption characteristics of the metal-supported hydroxylated Al-silicate films fit well into what is known about zeolite surfaces. Therefore, such films may be considered as well-defined model systems to study the chemistry of zeolites using surface-science tools.

Another example is water adsorption on Fe-silicate films. It turned out that hydroxylation reaction bears close similarities to that of pure silicate films. This is not surprising since the topmost layers in both, pure and Fe-silicate, films are virtually identical (Chapter 6). As far as mixed Fe–Al-silicate films are concerned, IRAS measurements only showed a weak band at 2764 cm^{-1} which can straightforwardly be assigned to terminal silanols.[25] Also STM images of the hydroxylated films revealed atomic size features which can be assigned to isolated hydroxyls located on top of the Si atoms in full agreement with IRAS data. Interestingly, these hydroxyls are only found on the Fe-silicate domains present in the films. To date, no signature for the presence of Si–OH–Fe linkages in such films has been found, which would resemble bridging hydroxyls sometimes reported for Fe-containing zeolites.

References

1. Thiel, P. A., and T. E. Madey, The interaction of water with solid surfaces: Fundamental aspects. *Surface Science Reports*, 1987. **7**(6–8): pp. 211–385.
2. Henderson, M. A., The interaction of water with solid surfaces: Fundamental aspects revisited. *Surface Science Reports*, 2002. **46**(1–8): pp. 1–308.
3. Zhuravlev, L. T., The surface chemistry of amorphous silica. Zhuravlev model. *Colloids and Surfaces A: Physicochemical and Engineering Aspects*, 2000. **173**(1–3): pp. 1–38.
4. Rimola, A., et al., Silica surface features and their role in the adsorption of biomolecules: Computational modeling and experiments. *Chemical Reviews*, 2013. **113**(6): pp. 4216–4313.
5. de Leeuw, N. H., F. M. Higgins, and S. C. Parker, Modeling the surface structure and stability of α-quartz. *The Journal of Physical Chemistry B*, 1999. **103**(8): pp. 1270–1277.
6. Chen, Y.-W., et al., Water thin film-silica interaction on alpha-quartz (0001) surfaces. *Physical Review B*, 2011. **84**(15): p. 155444.
7. Rozanska, X., F. Delbecq, and P. Sautet, Reconstruction and stability of β-cristobalite 001, 101, and 111 surfaces during dehydroxylation. *Physical Chemistry Chemical Physics*, 2010. **12**(45): pp. 14930–14940.
8. Odelius, M., M. Bernasconi, and M. Parrinello, Two dimensional ice adsorbed on mica surface. *Physical Review Letters*, 1997. **78**(14): pp. 2855–2858.
9. Hu, J., et al., Imaging the condensation and evaporation of molecularly thin films of water with nanometer resolution. *Science*, 1995. **268**(5208): pp. 267–269.
10. Yamamoto, S., et al., Surface structure of natural crystals of mordenite as imaged by atomic force microscopy. *Chemical Physics Letters*, 1996. **260**(1): pp. 208–214.
11. Wendt, S., et al., The interaction of water with silica thin films grown on Mo(112). *Surface Science*, 2004. **565**(2–3): pp. 107–120.
12. Kaya, S., et al., Formation of an ordered ice layer on a thin silica film. *The Journal of Physical Chemistry C*, 2006. **111**(2): pp. 759–764.
13. Bensebaa, F., and T. H. Ellis, Water at surfaces: What can we learn from vibrational spectroscopy? *Progress in Surface Science*, 1995. **50**(1): pp. 173–185.

14. Michaelides, A., and K. Morgenstern, Ice nanoclusters at hydrophobic metal surfaces. *Nature Materials*, 2007. **6**: p. 597.

15. Maier, S., et al., Growth and structure of the first layers of ice on Ru(0001) and Pt(111). *Journal of the American Chemical Society*, 2016. **138**(9): pp. 3145–3151.

16. Tosoni, S., B. Civalleri, and P. Ugliengo, Hydrophobic behavior of dehydroxylated silica surfaces: A B3LYP periodic study. *The Journal of Physical Chemistry C*, 2010. **114**(47): pp. 19984–19992.

17. Yang, B., S. Shaikhutdinov, and H.-J. Freund, Tuning spatial distribution of surface hydroxyls on a metal-supported single-layer silica. *The Journal of Physical Chemistry Letters*, 2014. **5**(10): pp. 1701–1704.

18. Yang, B., et al., Hydroxylation of metal-supported sheet-like silica films. *The Journal of Physical Chemistry C*, 2013. **117**(16): pp. 8336–8344.

19. Yu, X., et al., Electron stimulated hydroxylation of a metal supported silicate film. *Physical Chemistry Chemical Physics*, 2016. **18**(5): pp. 3755–3764.

20. Smith, R. S., et al., Thermal and nonthermal physiochemical processes in nanoscale films of amorphous solid water. *Accounts of Chemical Research*, 2011. **45**(1): pp. 33–42.

21. Klyachko, D., P. Rowntree, and L. Sanche, Oxidation of hydrogen-passivated silicon surfaces induced by dissociative electron attachment to physisorbed H_2O. *Surface Science*, 1996. **346**(1–3): pp. L49–L54.

22. Bordiga, S., et al., Probing zeolites by vibrational spectroscopies. *Chemical Society Reviews*, 2015. **44**(20): pp. 7262–7341.

23. Boscoboinik, J. A., et al., Modeling zeolites with metal-supported two-dimensional aluminosilicate films. *Angewandte Chemie International Edition*, 2012. **51**(24): pp. 6005–6008.

24. Boscoboinik, J. A., et al., Interaction of probe molecules with bridging hydroxyls of two-dimensional zeolites: A surface science approach. *The Journal of Physical Chemistry C*, 2013. **117**(26): pp. 13547–13556.

25. Tissot, H., et al., Preparation and structure of Fe-containing aluminosilicate thin films. *Physical Chemistry Chemical Physics*, 2016. **18**(36): pp. 25027–25035.

Chapter 9

Silicatene *una* Graphene

Single-layer materials have recently received enormous attention owing to their two-dimensional (2D) character leading to peculiar physical and chemical properties very different from or even missing in their three-dimensional (3D) counterparts. The most famous one—graphene[1]—triggered a quest for other truly 2D materials. In fact, free-standing single layers of Si and Ge had been predicted on theoretical grounds a while ago.[2] Later they were discovered experimentally, so far only supported on metal substrates, and received the names "silicene" and "germanene," respectively.[3–6] The studies have been extended to a Sn single layer called "stanene."[7] Note, however, that possible incorporation and mixing with substrate atoms in these layers remains the subject of debate.[8,9]

From purely a geometrical point of view, the silicate layer discussed in the previous chapters may be viewed as graphene composed of SiO_4 entities instead of the C atoms. Even the defect structures in silicate layers bear close similarities to those observed on graphene. Solely on this ground, the single silicate layer merits its place in the growing family of 2D materials and as such be named "silicatene" in the same manner as for other analogs of graphene, albeit the suffix "-ene" used for such materials is somewhat

Introduction to Ultrathin Silica Films: Silicatene and Others
Shamil Shaikhutdinov

ISBN 978-981-4877-28-2 (Hardcover), 978-1-003-08862-2 (eBook)
www.jennystanford.com

confusing as it implies compounds having double bonds, according to the IUPAC's (International Union of Pure and Applied Chemistry) standards and regulations. Structural diversity of silicatene and differences compared to single layers of individual element can reasonably be attributed to the relatively high flexibility of the Si–O bonds constituting a SiO_4 unit, thus providing additional degree of freedom. This latter may explain why periodic defect structures theoretically predicted for graphene were not yet found on graphene and related carbon nanomaterials, but were observed on silicatene.

Similar to silicene and germanene, silicatene does not exist as a free-standing layer and needs a suitable substrate to coordinate otherwise unsaturated O atom out of four in SiO_4 tetrahedra. Its bonding to the metal surface can be strong, as in case of Mo and other 3d-metals, or relatively weak, as on Ru and Pd. In the bilayer film, the two silicatene layers are linked together via Si–O–Si bridges thus forming a mirror plane, and all Si–O bonds are saturated.

Inspired by potential applications of single-layer materials in nanotechnologies, one might even think about combining graphene and silicatene, for example, in sandwich-like structures which could potentially result in interesting, yet unknown functional properties of such composites. According to the theoretical calculations,[10] graphene does not form covalent bonds with a bilayer silicatene, albeit shows some stabilizing effects on its structure.

Actually, interaction of graphene with silica has only recently received attention in the course of fabrication and development of graphene-based electronic devices (in which silica is a common gate dielectric). In the early studies, silica has been used solely as a support for the optical visualization and investigation of graphene layers themselves. There were few attempts to elucidate the structure of graphene on SiO_2 with atomic resolution.[11] Thin graphite flakes were deposited using the mechanical exfoliation ("rubbing") technique on thermally grown SiO_2 films about 300 nm in the thickness. For imaging with an STM microscope integrated into standard electron microscope for coarse positioning, the e-beam lithography was used for electrodes deposition. It has turned, however, that the uncontrollable residue remains covering all graphene devices fabricated using the conventional photoresist process. The residue could only be removed by prolonged annealing in H_2/Ar atmosphere. It was concluded that the graphene corrugation, which is relevant for interpreting of device performance, is largely due to partial conformation of the graphene to the SiO_2. The latter finding was further supported by comparative STM studies of graphene on *h*-BN and SiO_2 substrates which revealed a much smoother surface corrugation on *h*-BN because of the graphene conforming to the atomically flat surface of *h*-BN.[12] It is believed that the surface roughness may be critical for graphene-based devices because local curvature can alter electronic properties. Some experimental results pointed out that the morphology of a substrate plays a crucial role in the extrinsic strain-induced disorder in graphene and its effect on the transport properties.[13] Accordingly, the nanoscale transport properties of graphene may, to some extent, be controlled by substrate morphology. Therefore, it is not straightforward that unique physical properties of graphene persist on supported systems. In addition, they may alter under ambient atmosphere.

The interaction between graphene and SiO_2 has also been studied quite intensively by density functional theory (DFT), although the results may deviate depending on the structural models and the DFT level used.[14] Not surprisingly, the electronic structure of graphene on silica depends on the SiO_2 surface structure. The interface structures computed for some surface terminations are depicted in Fig. 9.1. Surface dangling bonds mediate the coupling to graphene and can induce hole or electron doping via charge transfer even in the absence of extrinsic impurities in SiO_2. In an interface geometry, where graphene is weakly bonded to an O-terminated silica surface, graphene is *p*-doped, whereas *n*-doping takes place on a Si-terminated surface.

A more recent DFT study[15] used α-quartz and cristobalite derived surface structures to systematically examine the interaction of graphene with several surfaces, namely: the Si-polar surface with two dangling bonds per surface Si atom, the Si-polar surface with one dangling bond per surface Si atom, the O-polar surface, the Si-adsorbed surface, the O-adsorbed surface, the reconstructed O-polar surface, and also a surface with defects. The interaction between the graphene and the silica surfaces is weak and dominated by dispersive forces which do not break the band structure of

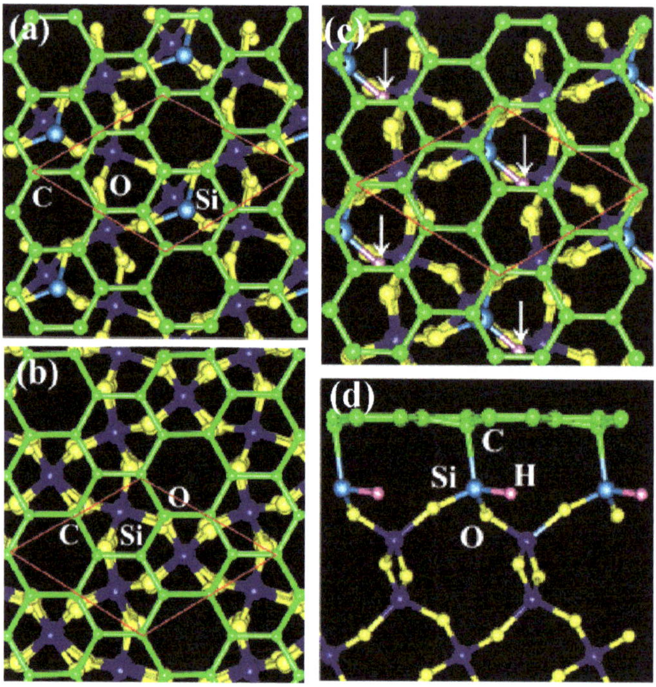

Figure 9.1 Top view of a single graphene layer on (a) Si-terminated and (b) O-terminated SiO_2 surfaces. Top (c) and side (d) views of graphene on the partially hydrogenated Si-terminated SiO_2 surface. The red lines represent the (2×2) graphene unit cell. Adapted with permission from ref. 14.

the graphene. The graphene layer keeps its planar hexagonal network. The binding energy to the Si-polar surfaces is smaller than that for the O-polar surface. H-passivated surfaces do not affect the electronic properties of the graphene irrespective of whether they are on the O-polar or Si-polar surfaces. Only surfaces with defects have a strong effect on the electronic properties of graphene. However, the cleaved SiO_2 can readily be reconstructed or hydroxylated (or hydrogenated) to eliminate the dangling bonds.

All these theoretical works highlight the point that the silica surface to be used as a substrate for fabrication of graphene/silica devices should be cleaved with an atomic level of flatness and passivated by hydrogen. In this respect, silicatene layer(s) may be a good option.

It is worth mentioning here a few studies addressing the interaction of graphene with Si ad-atoms. It appears that Si vapor-deposited onto graphene grown on a SiC(000-1) single crystal rapidly diffuses to the interface on heating.[16] DFT calculations predicted an intercalation mechanism where Si migrates through point defects in graphene.[17] It has been found that the intercalated Si atoms weaken the interaction of graphene with a metal substrate, and retain its superlative properties.[18] Another intercalation mechanism includes the penetration of Si through the graphene lattice facilitated by a Si–C exchange process.[19]

The Si layer formed underneath graphene can be oxidized in oxygen ambient, forming an amorphous silica between the graphene and the metal.[20] In order to make the interfacial silica film better ordered one can use annealing at higher temperatures. However, such treatments results in graphene disappearance from the surface due to its gasification into CO and CO_2 ("burning"), most notably observed at temperatures above 850 K. To minimize and even to avoid graphene gasification, one can, in principle, reverse the preparation process by starting with a well-ordered silica film as a substrate for subsequent graphene deposition.

To recall, the preparation of graphene layer(s) on metals commonly includes segregation of carbon impurities from the metal bulk or thermal decomposition of unsaturated hydrocarbons like ethylene.[21] Since metal supports used for the silica film growth are always thoroughly cleaned prior to the Si deposition, a carbon segregation route is hardly possible. Also, adventitious carbon readily burns out in oxygen environment during the silica film preparation. Exposure of the bilayer silicatene to ethylene at UHV-compatible pressures (typically, 10^{-6} mbar) at room temperature causes no visible effects on electronic structure and vibrational characteristics monitored by x-ray photoelectron spectroscopy (XPS) and infrared reflection absorption spectroscopy (IRAS). Increasing exposure temperature to those at which graphene is usually formed on pure metals, results in partial decomposition of the silica film accompanied by the substantial formation of reduced silica (Si^{+}) species. If the sample is exposed to elevated pressures of ethylene at relatively low temperatures, then substantial amounts of carbonaceous residues were found at the surface, yet not identified, which, however, do not alter the principal film structure as judged by IRAS. If, however, this sample is vacuum-annealed at around 1000 K, i.e., in the temperature range of graphene formation on Ru(0001)), two vibrational bands are observed which are strongly shifted compared to original film, i.e., from 1300 to 1264 cm^{-1} and from 696 to 802 cm^{-1}, as shown in Fig. 9.1a. The intensity of new bands depends on the ethylene exposure: the bands gain in intensity at the expense of the initial bands which attenuate and ultimately disappear. Such a spectral evolution has previously been observed on the films modified by 3d-metals (see Chapter 6) and suggests the formation of two coexisting structures until the new structure covers the entire surface. Interestingly, the resulted film becomes well-ordered (see low-energy electron diffraction (LEED) pattern in Fig. 9.2a) regardless of whether the original film was fully amorphous or partially crystalline. In good agreement with LEED, the scanning tunneling microscopy (STM) images (displayed in Fig. 9.2b) show domains with a honeycomb-like structure having a ~5.5 Å periodicity, like in the crystalline silicatene, which are separated by line defects typical for domain boundaries in hexagonal lattices.

Final piece of information came from XPS measurements (Fig. 9.2c) showing that the amount of oxygen in C-modified films decreases by a factor of 2, and no more O ad-atoms remain adsorbed onto Ru, whereas the amount of Si remains the same in total, but splits in two, almost equally populated states. The high binding energy (BE) signal at around 104 eV is shifted by about 1 eV from the one measured before the treatment. However, such a shift is also observed for the O 1s level, suggesting that it has the same origin as previously found for gas adsorption at the silica/metal interface and attributed to the band re-alignment (Chapter 7). Therefore, the high BE state can safely be assigned to the Si ions only having bonds to oxygen. Accordingly, the low BE signal at 101 eV could be indicative of sub-stoichiometric, reduced silica phase. On the other hand, it is close to the values reported for silicon carbides.[23] Therefore, the XPS results can be well explained by that about a half of oxygen ions in the film are replaced by carbon. (The carbon C 1s signal overlaps with the much stronger Ru 3d signals, and hence cannot be determined precisely.)

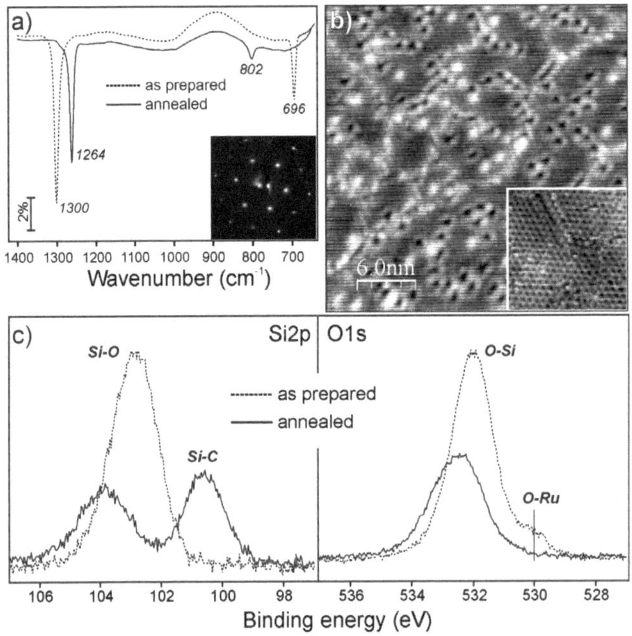

Figure 9.2 (a) Comparison of IRAS-spectra of a bilayer silicate film before (dash line) and after exposure to 10 mbar of C_2H_4 at 450 K and subsequent annealing at 1100 K in UHV (solid line). Inset shows LEED pattern demonstrating good crystallinity of the C-modified film, which is imaged by STM in panel (b). Si 2p and O 1s photoelectron spectra of the samples are shown in panel (c). Adapted with permission from ref. 22.

Random distribution of C and O atoms as well as substitution only of O atoms in the Si-O–Si interlayer linkage should result in electronically equivalent O–Si–C species and hence a single Si 2p state, which is at variance to the XPS results showing two states of Si. Accordingly, carbon only substitutes the O ions ether in the top or in the bottom silicatene layer. Taking into account that C atoms formed by ethylene decomposition only adsorb on the metal surface, carbon most likely substitutes the oxygen atoms in the bottom layer. If we further recall that silicatene layers can be formed on the SiC(0001) and (000-1) surfaces (Chapter 3), all experimental findings can be explained by the structure where a silicate layer is placed on top of a silicon carbide layer. Carbon atoms in the bottom layer most

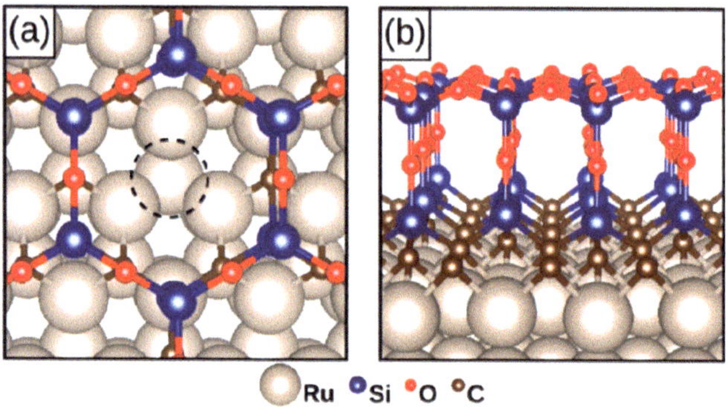

Figure 9.3 Structure of a carbon substituted silica film with a $SiC_{0.75}O_{1.25}$ composition on Ru(0001) in top (a) and side (b) views. Dashed circle indicates the Ru-hcp site. Reproduced with permission from ref. 24.

likely form strong bonds with the Ru(0001) surface, which can also explain long-range ordering in the resulted films. As we have shown before (Chapter 4), crystallinity of silicatene is primarily governed by strong interfacial bonds forcing an epitaxial relationship to a metal surface.

The proposed structure was fully supported by the DFT calculations (Fig. 9.3).[24] The adhesion energies for the silicatene/silicon-carbide film are more than one order of magnitude larger than for the pure bilayer silicatene (−510 vs. −33 meV $Å^{-2}$) due to the formation of chemical bonds to the metal surface atoms reflected in the short (2 Å) distances between C and Ru. On the basis of adhesion energies computed for various film orientations with respect to the Ru(0001) surface, the film is oriented such that the center of a silica hexagon is located over the threefold hollow site on the Ru(0001) surface as shown in Fig. 9.3a. Apparently, the hollow sites provide the best approximation for C atoms in the tetrahedral symmetry and hence indicate a preference of sp^3 hybridization over sp^2 in this system. In addition, the calculations revealed partial electron transfer from Ru to C, and also the Si atoms connected to C were less positive than the Si atoms only bonded to O.

The calculated (scaled) vibrational frequencies of IRAS active bands were compared for two systems. The principal band at 1289 cm^{-1} associated with vertical Si–O–Si bonds in the silicatene/silicon-carbide bilayer is only 3 cm^{-1} red-shifted with respect to that of the bilayer silicatene (1292 cm^{-1}), whereas 36 cm^{-1} is found in experiment. However, the second band (at 744 cm^{-1}) is found strongly blue-shifted (by 120 cm^{-1}) with respect to the bilayer-silicatene (624 cm^{-1}), in good agreement with experiment ($\Delta\nu = 106$ cm^{-1}). The deviation from experiment may be caused by high defectness of the prepared films, as imaged by STM, whereas the calculations were carried out for perfect films.

Even though the above-described experiments have not yet been performed on other silicatene systems, it appears that the fabrication of layered graphene-silicatene structures on a metal substrate is hardly possible by synthesis based on the "intercalation" approach. Indeed, either oxidation of Si underneath graphene results in its gasification or, alternatively, annealing of carbon underneath silicatene in oxygen-free atmosphere leads to the formation of a silicon carbide layer. Although further elaborative studies have to be done to see whether silicatene/silica carbide/metal heterostructures would be potentially interesting for some applications, it appears that methods based on precipitation of graphene and silicatene "flakes" would be a more promising approach for such purposes.

In principle, the bilayer silicatene weakly binds to a metal substrate. In addition, the interface between the silica and the metal may readily be intercalated by gas molecules in the ambient. This, in turn, results in lifting of the silicate layer away from the metal surface and making the bonding even weaker. These findings allow one to believe that the films can, in principle, be stripped or exfoliated from a substrate to be further handled *à la* graphene. However, the exfoliation of bilayer silicatene remains challenging. There is only one successful approach[25] reported thus far showing transfer of the bilayer silicatene from one to another metal substrate using poly-methylmethacrylate coating, the technique also used for manipulation with graphene.

Obviously, the situation becomes even more involved for silica monolayer films having a support, which is mandatory for saturation of all the bonds in the silicate layer. This is obvious for the ideal silicate layer, where all unsaturated apical oxygen atoms are on the one side of the silicate plane. On the other hand, from topological

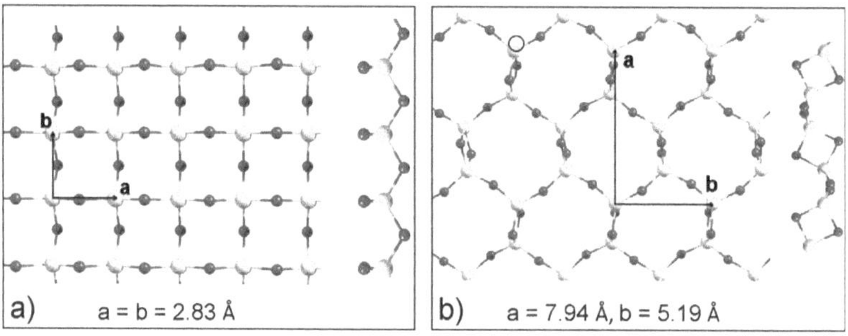

Figure 9.4 Proposed structures of 2D silica polymorphs in top and side views (**a** and **b** are the primitive translation vectors). Adapted with permission from ref. 26.

considerations and bearing in mind certain flexibility of silica bonds, it is possible to create "self-saturated" structures composed of silica tetrahedra as well as octahedra. In the lack of experimental findings, a new playground was opened up for theoreticians to search for free-standing single-layer silica structures.

Interestingly, even before substantial breakthrough in synthesis of silica monolayers on metals, Chernozatonskii analyzing nanotube structures of MO_2 dioxides proposed a "square" silica,[26,27] shown in Fig. 9.4a, which clearly differs from the hexagonal lattices commonly considered for graphene-like 2D-structures. The hypothetical nanotubes were rolled from silica monolayer showing a square unit cell with lattice constant of 2.83 Å. The Si–O bonds in this structure (1.65 Å) are somewhat stretched as compared to 1.60 Å in quartz. The Si–O–Si angle is 149° vs. 144° in quartz. However, the O–Si–O angle is by 40° larger than 109.5° in four-coordinated silica polymorphs. Therefore, bond angles around Si nuclei in this structure are quite strongly distorted. As a result, the calculated binding energy per SiO_2 unit for this "tetragonal" silica sheet is 0.75 eV higher than for quartz.

A more typical for silicas "trigonal" structure is shown in Fig. 9.4b. However, the layer has a two-membered ring in the structure. The Si–O bonds in two-membered rings are stretched to 1.67 Å, and the Si–O–Si angle is 89°, which again implies a strong deviation from tetrahedral geometry around Si nuclei. The calculated energy is 0.65 eV higher than for α-quartz.

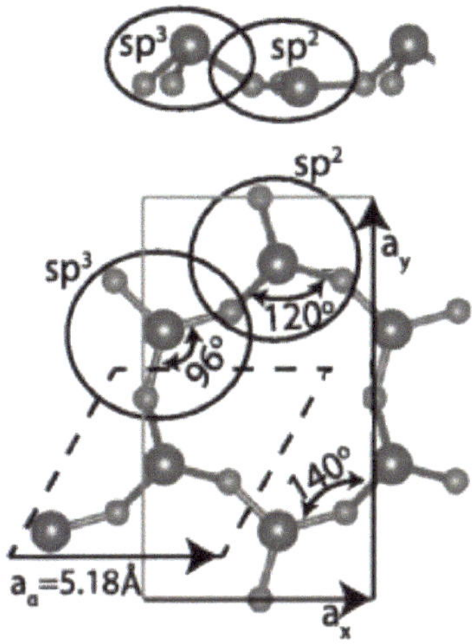

Figure 9.5 Optimized structure of *hα* silica with its hexagonal and rectangular unit cells. Two types of Si atoms, i.e., like sp^3 - and sp^2–bonded, are indicated. Reproduced with permission from ref. 28.

Another hypothetical structure, named *hα* silica, was derived from the oxidation of silicene resulting in a single silica layer with a Si_2O_3 stoichiometry having alternating sp^3- and sp^2–like bonds (Fig. 9.5). The 2D silica showed a wide band gap, a negative Poisson's ratio, and a high piezoelectric coefficient.[28] Although the layer was found thermodynamically stable, high reactivity of dangling Si bonds renders its fabrication hardly possible.

The follow-up first-principles calculations[29] examined various free-standing 2D silica monolayers with tetrahedral (*T*-silica) and octahedral (*O*-silica) building blocks. *T*-silica is formed by corner-sharing SiO_4 tetrahedra in a rectangular network, and *O*-silica consists of edge-sharing SiO_6 octahedra. *O*-silica is the first proposed 2D silica with octahedral-coordinated building blocks. The results for three 2D structures are shown in Fig. 9.6. The calculated formation energies of *O*-silica and *T*-silica are −2.77 and −2.70 eV/atom, respectively, which are very close to the value −2.82 eV/atom computed for the bilayer silicatene also shown in Fig. 9.6. For comparison, the above-mentioned *hα* silica only yielded −1.94 eV at the minimum, thus suggesting its considerably lower stability.

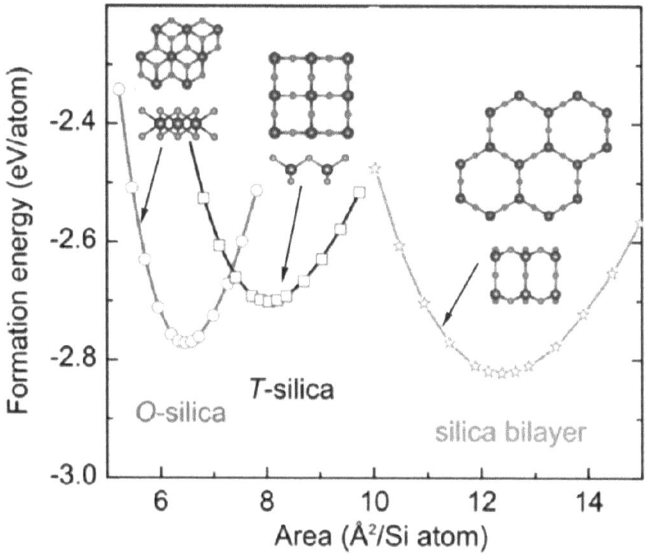

Figure 9.6 Calculated formation energy vs. area/Si atom for several 2D silica configurations schematically shown in top and side views. Large blue and small red balls in the represent silicon and oxygen atoms, respectively. Adapted with permission from ref. 29.

Considering that the sixfold-coordinated silica polymorphs (e.g., stishovite) exists at high pressure in the bulk phase, the prediction of a small energy difference of *O*-silica with the synthesized silica bilayer, together with the thermal stability at 1000 K, suggests that synthesis of *O*-silica can be achieved in experiments.

More recently, three new hypothetical 2D silica structures were reported, denoted by the authors as β-, γ-, and δ-2D silica (all shown in Fig. 9.7), from calculations using evolutionary algorithm.[30] Note, however, that the δ-2D structure, in fact, reproduces the one previously proposed (see Fig. 9.4a). All structures exhibit rectangular symmetry in the top view, and β- and γ-2D layers are strongly buckled in terms of the Si atom position in the side view, and have three different Si–O bond lengths in the structure. The calculations predicted extremely large in-plane negative Poisson's ratios in all these 2D layers. In addition, these materials are superhard, suggesting superior thermal conductivities compared

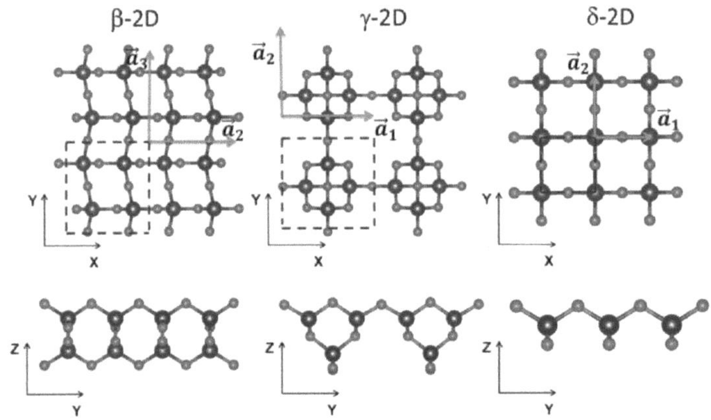

Figure 9.7 Equilibrium 2D silica structures in top and side views. Large blue and small red balls in the represent silicon and oxygen atoms, respectively. Adapted with permission from ref. 30.

to bulk silica, and exhibit large electronic band gaps as compared to other 2D materials. Such "exotic" 2D silica may find interesting applications in nanomechanics and nanoelectronics.

To sum up, a growing up body of theoretical results favors the possibility that low-energy 2D silica structures do exist in nature and can be prepared in laboratories. In principle, they all may be referred to as silicatenes, implying that they are single-layer silicas. The fact that only a few these silica structures have been experimentally identified suggests that their existence is determined by the kinetics rather than by the energetics of the final structure. It is also possible that there remain a large number of 2D silica structures yet to be discovered.

References

1. Geim, A. K., and K. S. Novoselov, The rise of graphene. *Nature Materials*, 2007. **6**(3): pp. 183–191.
2. Takeda, K., and K. Shiraishi, Theoretical possibility of stage corrugation in Si and Ge analogs of graphite. *Physical Review B*, 1994. **50**(20): pp. 14916–14922.

3. Vogt, P., et al., Silicene: Compelling experimental evidence for graphene-like two-dimensional silicon. *Physical Review Letters*, 2012. **108**(15): p. 155501.
4. Guzmán-Verri, G. G., and L. C. Lew Yan Voon, Electronic structure of silicon-based nanostructures. *Physical Review B*, 2007. **76**(7): p. 075131.
5. Dávila, M. E., et al., Germanene: A novel two-dimensional germanium allotrope akin to graphene and silicene. *New Journal of Physics*, 2014. **16**(9): p. 095002.
6. Cahangirov, S., et al., Two- and one-dimensional honeycomb structures of silicon and germanium. *Physical Review Letters*, 2009. **102**(23): p. 236804.
7. Zhu, F.-F., et al., Epitaxial growth of two-dimensional stanene. *Nature Materials*, 2015. **14**: p. 1020.
8. Cantero, E. D., et al., Growth of germanium on Au(111): Formation of germanene or intermixing of Au and Ge atoms? *Physical Chemistry Chemical Physics*, 2017. **19**(28): pp. 18580–18586.
9. Junji, Y., et al., Large area planar stanene epitaxially grown on Ag(1 1 1). *2D Materials*, 2018. **5**(2): p. 025002.
10. Huang, P. Y., et al., Direct imaging of a two-dimensional silica glass on graphene. *Nano Letters*, 2012. **12**(2): pp. 1081–1086.
11. Ishigami, M., et al., Atomic structure of graphene on SiO_2. *Nano Letters*, 2007. **7**(6): p. 1643–1648.
12. Xue, J., et al., Scanning tunnelling microscopy and spectroscopy of ultra-flat graphene on hexagonal boron nitride. *Nature Materials*, 2011. **10**: p. 282.
13. Gao, N., and X. Fang, Synthesis and development of graphene–inorganic semiconductor nanocomposites. *Chemical Reviews*, 2015. **115**(16): pp. 8294–8343.
14. Kang, Y.-J., J. Kang, and K. J. Chang, Electronic structure of graphene and doping effect on SiO_2. *Physical Review B*, 2008. **78**(11): p. 115404.
15. Fan, X. F., et al., Interaction between graphene and the surface of SiO_2. *Journal of Physics: Condensed Matter*, 2012. **24**(30): p. 305004.
16. Wang, F., et al., Silicon intercalation into the graphene-SiC interface. *Physical Review B*, 2012. **85**(16): p. 165449.
17. Kaloni, T. P., et al., Mechanism of Si intercalation in defective graphene on SiC. *Journal of Materials Chemistry*, 2012. **22**(44): pp. 23340–23343.

18. Mao, J., et al., Silicon layer intercalation of centimeter-scale, epitaxially grown monolayer graphene on Ru(0001). *Applied Physics Letters*, 2012. **100**(9): p. 093101.
19. Cui, Y., et al., An exchange intercalation mechanism for the formation of a two-dimensional Si structure underneath graphene. *Nano Research*, 2012. **5**(5): pp. 352–360.
20. Lizzit, S., et al., Transfer-free electrical insulation of epitaxial graphene from its metal substrate. *Nano Letters*, 2012. **12**(9): pp. 4503–4507.
21. Tetlow, H., et al., Growth of epitaxial graphene: Theory and experiment. *Physics Reports*, 2014. **542**(3): pp. 195–295.
22. Yang, B., S. Shaikhutdinov, and H.-J. Freund, Ultrathin silicatene/silicon-carbide hybrid film on a metal substrate. *Surface Science*, 2015. **632**(0): pp. 9–13.
23. Kusunoki, I., and Y. Igari, XPS study of a SiC film produced on Si(100) by reaction with a C_2H_2 beam. *Applied Surface Science*, 1992. **59**(2): pp. 95–104.
24. Schlexer, P., and G. Pacchioni, Modelling of an ultra-thin silicatene/silicon-carbide hybrid film. *Journal of Physics: Condensed Matter*, 2016. **28**(36): p. 364005.
25. Büchner, C., and M. Heyde, Two-dimensional silica opens new perspectives. *Progress in Surface Science*, 2017. **92**(4): pp. 341–374.
26. Artyukhov, V. I., and L. A. Chernozatonskii, Theoretical study of two-dimensional silica films. *The Journal of Physical Chemistry C*, 2010. **114**(21): pp. 9678–9684.
27. Chernozatonskii, L. A., A new class of MO_2 dioxide nanotubes (M=Si, Ge, Sn, Pb) composed of "square" lattices of atoms: Their structure and energy characteristics. *Journal of Experimental and Theoretical Physics Letters*, 2004. **80**(10): pp. 628–632.
28. Özçelik, V. O., S. Cahangirov, and S. Ciraci, Stable single-layer honeycomblike structure of silica. *Physical Review Letters*, 2014. **112**(24): p. 246803.
29. Wang, G., et al., Novel two-dimensional silica monolayers with tetrahedral and octahedral configurations. *The Journal of Physical Chemistry C*, 2015. **119**(27): pp. 15654–15660.
30. Gao, Z., et al., Novel two-dimensional silicon dioxide with in-plane negative Poisson's ratio. *Nano Letters*, 2017. **17**(2): pp. 772–777.

Index